Cortex-M4
이론과 실무(기초 활용편)

ARM Core STM 32 마스터하기

정금섭 저

光文閣
www.kwangmoonkag.co.kr

본 교재는 32비트 마이크로컨트롤러 유닛(MCU)의 범용 기능들을 쉽게 사용할 수 있도록 이론적인 배경은 배제하고 쉽게 따라 할 수 있도록 구성하였다. 일반적으로 32비트 프로세서를 활용해서 프로젝트를 구현하려면 관련된 많은 지식과 값비싼 컴파일러가 요구된다. 또한, 복잡한 환경 구축이 필수이며 이를 체계적으로 정리하지 않으면 매번 다시 시작해야 하는 수고로움이 동반되고, 하고자 하는 사기를 떨어뜨리며, 경우에 따라서는 포기하게 만든다. 따라서 본서는 컴파일러와 개발 툴이 결합되어 있고 무료로 다운로드해서 사용할 수 있는 ST사의 STM32CubeIDE(STM32CubeMX + TureSTUDIO) 개발 툴을 사용하며 내용은 간략하게 기술하고 따라 할 수 있도록 쉽게 기술하고자 하였다. 마이크로컨트롤러 유닛을 활용하는 기술자나 학생들의 목표는 32비트 ARM Core를 마음대로 활용하는 것이 아닐까 한다. 이 책을 접하시는 분들에게 조금이나마 도움이 되었으면 한다. 책의 주요 내용은 다음과 같다.

1. 최소한의 STM32의 지식: 1,2장
   개요 및 개발 환경
2. 필수 내용: 3장~11장
   GPIO, 시리얼통신, EXTI, Timer, ADC, 디스플레이 장치, FreeRTOS
3. 응용 분야: 12장
   딜레이 함수 만들기, 초음파센서, 압력센서, Load Cell, 서미스터

목차

CHAPTER 1

STM32 개요

CHAPTER 01

Cortex-M4 기초 활용(ARM Core STM32)

STM32 개요

Arm® Cortex®-M 프로세서를 기반으로 하는 STM32 32비트 마이크로컨트롤러 제품군[그림 1.1]으로 통합과 개발 용이성을 지원하면서 매우 강력한 성능, 실시간 처리, 디지털 신호 처리, 저전력/저전압 작동, 연결성을 겸비한 제품 라인으로 구성되어 있습니다. 업계 표준 코어를 기반으로 하는 다양한 종류의 STM32 마이크로컨트롤러는 개발 작업에 따라 선택해 사용할 수 있는 다양한 도구와 소프트웨어를 지원하므로, 소규모 프로젝트와 엔드 투 엔드 플랫폼 모두에 이상적입니다.

1. STM32 MCU 제품

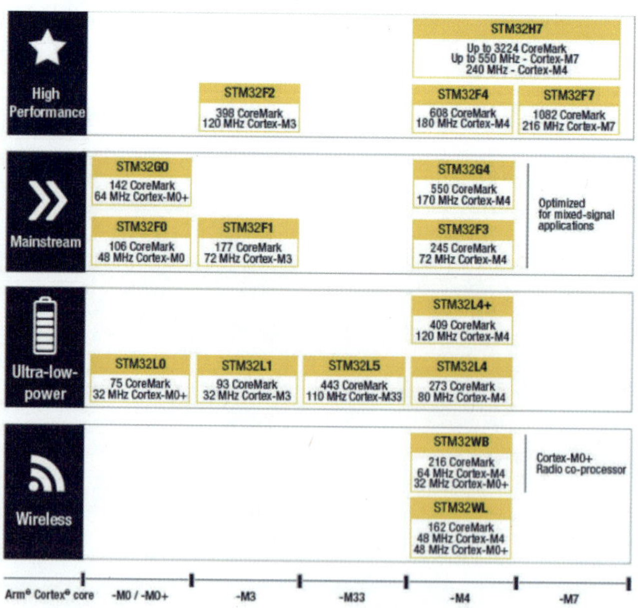

[그림 1.1] MCU 제품군

2. STM32 Ecosystem

STM32 Cube Ecosystem[그림 1.2], [표 1.1]은 STM32 MCU 및 MPU를 위한 소프트웨어 솔루션으로 무료이며 다운로드하여 사용할 수 있다.

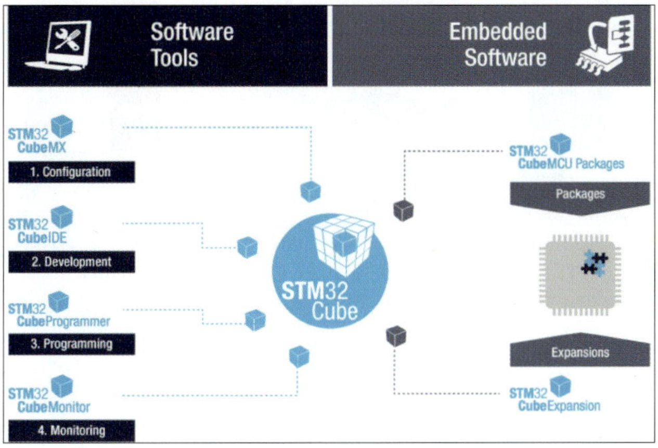

[그림 1.2] STM32 Ecosystem

Part Number	Configuration	Initialization code Generation	Code Edition	Code Building	Debugging	Binary Programming	Monitoring
STM32CubeMX	✔	✔					
STM32CubeIDE	✔ (Integrates STM32CubeMX)	✔ (Integrates STM32CubeMX)	✔	✔	✔	✔	
STM32CubeProgrammer						✔	
STM32CubeMonitor							✔

[표 1.1] STM32 툴들의 기능

▷ STM32CubeMX: 그래픽 기반 소프트웨어 구성 도구로 MCU, GPIO, 클럭, 타이머 카운터, 통신, 인터럽트, ADC 등 시스템을 구성하는 마법사 기능 툴이다. 이 경우 별도의 컴파일러(Keil or IAR, TRUESTUDIO)가 필요하다.

▷ STM32CubeIDE: 통합 개발 환경 툴로서 컴파일 기능과 향상된 디버깅 기능이 포함되어 있어 시스템의 구성과 컴파일 디버깅을 할 수 있는 툴이다. 본 교재에서는 이 툴을 사용한다.

3. 여러 가지 Board 소개

본 교재에서 사용할 보드 및 시중에서 쉽게 구할 수 있는 보드를 소개하고자 한다.

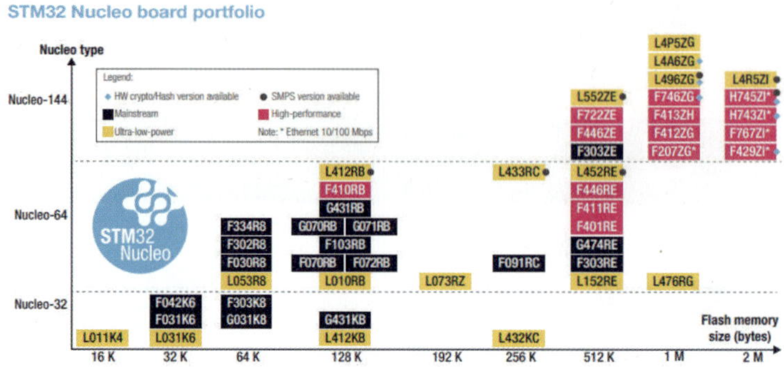

[그림 1.3] STM32 NUCLEO Board Portfolio

현재 출시된 Nucleo 보드[그림 1.3]는 MCU의 핀 수에 따라서 Nucleo-32, Nucleo-64, Nucleo-144 보드가 시판되고 있어 성능과 기능 핀 수를 감안해서 선정하여 사용하면 된다. Nucleo 보드의 특징 중 하나는 ST-Link/V2 가 포함되어 있어 별도의 장치 없이 USB 단자만 있으면 MCU를 Erase/program 및 dubugging을 할 수가 있다.

▷ Nucleo-32(Nucleo-L432KC) (여러 종류가 있음): 특징

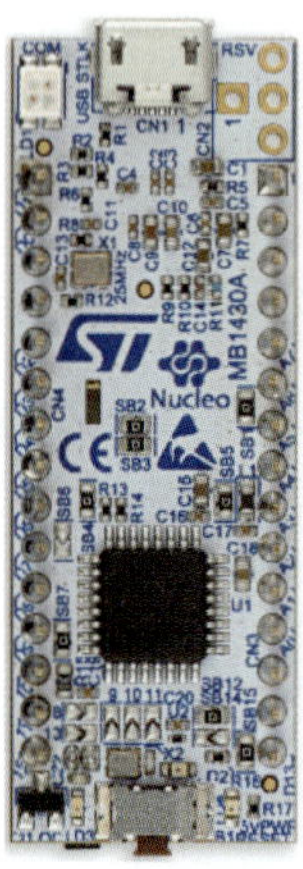

Features

- Common features
  - STM32 microcontroller in 32-pin package
  - 1 user LED
  - 1 reset push-button
  - Board connectors:
    - Arduino™ Nano V3 expansion connector
    - Micro-AB USB connector for the ST-LINK
  - Flexible power-supply options: ST-LINK, USB V_{BUS} or external sources
  - On-board ST-LINK debugger/programmer with USB re-enumeration capability: mass storage, Virtual COM port and debug port
  - Comprehensive free software libraries and examples available with the STM32Cube MCU Package
  - Support of a wide choice of Integrated Development Environments (IDEs) including IAR™, Keil® and GCC-based IDEs
- Board-specific features
  - 24 MHz crystal oscillator
  - Arm® Mbed Enabled™ compliant

- 구매 사이트 및 보드 가격: 엘레파츠, 가격(부가세 별도): 13,890원(내외)

▷ Nucleo-64(Nucleo-G474RE) (여러 종류가 있음): 특징

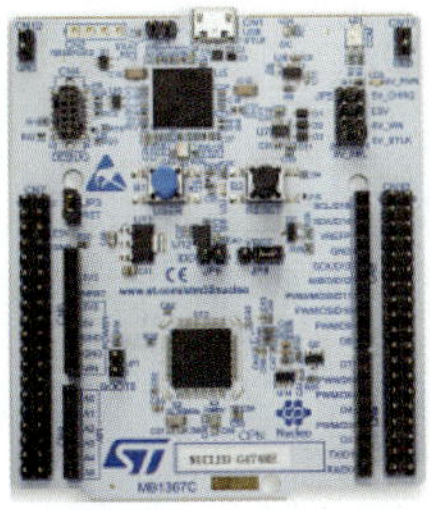

NUCLEO-G474RE example. Boards with different references show different layouts. Picture is not contractual.

Features

- Common features
  - STM32 microcontroller in LQFP64 package
  - 1 user LED shared with ARDUINO®
  - 1 user and 1 reset push-buttons
  - 32.768 kHz crystal oscillator
  - Board connectors:
    - ARDUINO® Uno V3 expansion connector
    - ST morpho extension pin headers for full access to all STM32 I/Os
  - Flexible power-supply options: ST-LINK, USB V_{BUS}, or external sources
  - On-board ST-LINK debugger/programmer with USB re-enumeration capability: mass storage, Virtual COM port and debug port
  - Comprehensive free software libraries and examples available with the STM32Cube MCU Package
  - Support of a wide choice of Integrated Development Environments (IDEs) including IAR Embedded Workbench®, MDK-ARM, and STM32CubeIDE
- Board-specific features
  - External SMPS to generate V_{core} logic supply
  - 24 MHz HSE
  - Board connectors:
    - External SMPS experimentation dedicated connector
    - Micro-AB or Mini-AB USB connector for the ST-LINK
    - MIPI® debug connector
  - Arm® Mbed Enabled™ compliant

- 구매 사이트 및 보드 가격: 엘레파츠, 가격(부가세 별도): NUCLEO F401RE: 16,370원(내외)

▷ Nucleo-144(Nucleo-H755ZI, F429ZI) (여러 종류가 있음): 특징

CLEO-H755ZI-Q example. Boards with erent references show different outs. Picture is not contractual.

Features

- Common features
  - STM32 microcontroller in LQFP144 package
  - 3 user LEDs
  - 2 user and reset push-buttons
  - 32.768 kHz crystal oscillator
  - Board connectors:
    - SWD
    - ST Zio expansion connector including ARDUINO® Uno V3
    - ST morpho expansion connector
  - Flexible power-supply options: ST-LINK, USB V_{BUS} or external sources
  - On-board ST-LINK debugger/programmer with USB re-enumeration capability: mass storage, Virtual COM port, and debug port
  - Comprehensive free software libraries and examples available with the STM32Cube MCU Package
  - Support of a wide choice of Integrated Development Environments (IDEs) including IAR™, Keil®, and STM32CubeIDE
- Board-specific features
  - External or internal SMPS to generate V_{core} logic supply
  - Ethernet compliant with IEEE-802.3-2002
  - USB OTG full speed or device only
  - Board connectors:
    - USB with Micro-AB or USB Type-C™
    - Ethernet RJ45
  - Arm® Mbed Enabled™ compliant

- 구매 사이트 및 보드 가격: 엘레파츠, 가격(부가세 별도): NUCLEO-H743ZI2: 33,770원(내외)

▷ **XNucleo-64(XNucleo-F411RE)**

ST사의 Nucleo-64와 차이점은 제조사가 Waveshare이며 ST-Link/V2 장치가 분리되어 있으며, 외부 8Mhz 크리스탈과 RTC 발진자가 연결되어 있고 LED가 4개 연결되어 있다. 가격은 조금 더 비싼 편이다. 정확한 타이머 가능과 LED를 이용할 수 있어 편리하다. 보드를 구매하면 USB 케이블과 mini ST-Link/V2가 포함되어 있다.

▷ **STM32F4DISCOVERY Board(STM32F407G-DISC1)**

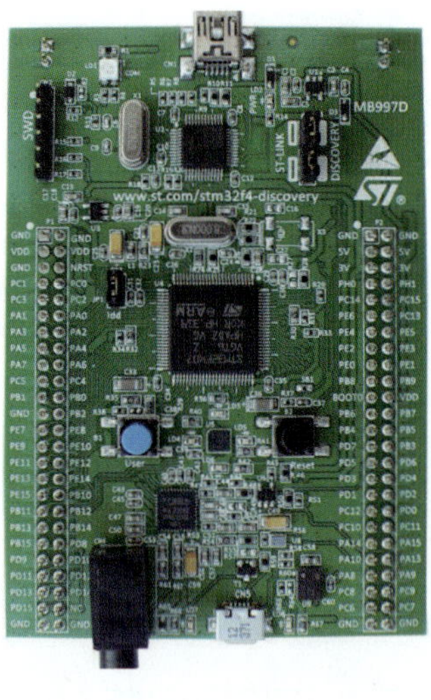

- 프로그램 가능한 LED가 4개 스위치가 1개 있으며, 3축 모션센서 모듈 오디오 미니 잭이 탑재되어 있다. 가격은 3만 원대 후반이다.

4. 실습 보드(NUCLEO-F411RE)

실습 보드[그림 1.4]~[그림 1.5]는 NUCLEO-64(NUCLE-411RE)로 선정하였다. 이 보드는 ST-Link/V2가 내장되어 있으며, 유저가 프로그램 가능한 LED 1개, 입력 스위치 1개 등으로 시스템이 구성되어 있어 별도의 하드웨어를 구성하지 않고도 다양한 실습을 할 수 있어 실습 보드로 사용한다. 가격은 2만 원 내외다.

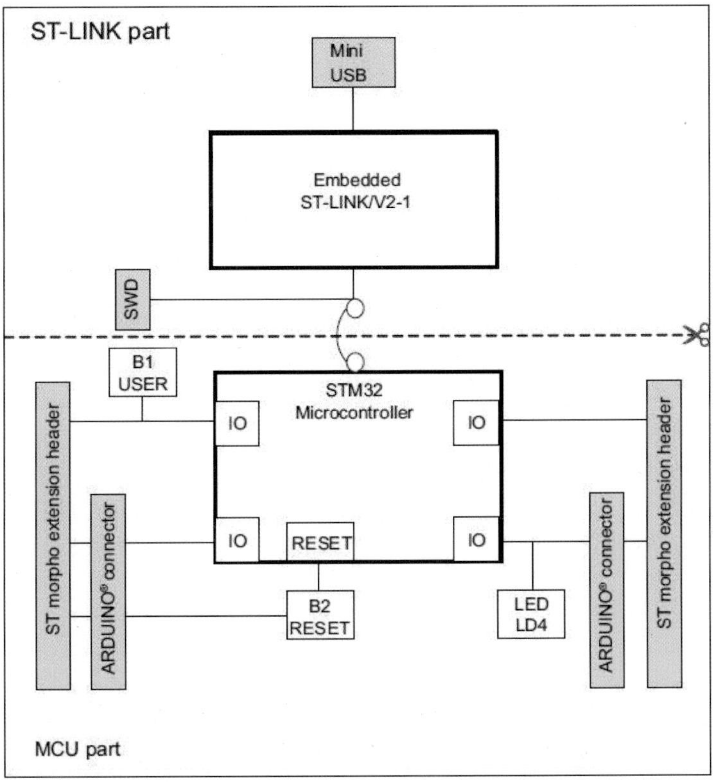

[그림 1.4] STM32-F411RE Block Diagram

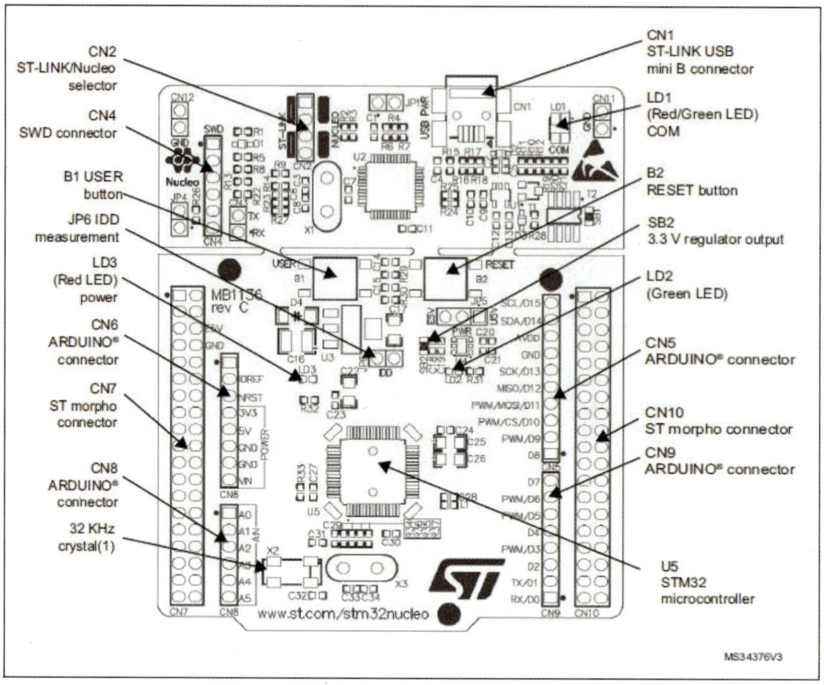

[그림 1.5] Board Top Layout

프로그램과 디버거를 위해서는 [그림 1.6]과 같은 점퍼 조건이 되어야 한다.

▷ **프로그램 및 디버그 점퍼 조건**:

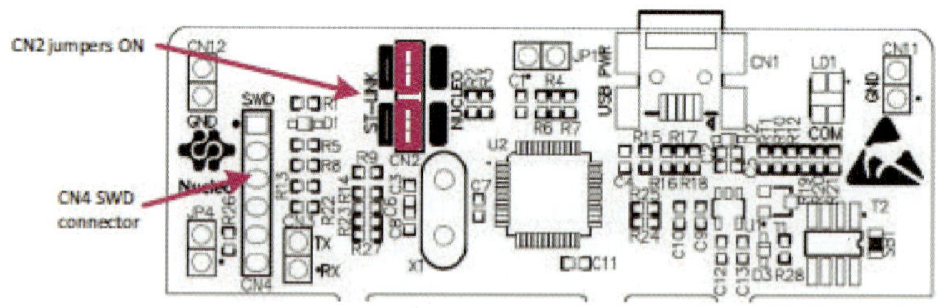

[그림 1.6] 프로그램과 디버그를 위한 점퍼 조건

▷ **보드 파워 설정**

USB와 연결된 파워를 이용하려면 아래 [표 1.2]의 첫 번째와 같은 그림으로 JP5 점퍼를 설정하고, 외부 5V 전원을 연결해서 사용하려면 [표 1.2]의 중간 형태로 점퍼를 설정하며, 외부 전원 7V~12V를 연결해서 사용하려면 [표 1.2]의 가장 아래와 같이 JP3를 설정하여야 한다. 이와 관련된 자료는 [표 1.3]과 같다.

Jumper	Description
JP5	U5V (ST-LINK VBUS) is used as a power source when JP5 is set as shown below (Default setting) 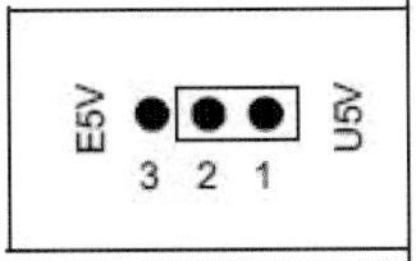
	VIN or E5V is used as a power source when JP5 is set as shown below. 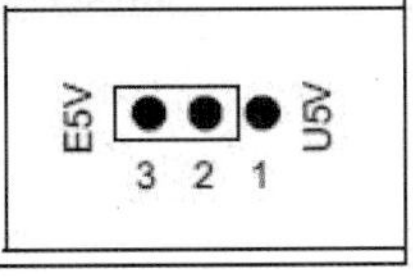

[표 1.2] Power 연결 점퍼 조건

Input power name	Connectors pins	Voltage range	Max current	Limitation
VIN	CN6 pin 8 CN7 pin 24	7 V to 12 V	800 mA	From 7 V to 12 V only and input current capability is linked to input voltage: 800 mA input current when Vin = 7 V 450 mA input current when 7 V < Vin <= 9 V 250 mA input current when 9 V < Vin <= 12 V
E5V	CN7 pin 6	4.75 V to 5.25 V	500 mA	-

[표 1.3] External Power Source 사용 조건

▷ **Board Pinout**

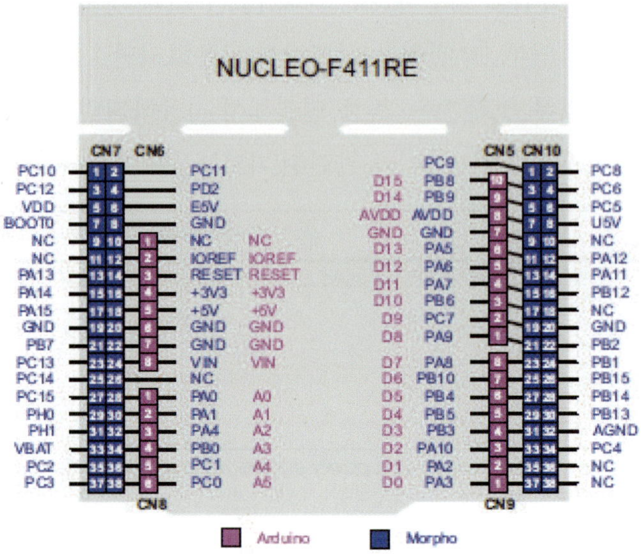

5. 실습 보드(NUCLEO-F411RE) 외부 장치

외부장치는 LED(LD2), Push-botton(B1), 외부 클럭, 시리얼통신, USB 및 시리얼 통신 등이 연결되어 있으므로 해당 포트는 다른 용도로 사용하지 말 것을 권한다.

▶ LED

LD1: USB 통신 기능

LD2: PA5, Yellow Color, User program 가능

LD3: Power 기능

▶ Push-buttons

B1: PC13, User Program 가능

B2: Reset 스위치

▶ OSC Clock

HSE(외부 하이스피드 클럭) 4가지

-.MCO from ST-LINK: ST-LINK-MCU의 출력 MCO 8Mhz

-.HSE oscillator on-board from X3 crystal (not provided)

-.Oscillator from external PF0/PD0/PH0

-.HSE not used :PF0/PD0/PH0 and PF1/PD1/PH1를 GPIO로 사용, 이를 위해서는 아래의 조치가 필요하다.

- SB54 and SB55 ON

- SB16 and SB50 (MCO) OFF

- R35 and R37 제거

OSC 32kHz 설정 3가지

-.On-board oscillator: X2: 32.768kHz

-.Oscillator from external PC14

-.LSE not used :PC14,PC15는 GPIO로 사용 가능, 이를 위해서는 아래의 조치가 필요하다.

-. SB48 and SB49 ON

-. R34 and R36 제거

▶ USART 통신

USART2 TX: PA2

USART2 RX: PA3

▶ SOLDER Bridges

Bridge	State[1]	Description
SB45 (VBAT/VLCD)	ON	VBAT or VLCD on STM32 is connected to VDD.
	OFF	VBAT or VLCD on STM32 is not connected to VDD.
SB57 (VDDA/VREF+)	ON	VDDA/VREF+ on STM32 is connected to VDD.
	OFF	VDDA/VREF+ on STM32 is not connected to VDD and can be provided from pin 8 of CN5 (Used for external VREF+ provided by ARDUINO® shield)
SB62, SB63 (USART)	**OFF**	PA2 and PA3 on STM32 are disconnected to D1 and D0 (pin 2 and pin 1) on ARDUINO® connector CN9 and ST morpho connector CN10.
	ON	PA2 and PA3 on STM32 are connected to D1 and D0 (pin 2 and pin 1) on ARDUINO® connector CN9 and ST morpho connector CN10 as USART signals. Thus SB13 and SB14 must be OFF.
SB13, SB14 (ST-LINK-USART)	**ON**	PA2 and PA3 on STM32F103CBT6 (ST-LINK MCU) are connected to PA3 and PA2 on STM32 to have USART communication between them. Thus SB61, SB62, and SB63 must be OFF.
	OFF	PA2 and PA3 on STM32F103CBT6 (ST-LINK MCU) are disconnected to PA3 and PA2 on STM32.
SB16,SB50(MCO)[2]	OFF	MCO on STM32F103CBT6 (ST-LINK MCU) are disconnected to PF0/PD0/PH0 on STM32.
	ON	MCO on STM32F103CBT6 (ST-LINK MCU) are connected to PF0/PD0/PH0 on STM32.

Bridge	State[1]	Description
SB54, SB55 (X3 crystal)[2]	OFF	X3, C33, C34, R35, and R37 provide a clock as shown in electrical schematics PF0/PD0/PH0, PF1/PD1/PH1 are disconnected from CN7.
	ON	PF0/PD0/PH0 and PF1/PD1/PH1 are connected to CN7. (R35, R37, and SB50 must not be fitted).
SB3, SB5, SB7,SB9 (DEFAULT)	**ON**	Reserved, do not modify.
SB4,SB6, SB8,SB10 (RESERVED)	OFF	Reserved, do not modify.
SB48, SB49 (X2 crystal)[3]	OFF	X2, C31, C32, R34, and R36 deliver a 32 kHz clock. PC14, PC15 are not connected to CN7.
	ON	PC14 and PC15 are only connected to CN7. Remove only R34, R36.
SB17 (B1-USER)	**ON**	B1 push button is connected to PC13.
	OFF	B1 push button is not connected to PC13.
SB12 (NRST)	ON	The NRST signal of the CN4 connector is connected to the NRST pin of the STM32.
	OFF	The NRST signal of the CN4 connector is not connected to the NRST pin of the STM32.
SB15 (SWO)	ON	The SWO signal of the CN4 connector is connected to PB3.
	OFF	The SWO signal is not connected.
SB11 (STM_RST)	**OFF**	No incidence on STM32F103CBT6 (ST-LINK MCU) NRST signal.
	ON	STM32F103CBT6 (ST-LINK MCU) NRST signal is connected to GND.
SB1 (USB-5V)	OFF	USB power management is functional.
	ON	USB power management is disabled.
SB2 (3.3 V)	ON	Output of voltage regulator LD39050PU33R is connected to 3.3V.
	OFF	Output of voltage regulator LD39050PU33R is not connected.
SB21 (LD2-LED)	ON	Green user LED LD2 is connected to D13 of ARDUINO® signal.
	OFF	Green user LED LD2 is not connected.
SB56,SB51 (A4 and A5)	ON	PC1 and PC0 (ADC in) are connected to A4 and A5 (pin 5 and pin 6) on ARDUINO® connector CN8 and ST morpho connector CN7. Thus SB46 and SB52 must be OFF.
	OFF	PC1 and PC0 (ADC in) are disconnected to A4 and A5 (pin 5 and pin 6) on ARDUINO® connector CN8 and ST morpho connector CN7.
SB46,SB52 (I2C on A4 and A5)	OFF	PB9 and PB8 (I2C) are disconnected to A4 and A5 (pin 5 and pin 6) on ARDUINO® connector CN8 and ST morpho connector CN7.
	ON	PB9 and PB8 (I2C) are connected to A4 and A5 (pin 5 and pin 6) on ARDUINO® connector CN8 and ST morpho connector CN7 as I2C signals. Thus SB56 and SB51 must be OFF.

CHAPTER 2

개발 환경 구성하기

CHAPTER 02

Cortex-M$_4$ 기초 활용(ARM Core STM32)

개발 환경 구성하기

STM32CubeIDE는 STM32CubeMX와 TrueSTUDIO가 결합된 통합 개발 플래폼으로 시스템 구성, 코드 제너레이션과 컴파일 그리고 디버그 기능을 가지고 있다. 따라서 시스템 구성에 따른 코드가 자동 생성되고 컴파일 디버그 기능을 활용해서 펌웨어를 쉽게 개발할 수 있다. [그림 2.1]은 개발 툴의 특징을 나타낸다.

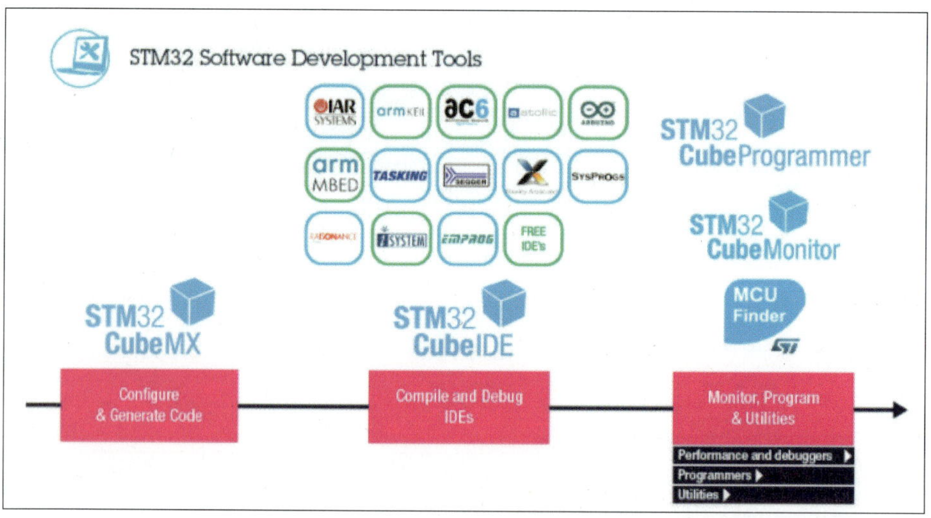

[그림 2.1] 소프트웨어 개발 툴 특징

1. STM32CubeIDE 다운로드하기

다운로드할 사이트를 방문하여 검색하거나 직접 주소로 다운로드 할 수 있다.

1) https://www.st.com/en/development-tools/stm32cubeide.html로 들어가서 [그림 2.2]와 같이 STM32CubeIDE-Win의 Get Software 선택 클릭한다.

Get Software

	Part Number ▲	General Description	Latest version	Download	All versions
+	STM32CubeIDE-DEB	STM32CubeIDE Debian Linux Installer	1.7.0	Get latest	Select version
+	STM32CubeIDE-Lnx	STM32CubeIDE Generic Linux Installer	1.7.0	Get latest	Select version
+	STM32CubeIDE-Mac	STM32CubeIDE macOS Installer	1.7.0	Get latest	Select version
+	STM32CubeIDE-RPM	STM32CubeIDE RPM Linux Installer	1.7.0	Get latest	Select version
+	STM32CubeIDE-Win	STM32CubeIDE Windows Installer	1.7.0	Get latest	Select version

[그림 2.2] STM32CubeIDE Win 선택 클릭

2) License Agreement 창이 [그림 2.3]과 같이 뜨면 우측 상단 ACCEPT 를 클릭한다.

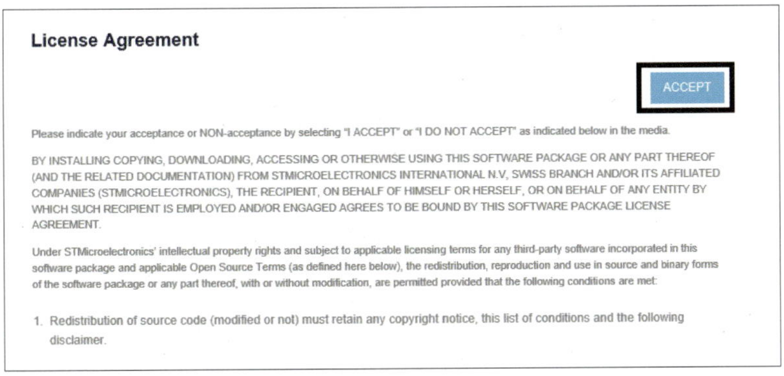
License Agreement
ACCEPT
Please indicate your acceptance or NON-acceptance by selecting "I ACCEPT" or "I DO NOT ACCEPT" as indicated below in the media.
BY INSTALLING COPYING, DOWNLOADING, ACCESSING OR OTHERWISE USING THIS SOFTWARE PACKAGE OR ANY PART THEREOF (AND THE RELATED DOCUMENTATION) FROM STMICROELECTRONICS INTERNATIONAL N.V, SWISS BRANCH AND/OR ITS AFFILIATED COMPANIES (STMICROELECTRONICS), THE RECIPIENT, ON BEHALF OF HIMSELF OR HERSELF, OR ON BEHALF OF ANY ENTITY BY WHICH SUCH RECIPIENT IS EMPLOYED AND/OR ENGAGED AGREES TO BE BOUND BY THIS SOFTWARE PACKAGE LICENSE AGREEMENT.
Under STMicroelectronics' intellectual property rights and subject to applicable licensing terms for any third-party software incorporated in this software package and applicable Open Source Terms (as defined here below), the redistribution, reproduction and use in source and binary forms of the software package or any part thereof, with or without modification, are permitted provided that the following conditions are met:
1. Redistribution of source code (modified or not) must retain any copyright notice, this list of conditions and the following disclaimer.

[그림 2.3] License Agreement 화면

3) [그림 2.4]와 같이 창이 뜨면, 이름과 성을 쓰고 E-mail 주소를 반드시 입력하고 개인정보 정책 동의란을 클릭하고 Download 를 클릭한다.

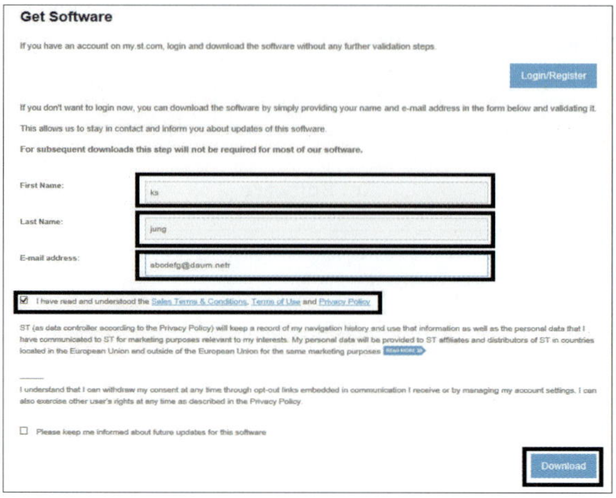
Get Software
If you have an account on my.st.com, login and download the software without any further validation steps.
Login/Register
If you don't want to login now, you can download the software by simply providing your name and e-mail address in the form below and validating it.
This allows us to stay in contact and inform you about updates of this software.
For subsequent downloads this step will not be required for most of our software.
First Name:
Last Name:
E-mail address:
Download

[그림 2.4] Get Software 화면

4) 위의 기록한 E-mail 주소로 이동하면 아래와 같이 ST사에서 메일이 한 개와 있으면 클릭해서 들어간다.

S
STMicroelectronics <onlinesu...
[받은 메일함] Start your software download

5) [그림 2.5]의 메일 중간쯤에 있는 Download now 를 클릭하면 프로그램이 다운로드된다.

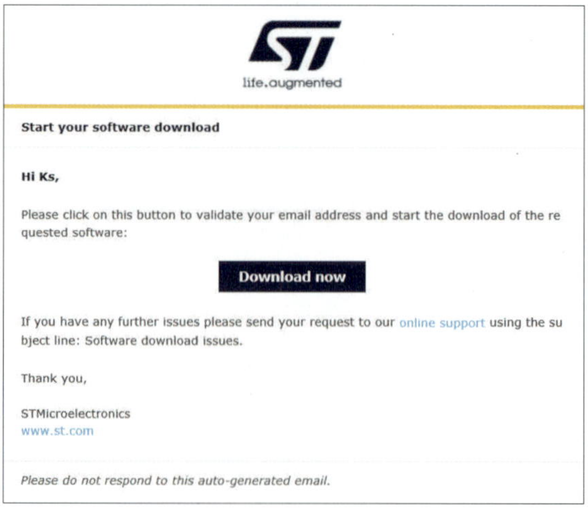
life.augmented
Start your software download
Hi Ks,
Please click on this button to validate your email address and start the download of the requested software:
Download now
If you have any further issues please send your request to our online support using the subject line: Software download issues.
Thank you,
STMicroelectronics
www.st.com
Please do not respond to this auto-generated email.

[그림 2.5] 받은 메일 내용

6) 다운로드가 완료된 파일은 아래 그림과 같다.

en.st-stm32cubeide_1.7.0_10852_20210715_0634_x86_64.exe_v1.7.0

2. STM32CubeIDE 설치하기

1) 다운로드한 파일을 압축 해제한 후에 실행하면 [그림 2.6] 화면이 나타난다. [Next >] 아이콘을 클릭한다.

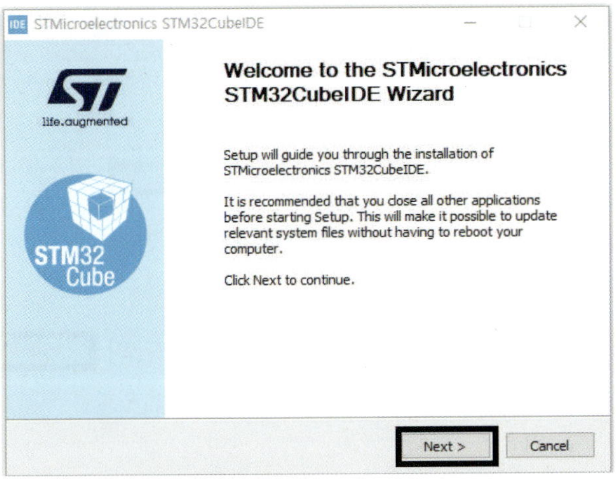

STMicroelectronics STM32CubeIDE
life.augmented
STM32 Cube
Welcome to the STMicroelectronics STM32CubeIDE Wizard
Setup will guide you through the installation of STMicroelectronics STM32CubeIDE.
It is recommended that you close all other applications before starting Setup. This will make it possible to update relevant system files without having to reboot your computer.
Click Next to continue.
Next >
Cancel

[그림 2.6] 설치 화면

2) [그림 2.7] 화면 하단의 [I Agree] 아이콘을 클릭한다.

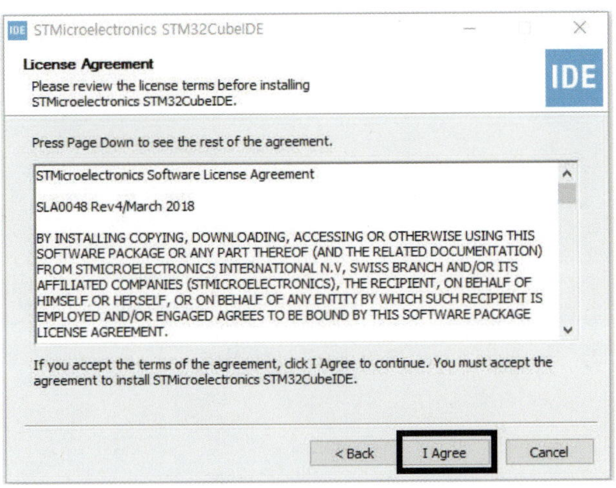

STMicroelectronics STM32CubeIDE
License Agreement
Please review the license terms before installing STMicroelectronics STM32CubeIDE.
IDE
Press Page Down to see the rest of the agreement.
STMicroelectronics Software License Agreement
SLA0048 Rev4/March 2018
BY INSTALLING COPYING, DOWNLOADING, ACCESSING OR OTHERWISE USING THIS SOFTWARE PACKAGE OR ANY PART THEREOF (AND THE RELATED DOCUMENTATION) FROM STMICROELECTRONICS INTERNATIONAL N.V, SWISS BRANCH AND/OR ITS AFFILIATED COMPANIES (STMICROELECTRONICS), THE RECIPIENT, ON BEHALF OF HIMSELF OR HERSELF, OR ON BEHALF OF ANY ENTITY BY WHICH SUCH RECIPIENT IS EMPLOYED AND/OR ENGAGED AGREES TO BE BOUND BY THIS SOFTWARE PACKAGE LICENSE AGREEMENT.
If you accept the terms of the agreement, click I Agree to continue. You must accept the agreement to install STMicroelectronics STM32CubeIDE.
< Back
I Agree
Cancel

[그림 2.7] 설치 화면

3) 설치 경로를 설정하거나 그림처럼 기본으로 하고 [그림 2.8] 화면 하단의 Next > 아이콘을 클릭한다.

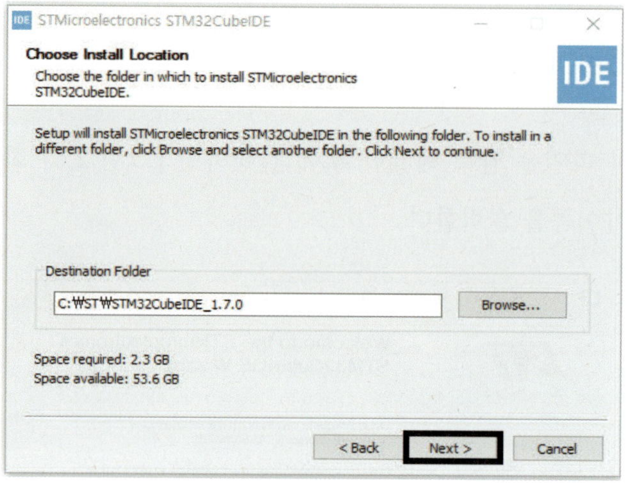
STMicroelectronics STM32CubeIDE
Choose Install Location
Choose the folder in which to install STMicroelectronics STM32CubeIDE.
IDE
Setup will install STMicroelectronics STM32CubeIDE in the following folder. To install in a different folder, click Browse and select another folder. Click Next to continue.
Destination Folder
C:₩ST₩STM32CubeIDE_1.7.0
Browse...
Space required: 2.3 GB
Space available: 53.6 GB
< Back
Next >
Cancel

[그림 2.8] 설치 화면

4) 드라이버들을 모두 선택하고 [그림 2.9] 화면 하단의 Install 아이콘을 클릭한다.

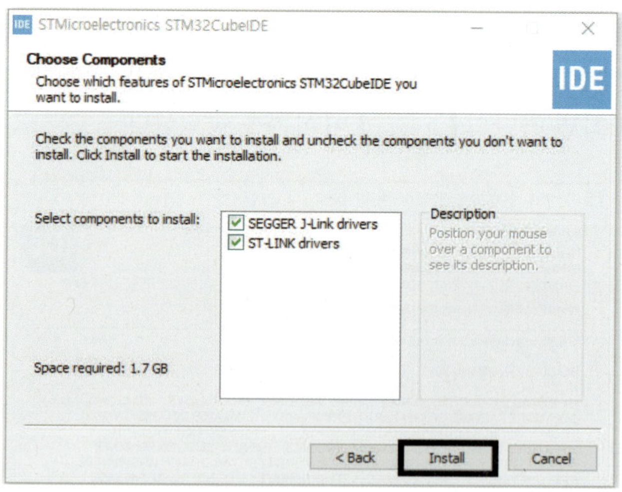
STMicroelectronics STM32CubeIDE
Choose Components
Choose which features of STMicroelectronics STM32CubeIDE you want to install.
IDE
Check the components you want to install and uncheck the components you don't want to install. Click Install to start the installation.
Select components to install:
SEGGER J-Link drivers
ST-LINK drivers
Description
Position your mouse over a component to see its description.
Space required: 1.7 GB
< Back
Install
Cancel

[그림 2.9] 설치 화면

5) [그림 2.10]은 프로그램 설치 중 화면이다.

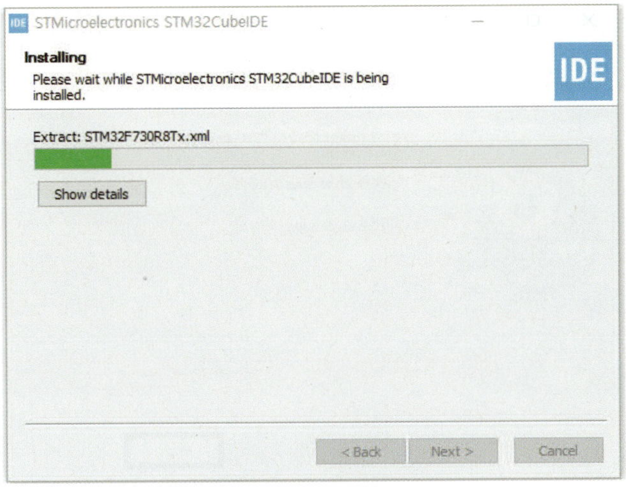

[그림 2.10] 설치 화면

6) [그림 2.11]의 하단의 Next > 아이콘을 클릭한다.

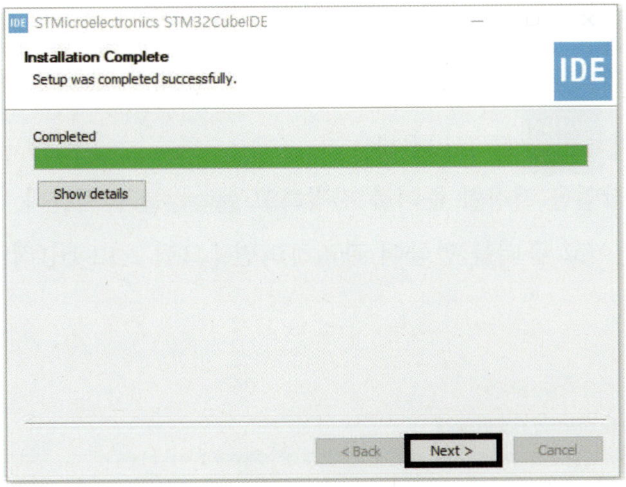

[그림 2.11] 설치 화면

7) [그림 2.12]는 설치가 완료된 화면으로 하단의 [Finish] 아이콘을 클릭한다.

[그림 2.12] 설치 화면

3. STM32CubeIDE 실행하기

설치가 완료되면 컴퓨터 바탕화면에 STM32CubeIDE 아이콘이 나타난다.

1) 바탕화면의  을 클릭한다.

2) 프로그램 작업을 저장할 폴더를 지정하고 [Launch]를 클릭한다. 디폴트로는 [그림 2.12_a]이고 폴더를 만들어 관리하려면 [그림 2.12_b](적극 추천)와 같이 사용한다.

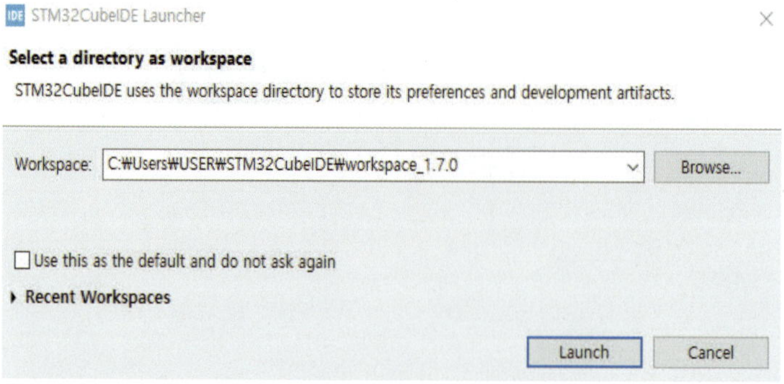

[그림 2.12_a] 설치 화면(디폴트 위치)

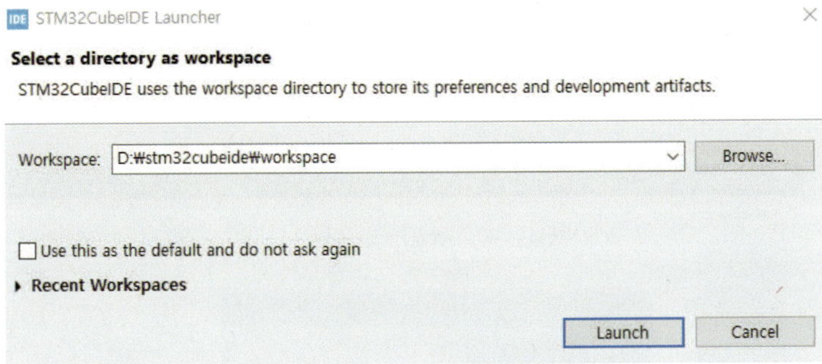
STM32CubeIDE Launcher
Select a directory as workspace
STM32CubeIDE uses the workspace directory to store its preferences and development artifacts.
Workspace: D:₩stm32cubeide₩workspace
Browse...
Use this as the default and do not ask again
Recent Workspaces
Launch
Cancel

[그림 2.12_b] 설치 화면(폴더 지정)

3) ST사 개인정보 정책 인지 화면 [그림 2.13_a]에 Yes 프로그램 설치나 실행 과정에서 [그림 2.13_b]와 같은 화면이 나타나면 액세스 허용(A) 클릭한다.

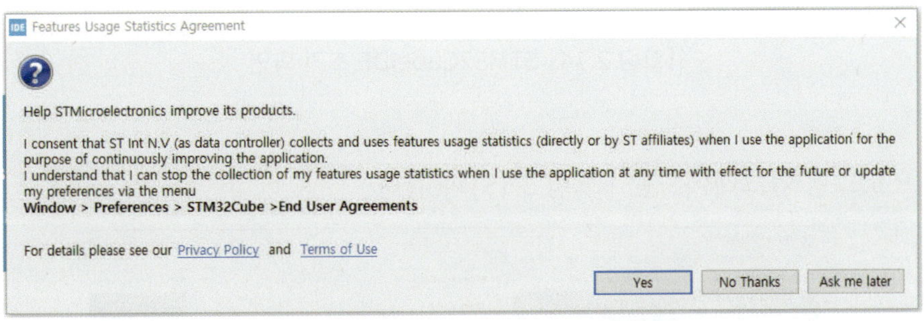
Features Usage Statistics Agreement
Help STMicroelectronics improve its products.
I consent that ST Int N.V (as data controller) collects and uses features usage statistics (directly or by ST affiliates) when I use the application for the purpose of continuously improving the application.
I understand that I can stop the collection of my features usage statistics when I use the application at any time with effect for the future or update my preferences via the menu
Window > Preferences > STM32Cube >End User Agreements
For details please see our Privacy Policy and Terms of Use
Yes
No Thanks
Ask me later

[그림 2.13_a] ST사 개인정보 정책 인지

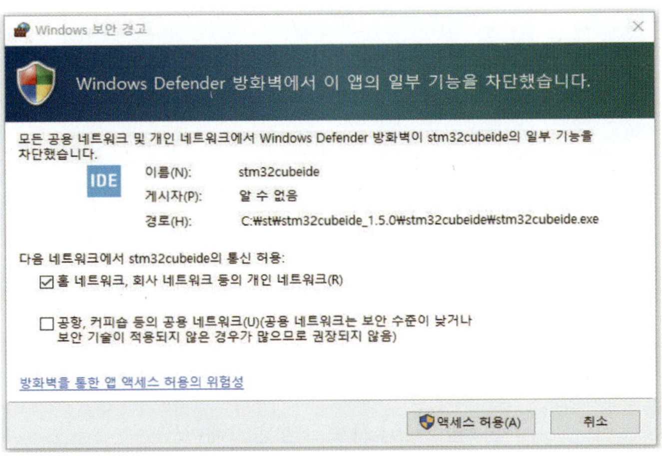
Windows 보안 경고
Windows Defender 방화벽에서 이 앱의 일부 기능을 차단했습니다.
모든 공용 네트워크 및 개인 네트워크에서 Windows Defender 방화벽이 stm32cubeide의 일부 기능을 차단했습니다.
IDE
이름(N): stm32cubeide
게시자(P): 알 수 없음
경로(H): C:₩st₩stm32cubeide_1.5.0₩stm32cubeide₩stm32cubeide.exe
다음 네트워크에서 stm32cubeide의 통신 허용:
홈 네트워크, 회사 네트워크 등의 개인 네트워크(R)
공항, 커피숍 등의 공용 네트워크(U)(공용 네트워크는 보안 수준이 낮거나 보안 기술이 적용되지 않은 경우가 많으므로 권장되지 않음)
방화벽을 통한 앱 액세스 허용의 위험성
액세스 허용(A)
취소

[그림 2.13_b] 방화벽 액세스

4) 실행이 완료되면 [그림 2.14]와 같이 초기 화면이 뜬다.

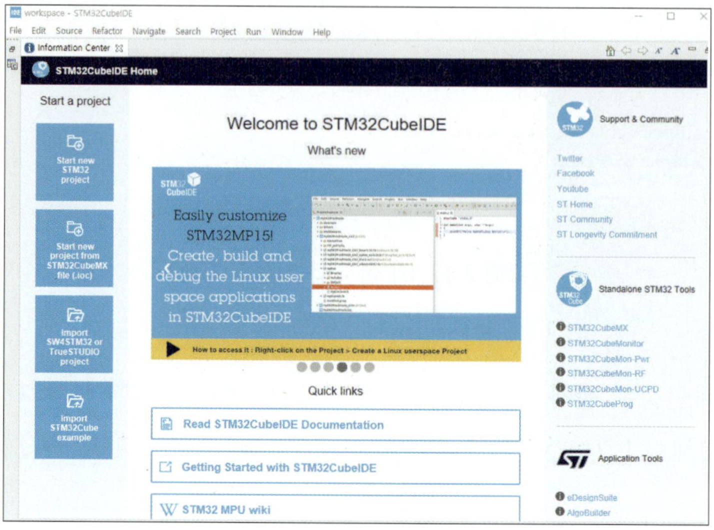

[그림 2.14] STM32CubeIDE 초기 화면

5) 프로그램 업그레이드는 [그림 2.15]와 같다.

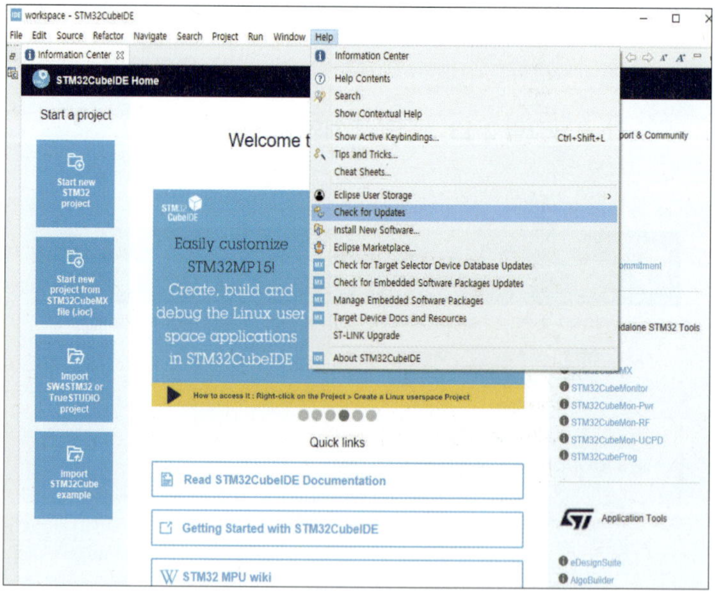

[그림 2.15] 프로그램 업그레이드

6) 디바이스 데이터베이스 업그레이드는 [그림 2.16_a], [그림 2.16_b]의 REFRESH 끝나면 OK 버튼을 눌러 주면 된다.

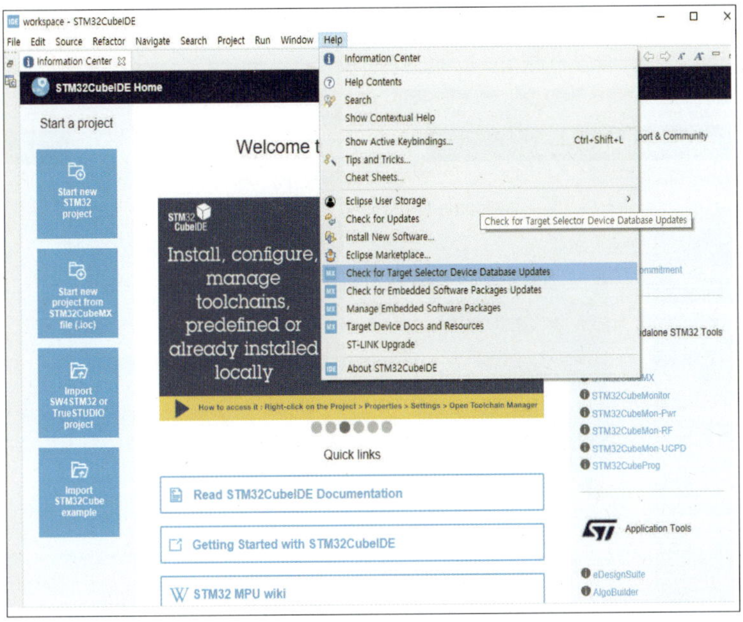

[그림 2.16_a] 프로그램 업그레이드

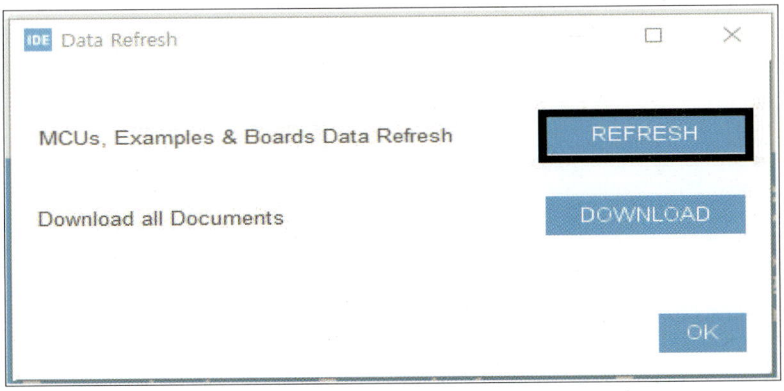

[그림 2.16_b] 디바이스 데이터베이스 업그레이드

7) 임베디드 소프트웨어 패키지 업그레이드는 업데이트할 파일이 있으면 Install Now 클릭하고 REFRESH 클릭하고 Close 버튼을 눌러 주면 된다[그림 2.17]~[그림 2.18].

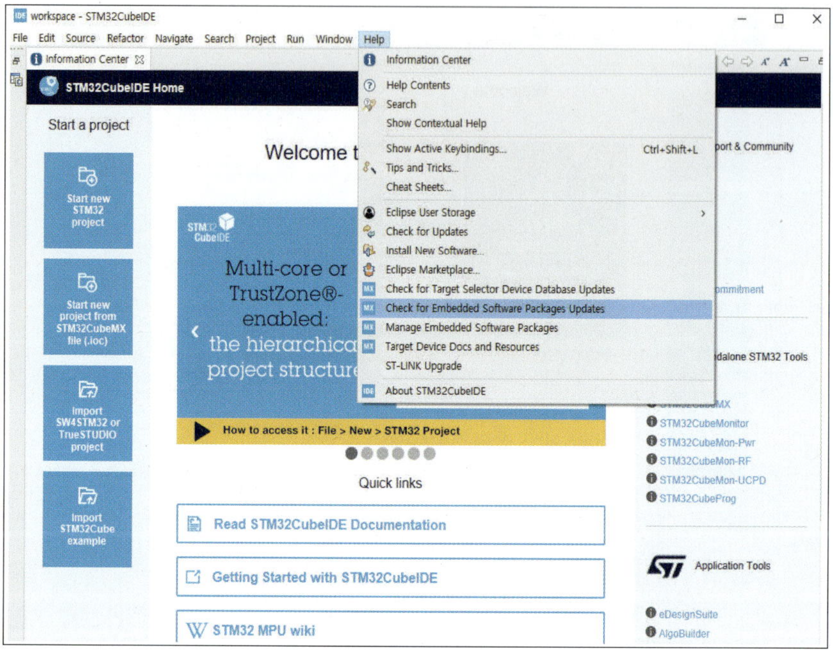

[그림 2.17] 임베디드 소프트웨어 패키지 업그레이드

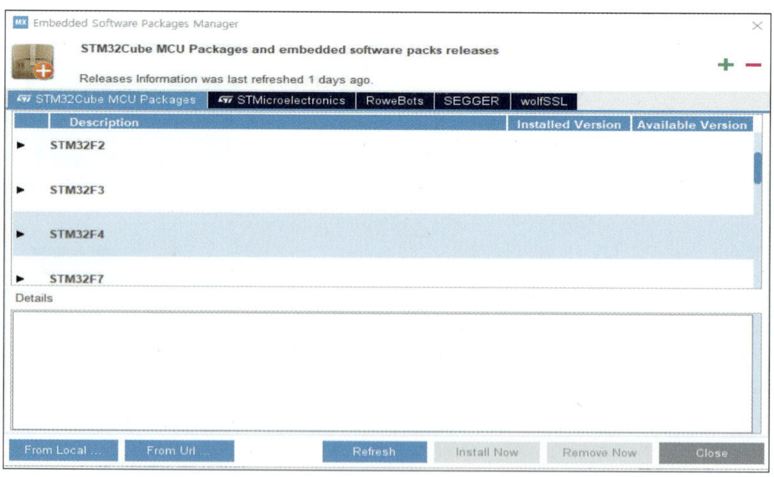

[그림 2.18] 임베디드 소프트웨어 패키지 업그레이드

8) ST-LINK 업그레이드 과정은 [그림 2.19]~[그림 2.21]과 같다.

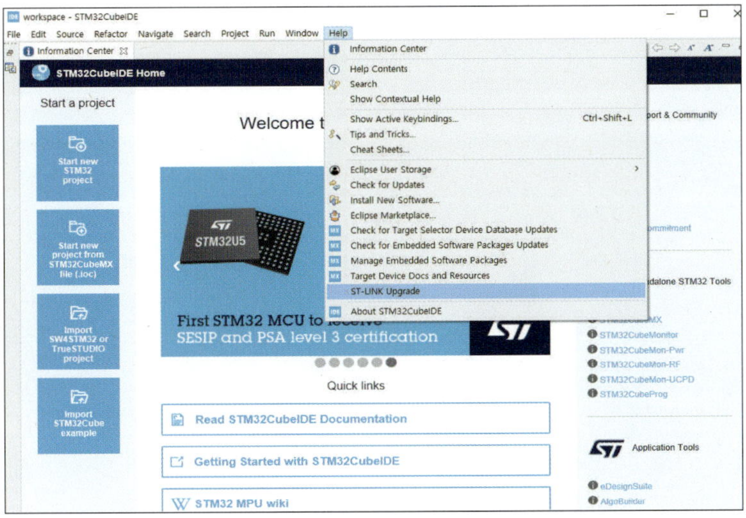

[그림 2.19] ST-LINK 업그레이드

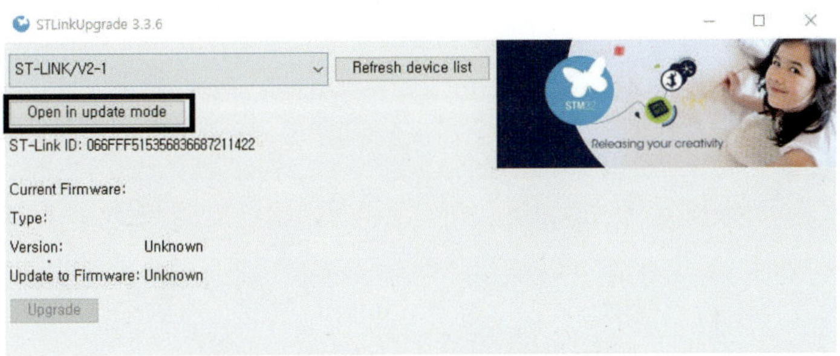

[그림 2.20] ST-WLINK 업그레이드

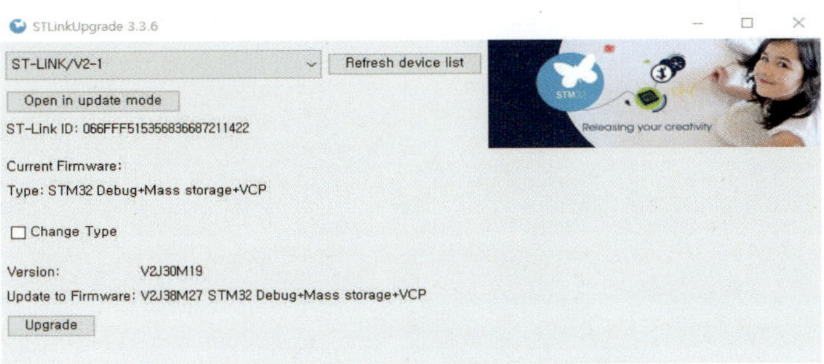

[그림 2.21] ST-LINK 업그레이드

Cortex-M4 기초 활용(ARM Core STM32)

CHAPTER 3

GPIO 제어하기

<GPIO 출력 기능 및 입력 기능>

1. STM32 프로젝트 시작
2. 보드 선정: NUCLEO-F411RE
3. 프로젝트 이름 결정 및 보드 구성 형태로 초기화: GPIO_LED
4. 실습 보드 LED 연결 확인
5. 클럭 설정
6. Pinout 초기화 설정
7. 디바이스 기본 설정
8. GPIO 설정
9. Generate Code 생성
10. Source Code 작성하기
11. 디버그 및 다운로드하기
12. 보드 프로그램 실행하기
13. LED 4개 제어하기

CHAPTER 03

Cortex-M4 기초 활용(ARM Core STM32)

GPIO 제어하기

<GPIO 출력 기능>

1. STM32 프로젝트 시작

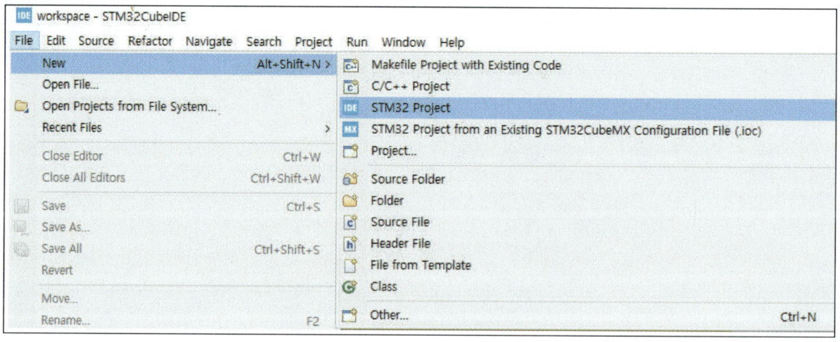

[그림 3.1] 프로그램 시작

2. 보드 선정: NUCLEO-F411RE

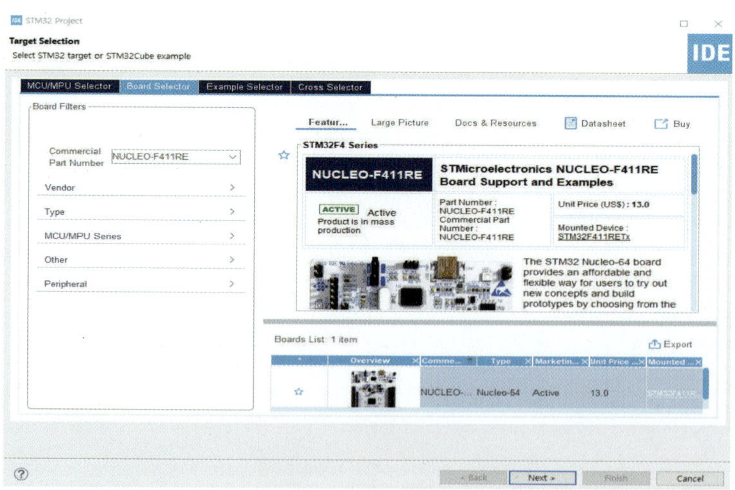

[그림 3.2] 보드 선정

3. 프로젝트 이름 결정 및 보드 구성 형태로 초기화: GPIO_LED

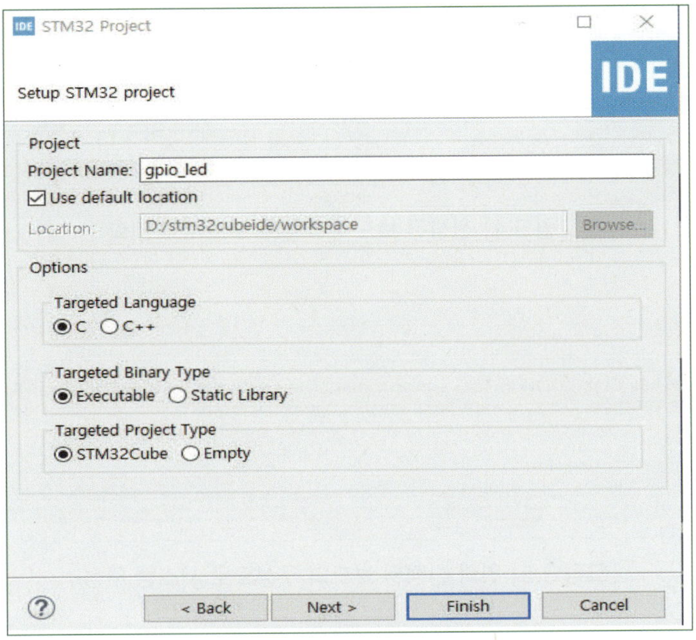

[그림 3.3] 프로젝트 이름 결정

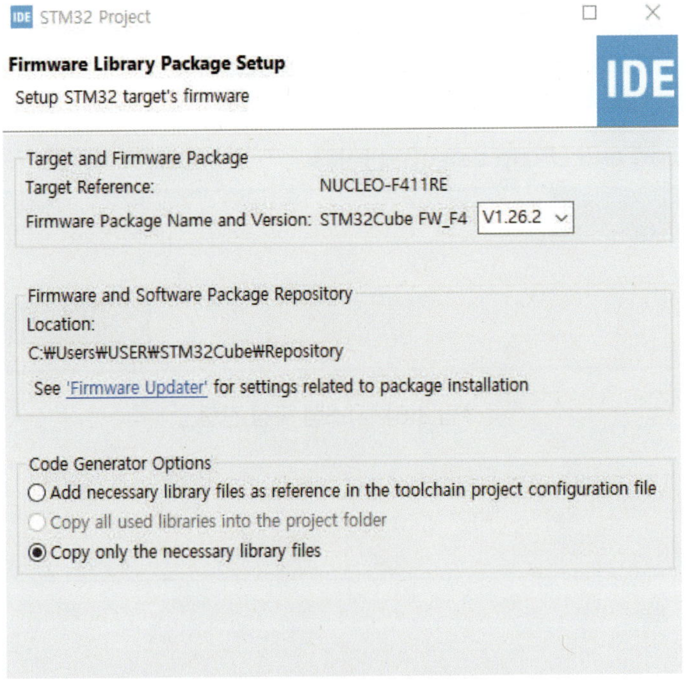

[그림 3.3_a] 펌웨어 패키지 설정

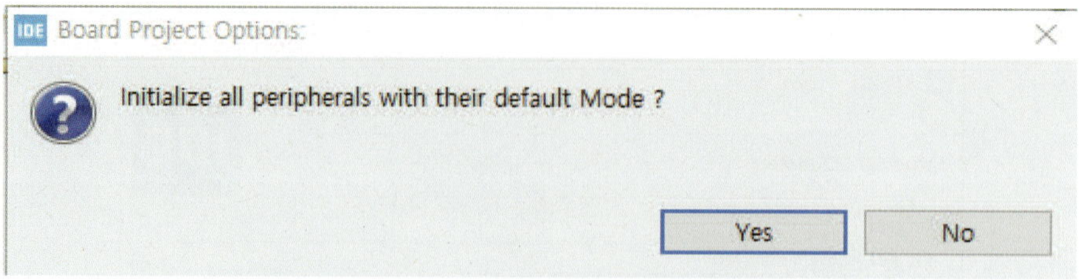
Board Project Options:
Initialize all peripherals with their default Mode ?
Yes
No

[그림 3.4] 선정된 보드의 디폴트 모드로 초기화

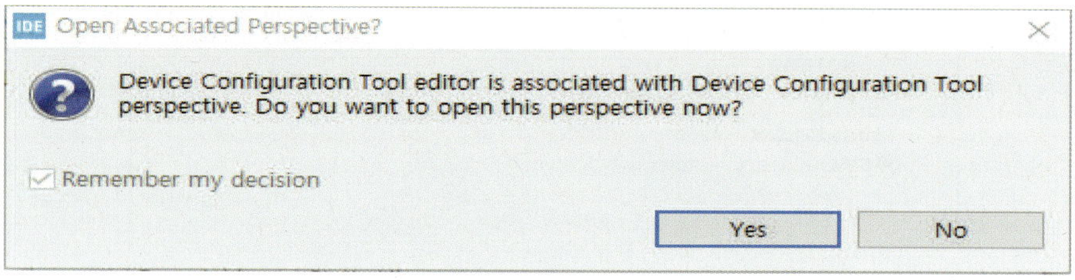
Open Associated Perspective?
Device Configuration Tool editor is associated with Device Configuration Tool perspective. Do you want to open this perspective now?
Remember my decision
Yes
No

[그림 3.5] 위의 선정된 보드의 구성으로 시스템 구성

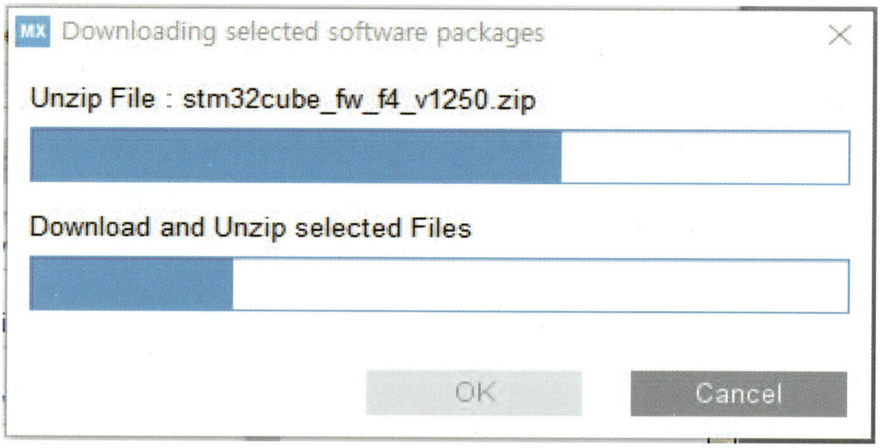
Downloading selected software packages
Unzip File : stm32cube_fw_f4_v1250.zip
Download and Unzip selected Files
OK
Cancel

[그림 3.6] 시스템 구성 진행

4. 실습보드 LED 연결 확인

[그림 3.7]에서 녹색은 디폴트로 핀 맵이 구성되어 있다는 것을 나타내고 있다. [그림 3.8]은 보드에서 사용한 디폴트 시스템을 나타내고 있다. 이번에는 LED on/off 실습이므로 Led가 GPIO에 어떻게 연결되어 있는지 확인한다.

```
> LED GPIO  출력: LD2 : PA5, Yellow color
```

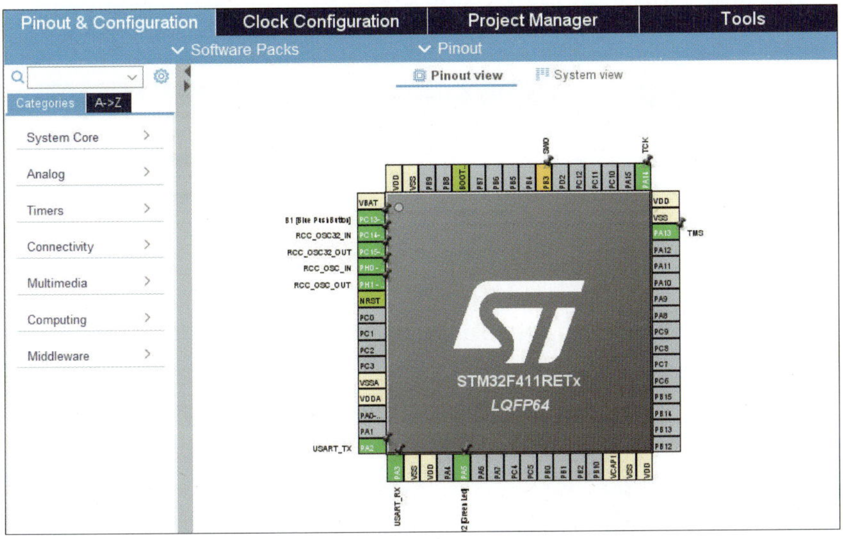

[그림 3.7] NUCLEO-F411RE 디폴트 보드 구성 핀 맵

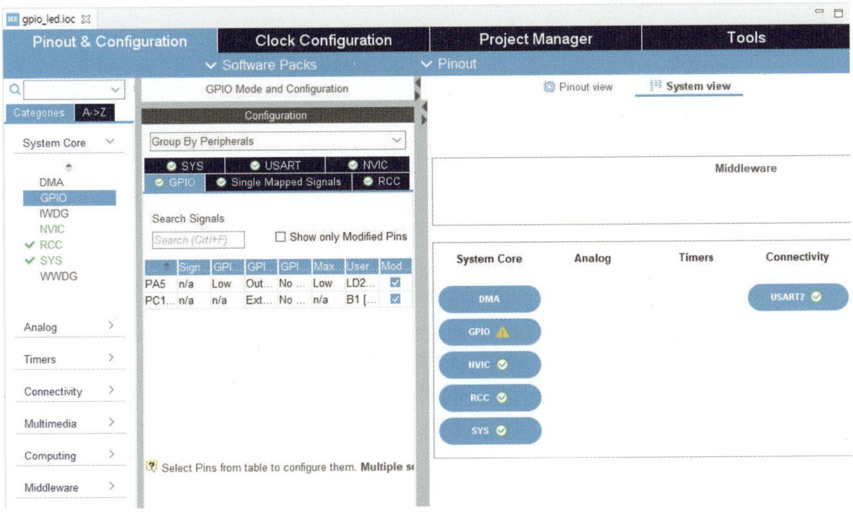

[그림 3.8] NUCLEO-F429ZI 디폴트 시스템 뷰

5. 클럭 설정

[그림 3.9]~[그림 3.10]은 디폴트로 클럭이 설정된 경우를 나타낸다. 메인 클럭은 외부 클럭(Crystal/Ceramic Resonator)과 내부 크럭(HSI RC/BYPASS Clock Source)둘 중에 하나를 이용하여야 한다. 내부 클럭보다는 외부 클럭이 정확하다. 디폴트로는 내부 클럭이 선정된다.

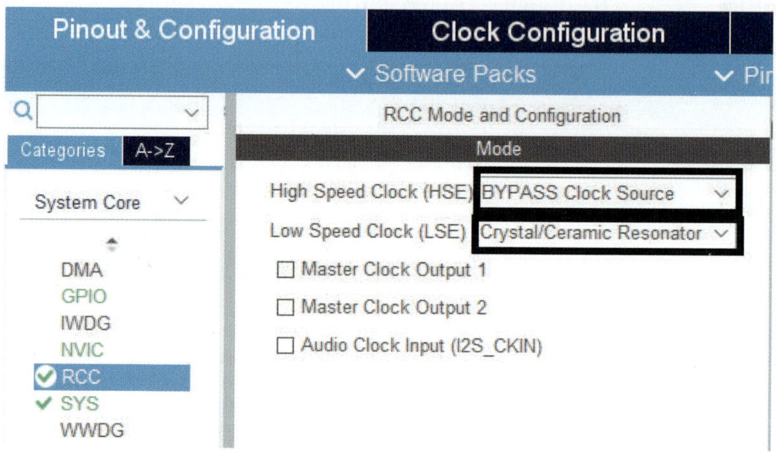

[그림 3.9] NUCLEO-F411RE 디폴트 클럭 설정

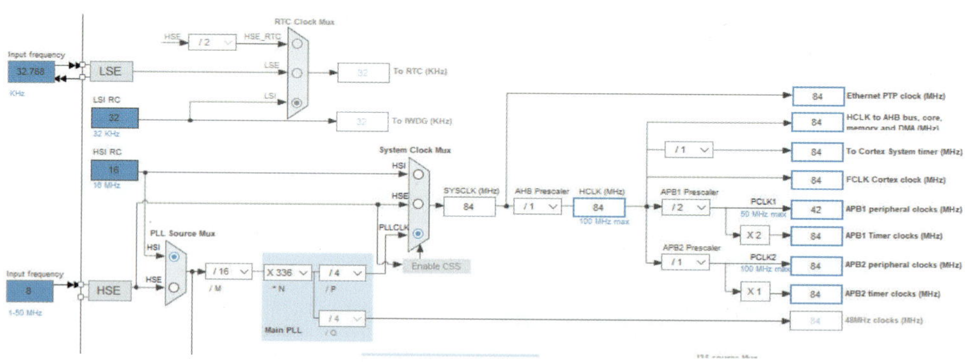

[그림 3.10] NUCLEO-F411RE 디폴트 클럭 설정

[그림 3.11]은 주변 장치의 클럭이 할당된 것을 알 수 있다. [그림 3.12]를 보면 APB1에 연결되어 있으면 42MHz, APB2에 연결되어 있으면 84MHz로 동작함을 알 수 있다.

▶APB1 Timer: 84MHz, Timer2(32b),3,4,5(32b),6,7,12,13,14

▶APB2 Timer: 84MHz, Timer1,8,9,10,11,

▶FCLK Clock: 84MHz

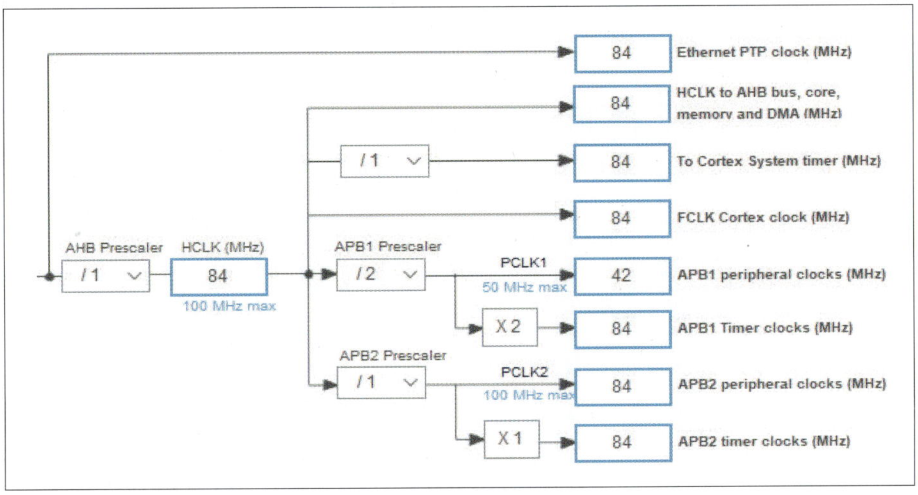

[그림 3.11] NUCLEO-F411RE 주변 장치 클럭

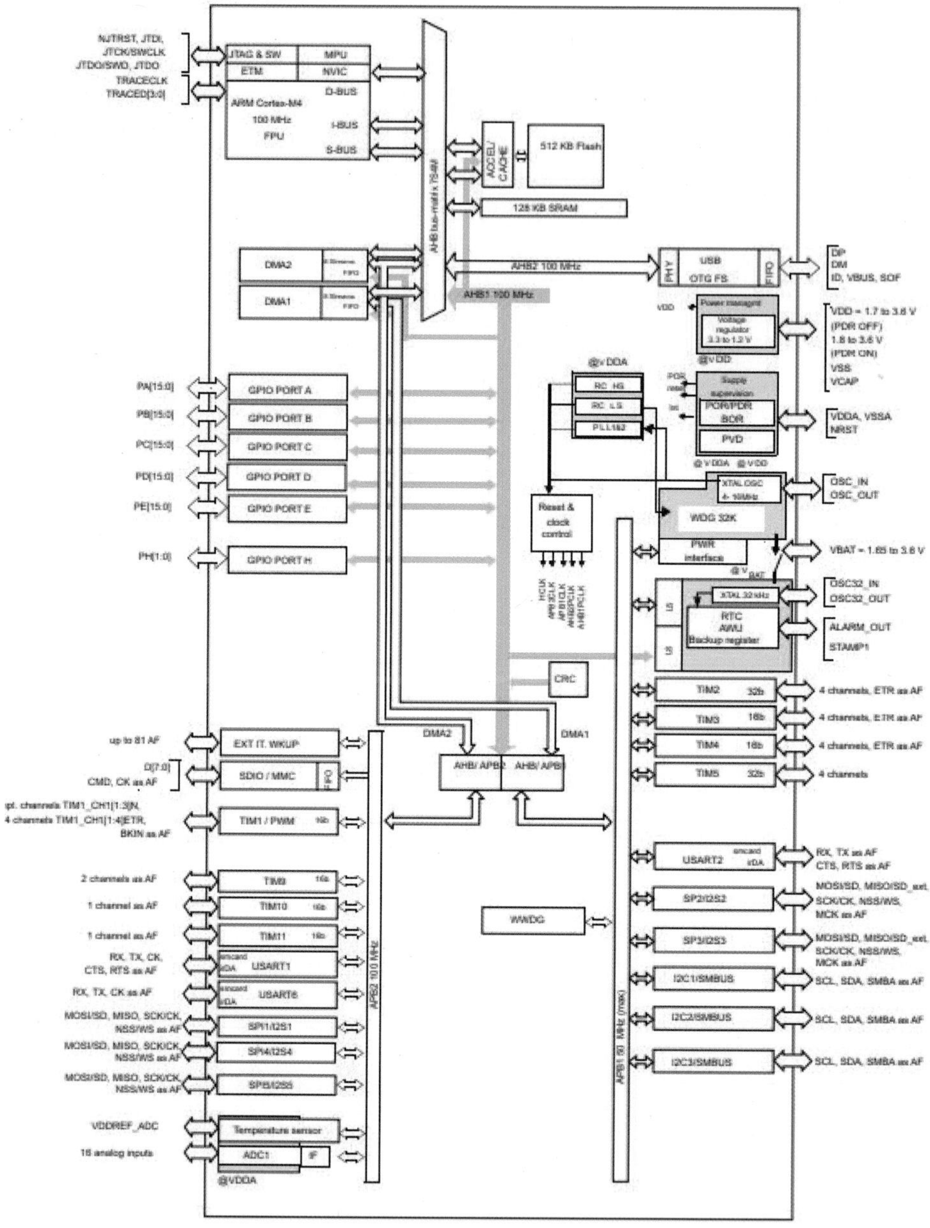

[그림 3.12] NUCLEO-F411RE APB1,APB2 블럭도

6. Pinout 초기화 설정

보드의 기능을 유지하면서 프로그램을 하는 경우의 [그림 3.7]의 경우이고 여기서는 사용자에 맞게 새로 환경 설정을 하는 경우에 해당 한다. 이 경우 [그림 3.13]과 같이하면 된다.

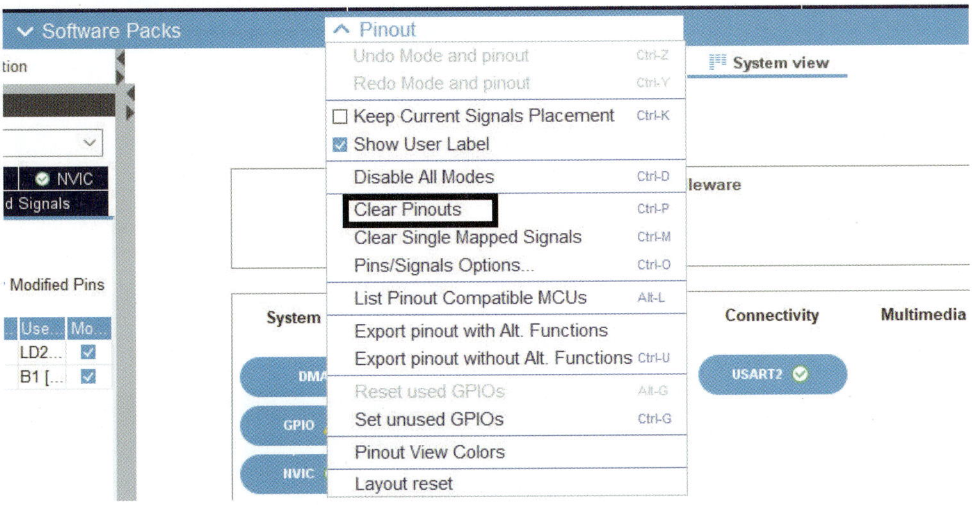

[그림 3.13] 디바이스 Pinout 설정

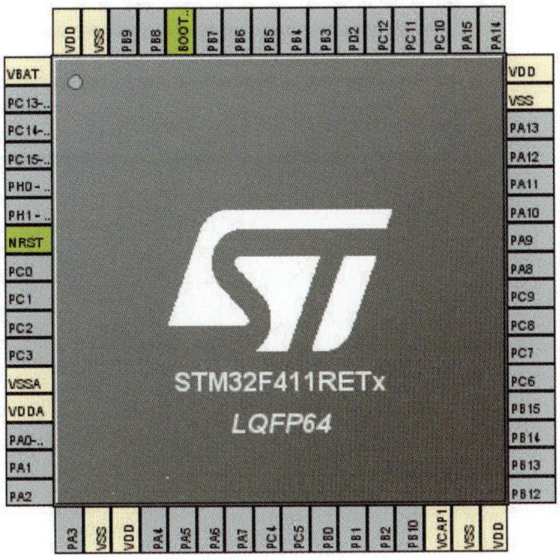

[그림 3.14] 디바이스 Pinout 초기화 화면

7. 디바이스 기본 설정

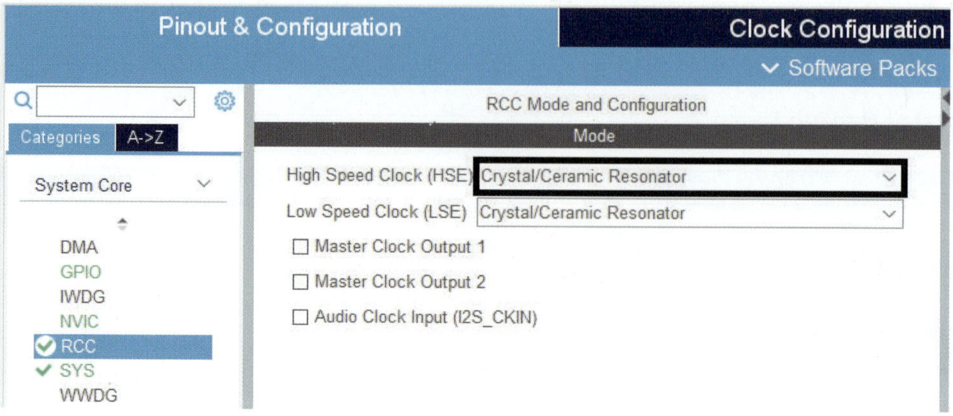

[그림 3.15_a] 클럭 설정

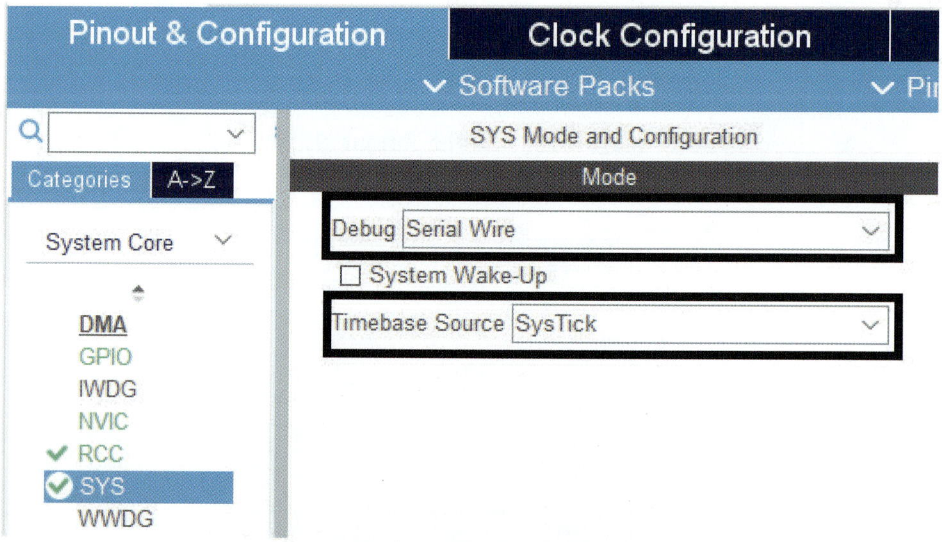

[그림 3.15_b] 시리얼 디버그 및 타임베이스 설정

8. GPIO 설정

LED가 PA5에 연결되어 있으므로 GPIO 출력 포트로 설정하면 된다. [그림 3.16], [그림 3.17], [그림 3.18]은 GPIO_Output으로 설정을 나타내고 있다.

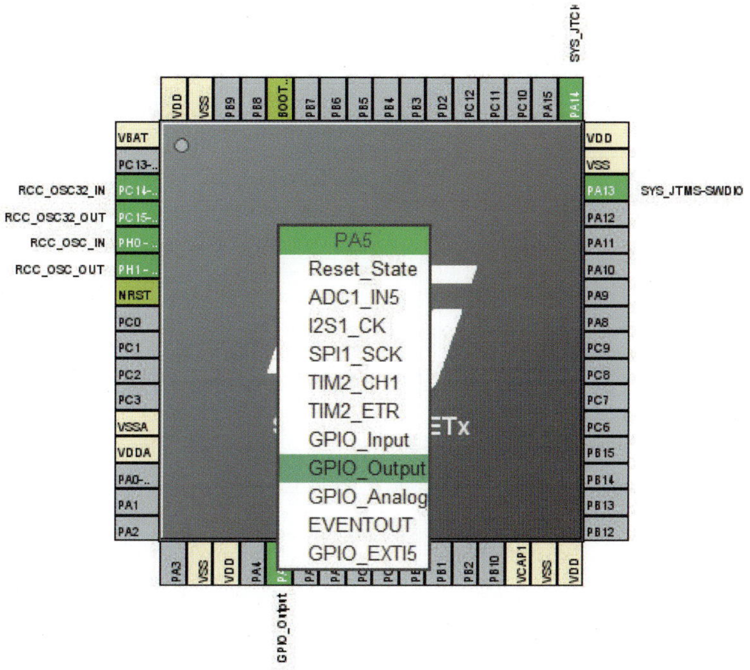

[그림 3.16] NUCLEO-F411RE PA5 출력 설정

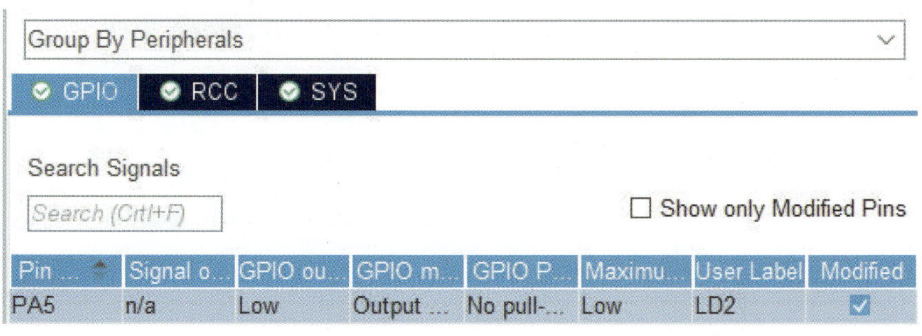

[그림 3.17] NUCLEO-F411RE PA5 출력 설정

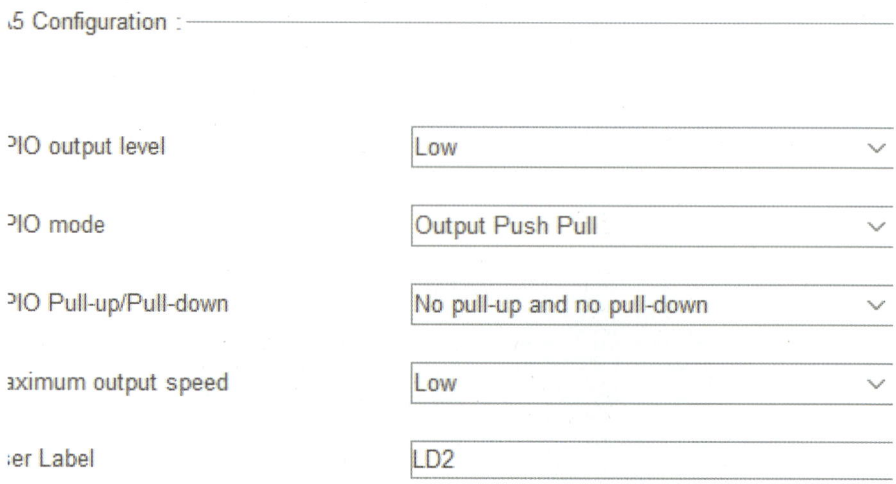

[그림 3.18] NUCLEO-F411RE PA5 출력 설정

9. Generate Code 생성

[그림 3.19]와 [그림 3.20]은 코드를 생성하기 위한 화면이다. 순차적으로 클릭한다.

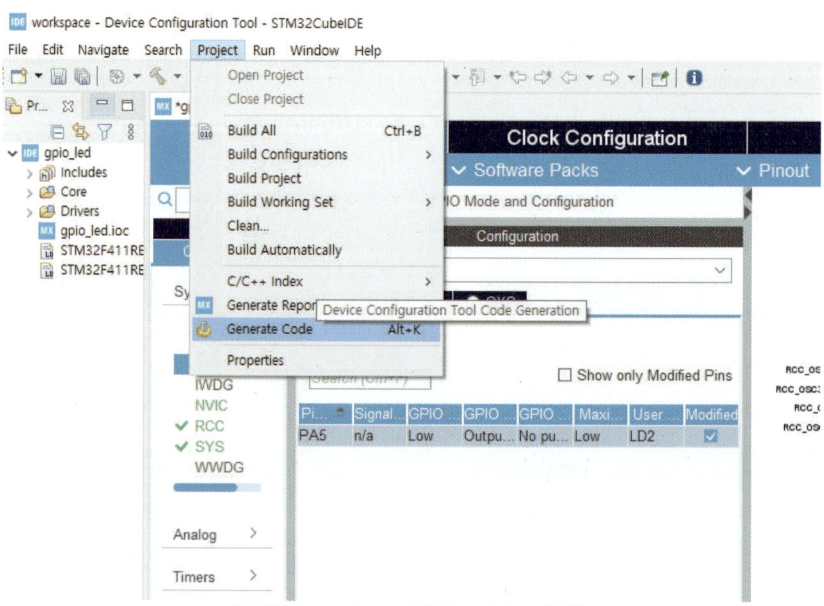

[그림 3.19] 프로젝트 코드를 생성

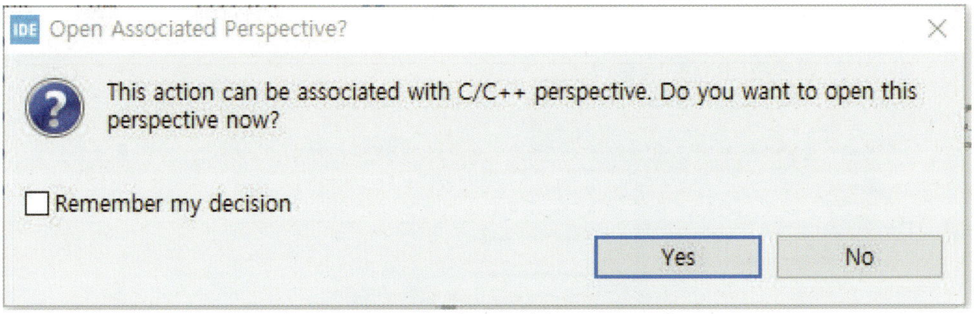
Open Associated Perspective?
This action can be associated with C/C++ perspective. Do you want to open this perspective now?
Remember my decision
Yes
No

[그림 3.20] 프로젝트 코드를 생성

10. Source Code 작성하기

소스 코드는 [그림 3.21]의 좌측 프로젝트 폴더에서 Core▷Src▷main.c에 작성한다. Main.c 파일을 위에서부터 아래까지 한 번 보면 함수 2개가 전역변수로 선언되었고, 메인에는 사용할 함수가 선언되었으며, 메인 아래에는 함수들이 작성되어 있다. 일반적으로 소스 코드는 while(1) 루프에 작성한다.

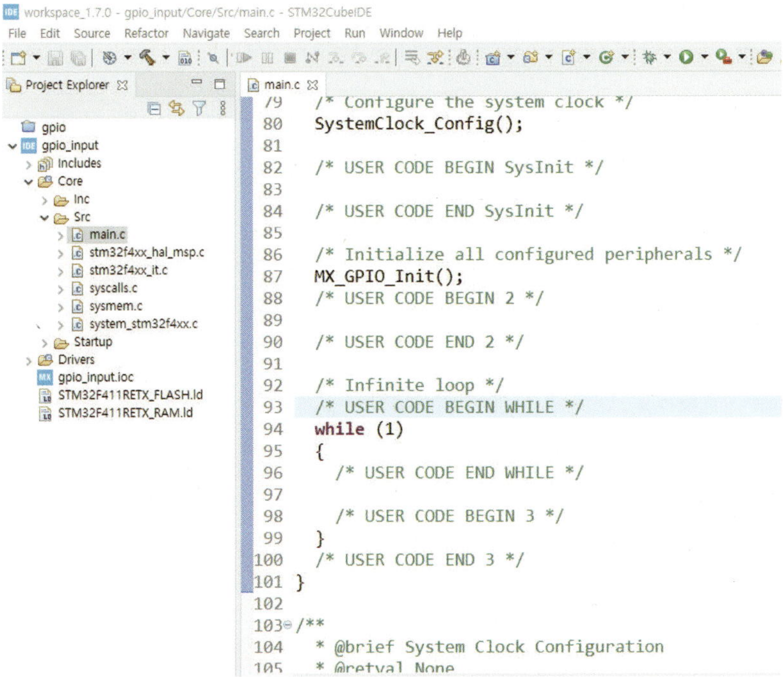
workspace_1.7.0 - gpio_input/Core/Src/main.c - STM32CubeIDE
File Edit Source Refactor Navigate Search Project Run Window Help
Project Explorer
gpio
gpio_input
Includes
Core
Inc
Src
main.c
stm32f4xx_hal_msp.c
stm32f4xx_it.c
syscalls.c
sysmem.c
system_stm32f4xx.c
Startup
Drivers
gpio_input.ioc
STM32F411RETX_FLASH.ld
STM32F411RETX_RAM.ld
main.c
/* Configure the system clock */
SystemClock_Config();
/* USER CODE BEGIN SysInit */
/* USER CODE END SysInit */
/* Initialize all configured peripherals */
MX_GPIO_Init();
/* USER CODE BEGIN 2 */
/* USER CODE END 2 */
/* Infinite loop */
/* USER CODE BEGIN WHILE */
while (1)
{
/* USER CODE END WHILE */
/* USER CODE BEGIN 3 */
}
/* USER CODE END 3 */
}
/**
* @brief System Clock Configuration
* @retval None

[그림 3.21] 소스 코드 작성 위치

▶ GPIO 명령 사용하기 1: [그림 3.22]와 같이 마우스로 MX_GPIO_Init를 클릭한 다음 단축키 F3를 누른다.

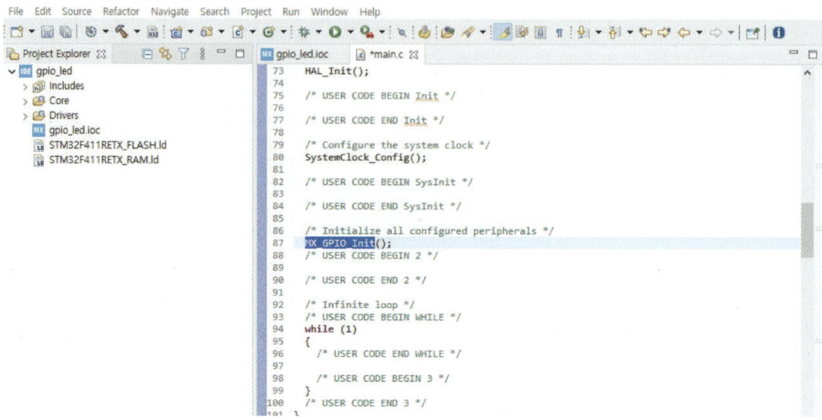

File Edit Source Refactor Navigate Search Project Run Window Help
Project Explorer
gpio_led
Includes
Core
Drivers
gpio_led.ioc
STM32F411RETX_FLASH.ld
STM32F411RETX_RAM.ld
gpio_led.ioc
*main.c
HAL_Init();
/* USER CODE BEGIN Init */
/* USER CODE END Init */
/* Configure the system clock */
SystemClock_Config();
/* USER CODE BEGIN SysInit */
/* USER CODE END SysInit */
/* Initialize all configured peripherals */
MX_GPIO_Init();
/* USER CODE BEGIN 2 */
/* USER CODE END 2 */
/* Infinite loop */
/* USER CODE BEGIN WHILE */
while (1)
{
/* USER CODE END WHILE */
/* USER CODE BEGIN 3 */
}
/* USER CODE END 3 */

[그림 3.22] GPIO 명령어 찾기

▶ GPIO 명령 사용하기 2: [그림3.23]과 같이 마우스로 HAL_GPIO_WritePin를 클릭한 다음 단축키 F3를 누른다.

```
/**
  * @brief  Sets or clears the selected data port bit.
  *
  * @note   This function uses GPIOx_BSRR register to allow atomic read/modify
  *         accesses. In this way, there is no risk of an IRQ occurring between
  *         the read and the modify access.
  *
  * @param  GPIOx where x can be (A..K) to select the GPIO peripheral for STM32F429X device or
  *                      x can be (A..I) to select the GPIO peripheral for STM32F40XX and STM32F427X devices.
  * @param  GPIO_Pin specifies the port bit to be written.
  *          This parameter can be one of GPIO_PIN_x where x can be (0..15).
  * @param  PinState specifies the value to be written to the selected bit.
  *          This parameter can be one of the GPIO_PinState enum values:
  *            @arg GPIO_PIN_RESET: to clear the port pin
  *            @arg GPIO_PIN_SET: to set the port pin
  * @retval None
  */
void HAL_GPIO_WritePin(GPIO_TypeDef* GPIOx, uint16_t GPIO_Pin, GPIO_PinState PinState)
{
  /* Check the parameters */
  assert_param(IS_GPIO_PIN(GPIO_Pin));
  assert_param(IS_GPIO_PIN_ACTION(PinState));

  if(PinState != GPIO_PIN_RESET)
  {
    GPIOx->BSRR = GPIO_Pin;
  }
  else
  {
    GPIOx->BSRR = (uint32_t)GPIO_Pin << 16U;
  }
}

/**
  * @brief  Toggles the specified GPIO pins.
  * @param  GPIOx Where x can be (A..K) to select the GPIO peripheral for STM32F429X device or
  *                      x can be (A..I) to select the GPIO peripheral for STM32F40XX and STM32F427X devices.
  * @param  GPIO_Pin Specifies the pins to be toggled.
  * @retval None
  */
void HAL_GPIO_TogglePin(GPIO_TypeDef* GPIOx, uint16_t GPIO_Pin)
{
```

[그림 3.23] GPIO 명령어 찾기

▶ GPIO 명령 사용하기 3: [그림3.24]는 HAL_GPIO_WritePin 함수를 나타내고 그 아래에는 HAL_GPIO_TogglePin 함수도 있다.

```
/**
  * @brief  Sets or clears the selected data port bit.
  *
  * @note   This function uses GPIOx_BSRR register to allow atomic read/modify
  *         accesses. In this way, there is no risk of an IRQ occurring between
  *         the read and the modify access.
  *
  * @param  GPIOx where x can be (A..K) to select the GPIO peripheral for STM32F429X device or
  *                      x can be (A..I) to select the GPIO peripheral for STM32F40XX and STM32F427X devices.
  * @param  GPIO_Pin specifies the port bit to be written.
  *          This parameter can be one of GPIO_PIN_x where x can be (0..15).
  * @param  PinState specifies the value to be written to the selected bit.
  *          This parameter can be one of the GPIO_PinState enum values:
  *            @arg GPIO_PIN_RESET: to clear the port pin
  *            @arg GPIO_PIN_SET: to set the port pin
  * @retval None
  */
void HAL_GPIO_WritePin(GPIO_TypeDef* GPIOx, uint16_t GPIO_Pin, GPIO_PinState PinState)
{
  /* Check the parameters */
  assert_param(IS_GPIO_PIN(GPIO_Pin));
  assert_param(IS_GPIO_PIN_ACTION(PinState));

  if(PinState != GPIO_PIN_RESET)
  {
    GPIOx->BSRR = GPIO_Pin;
  }
  else
  {
    GPIOx->BSRR = (uint32_t)GPIO_Pin << 16U;
  }
}

/**
  * @brief  Toggles the specified GPIO pins.
  * @param  GPIOx Where x can be (A..K) to select the GPIO peripheral for STM32F429X device or
  *                      x can be (A..I) to select the GPIO peripheral for STM32F40XX and STM32F427X devices.
  * @param  GPIO_Pin Specifies the pins to be toggled.
  * @retval None
  */
void HAL_GPIO_TogglePin(GPIO_TypeDef* GPIOx, uint16_t GPIO_Pin)
{
```

[그림 3.24] GPIO 명령어 찾기

▶ GPIO 명령 사용하기 4: GPIO 정리

```
GPIOx :x= (A..K)
GPIO_Pin : GPIO_PIN_x(x= (0..15))
PinState :  GPIO_PinState(GPIO_PIN_RESET: 0, GPIO_PIN_SET: 1)
>HAL_GPIO_WritePin(GPIO_TypeDef* GPIOx, uint16_t GPIO_Pin, GPIO_PinState PinState);
>PORT PB0을 0으로 출력하기 위한 설정
 ex)HAL_GPIO_WritePin(GPIOB, GPIO_PIN_0, GPIO_PIN_RESET);
>PORT PB0을 1으로 출력하기 위한 설정
 ex)HAL_GPIO_WritePin(GPIOB, GPIO_PIN_0, GPIO_PIN_SET);
```

>**HAL_GPIO_TogglePin**(GPIO_TypeDef* GPIOx, uint16_t GPIO_Pin);
>PORT PB0을 주기적으로 1과 0으로 출력하기 위한 설정
ex)**HAL_GPIO_TogglePin**(GPIOB, GPIO_PIN0);

▶ HAL_Delay() 함수 사용하기

HAL_D를 입력하고 <u>단축키 **CTRL+Space bar**</u>를 치면 관련 함수들이 나온다[그림 3.25]. 해당 함수를 선택하면 된다. **HAL_Delay**(uint32_t Delay)함수는 SysTick timer를 이용하고 기본 단위가 ms 이다. ex) 100ms , HAL_Delay(100)

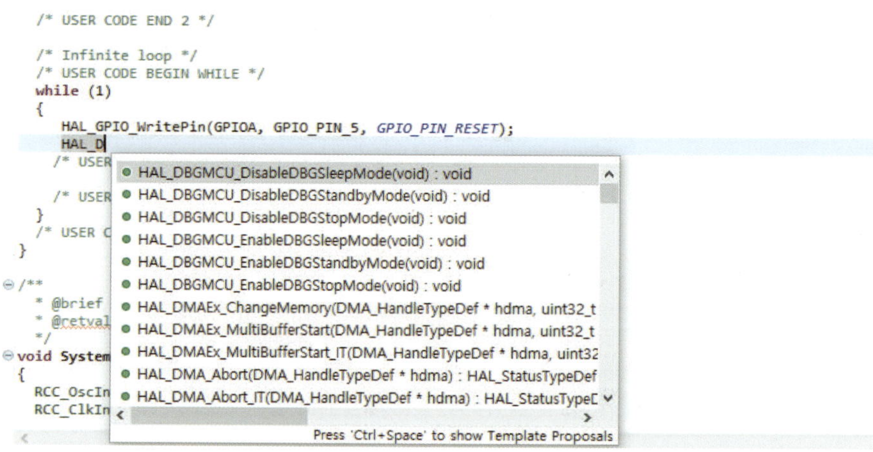

/* USER CODE END 2 */
/* Infinite loop */
/* USER CODE BEGIN WHILE */
while (1)
{
HAL_GPIO_WritePin(GPIOA, GPIO_PIN_5, GPIO_PIN_RESET);
HAL_D
HAL_DBGMCU_DisableDBGSleepMode(void) : void
HAL_DBGMCU_DisableDBGStandbyMode(void) : void
HAL_DBGMCU_DisableDBGStopMode(void) : void
HAL_DBGMCU_EnableDBGSleepMode(void) : void
HAL_DBGMCU_EnableDBGStandbyMode(void) : void
HAL_DBGMCU_EnableDBGStopMode(void) : void
HAL_DMAEx_ChangeMemory(DMA_HandleTypeDef * hdma, uint32_t
HAL_DMAEx_MultiBufferStart(DMA_HandleTypeDef * hdma, uint32_t
HAL_DMAEx_MultiBufferStart_IT(DMA_HandleTypeDef * hdma, uint32
HAL_DMA_Abort(DMA_HandleTypeDef * hdma) : HAL_StatusTypeDef
HAL_DMA_Abort_IT(DMA_HandleTypeDef * hdma) : HAL_StatusTypeD
Press 'Ctrl+Space' to show Template Proposals

[그림 3.25] GPIO 명령어 찾기

▶ Main 함수에 프로그램 작성하기

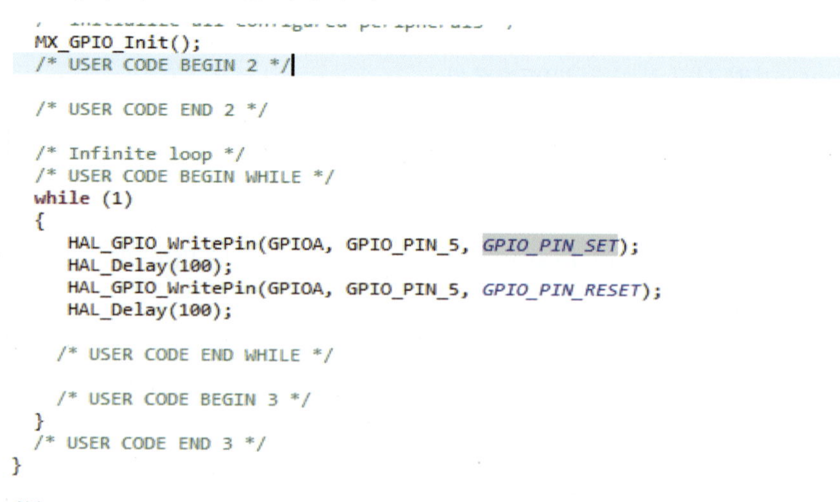

MX_GPIO_Init();
/* USER CODE BEGIN 2 */
/* USER CODE END 2 */
/* Infinite loop */
/* USER CODE BEGIN WHILE */
while (1)
{
HAL_GPIO_WritePin(GPIOA, GPIO_PIN_5, GPIO_PIN_SET);
HAL_Delay(100);
HAL_GPIO_WritePin(GPIOA, GPIO_PIN_5, GPIO_PIN_RESET);
HAL_Delay(100);
/* USER CODE END WHILE */
/* USER CODE BEGIN 3 */
}
/* USER CODE END 3 */
}

[그림 3.26] 메인 함수 작성

11. 디버그 및 다운로드하기

메인 함수에 프로그램이 작성되었다면 다음으로 디버그는 [그림 3.27]의  를 클릭하거나 단축키 F11을 누르면 디버깅과 프로그램 다운로드가 완료된다[그림 3.28], [그림 3.31].

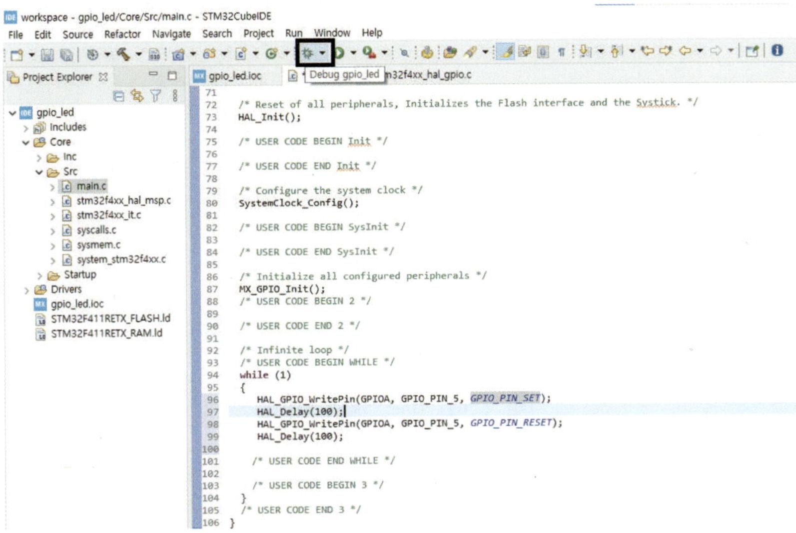

[그림 3.27] 프로그램 디버그

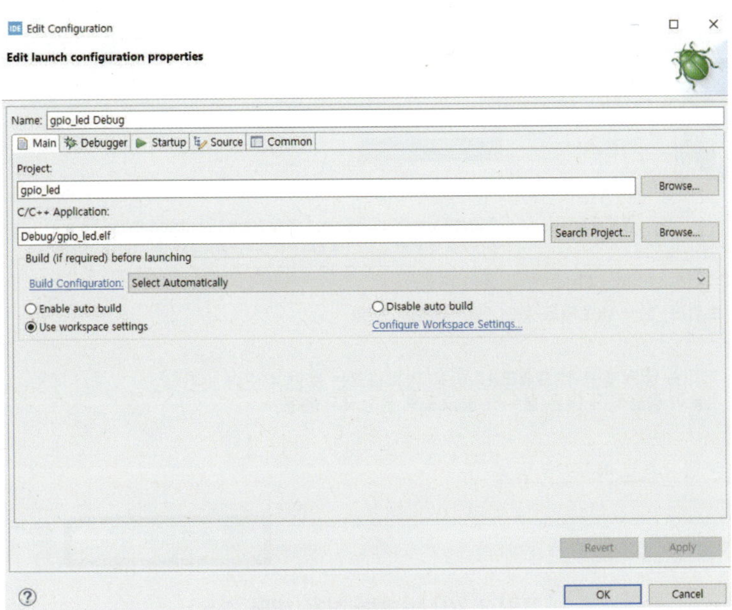

[그림 3.28] 프로그램 디버그

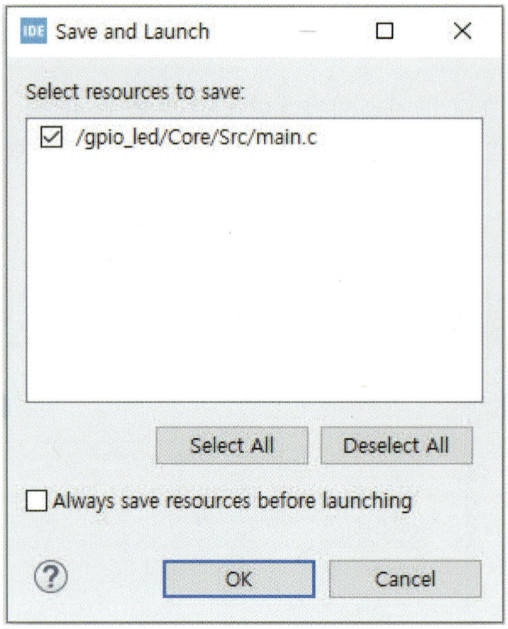

[그림 3.29] 프로그램 디버그

[그림 3.30] 방화벽 허용(해당 경우)

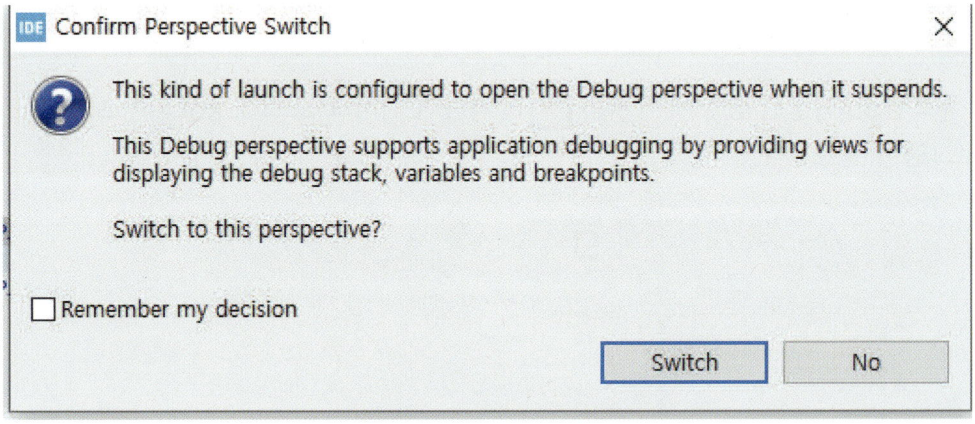
Confirm Perspective Switch
This kind of launch is configured to open the Debug perspective when it suspends.
This Debug perspective supports application debugging by providing views for displaying the debug stack, variables and breakpoints.
Switch to this perspective?
Remember my decision
Switch
No

[그림 3.31] 프로그램 디버그

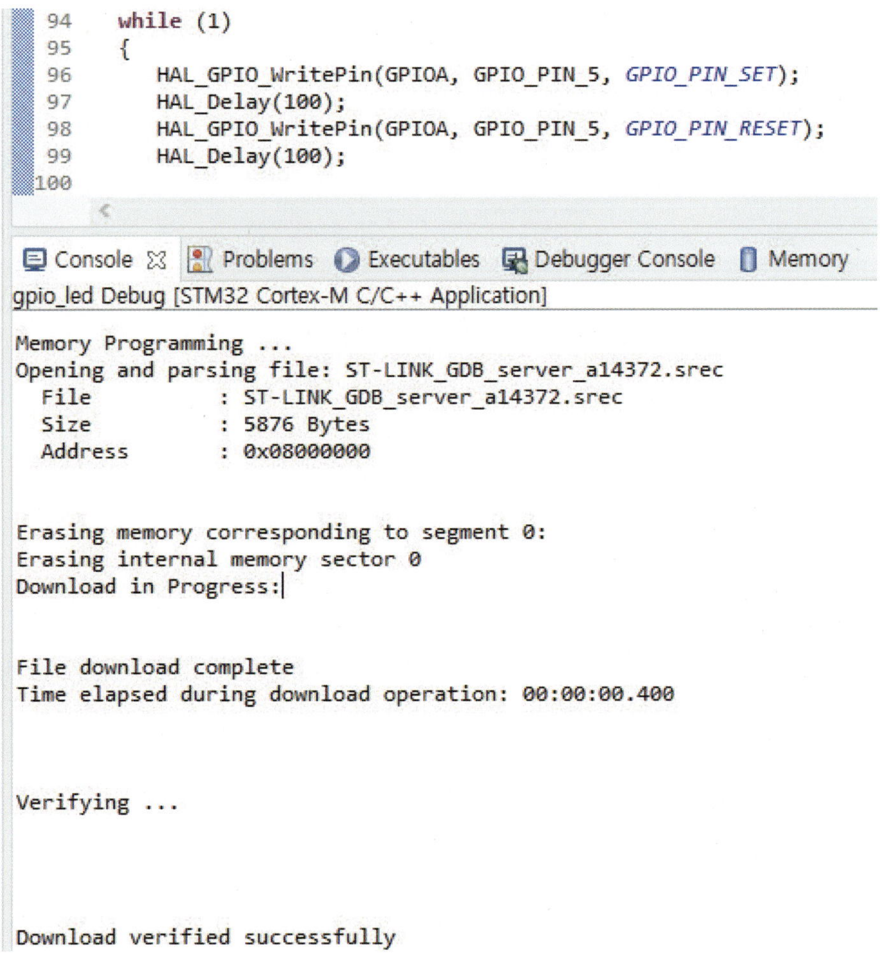
while (1)
{
HAL_GPIO_WritePin(GPIOA, GPIO_PIN_5, GPIO_PIN_SET);
HAL_Delay(100);
HAL_GPIO_WritePin(GPIOA, GPIO_PIN_5, GPIO_PIN_RESET);
HAL_Delay(100);
Console
Problems
Executables
Debugger Console
Memory
gpio_led Debug [STM32 Cortex-M C/C++ Application]
Memory Programming ...
Opening and parsing file: ST-LINK_GDB_server_a14372.srec
File : ST-LINK_GDB_server_a14372.srec
Size : 5876 Bytes
Address : 0x08000000
Erasing memory corresponding to segment 0:
Erasing internal memory sector 0
Download in Progress:
File download complete
Time elapsed during download operation: 00:00:00.400
Verifying ...
Download verified successfully

[그림 3.32] Console 창 내용

12. 보드 프로그램 실행하기

[그림 3.33]상단 중앙에 아이콘 클릭하면 디버그 모드가 종료되어 보드의 프로그램이 실행된다.

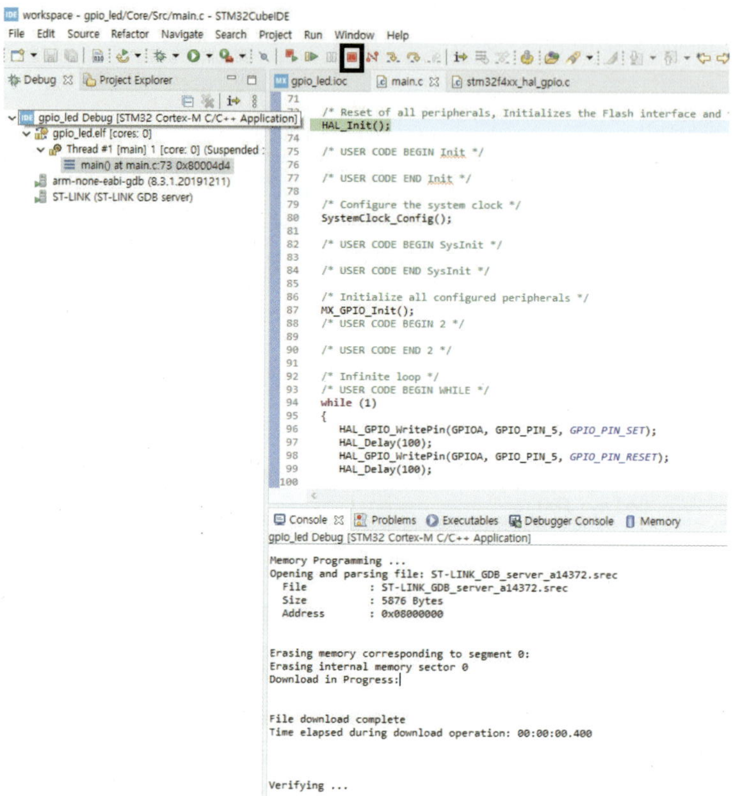
workspace - gpio_led/Core/Src/main.c - STM32CubeIDE
File Edit Source Refactor Navigate Search Project Run Window Help
Debug
Project Explorer
gpio_led.ioc
main.c
stm32f4xx_hal_gpio.c
gpio_led Debug [STM32 Cortex-M C/C++ Application]
gpio_led.elf [cores: 0]
Thread #1 [main] 1 [core: 0] (Suspended :
main() at main.c:73 0x80004d4
arm-none-eabi-gdb (8.3.1.20191211)
ST-LINK (ST-LINK GDB server)
/* Reset of all peripherals, Initializes the Flash interface and
HAL_Init();
/* USER CODE BEGIN Init */
/* USER CODE END Init */
/* Configure the system clock */
SystemClock_Config();
/* USER CODE BEGIN SysInit */
/* USER CODE END SysInit */
/* Initialize all configured peripherals */
MX_GPIO_Init();
/* USER CODE BEGIN 2 */
/* USER CODE END 2 */
/* Infinite loop */
/* USER CODE BEGIN WHILE */
while (1)
{
HAL_GPIO_WritePin(GPIOA, GPIO_PIN_5, GPIO_PIN_SET);
HAL_Delay(100);
HAL_GPIO_WritePin(GPIOA, GPIO_PIN_5, GPIO_PIN_RESET);
HAL_Delay(100);
Console
Problems
Executables
Debugger Console
Memory
gpio_led Debug [STM32 Cortex-M C/C++ Application]
Memory Programming ...
Opening and parsing file: ST-LINK_GDB_server_a14372.srec
File : ST-LINK_GDB_server_a14372.srec
Size : 5876 Bytes
Address : 0x08000000
Erasing memory corresponding to segment 0:
Erasing internal memory sector 0
Download in Progress:
File download complete
Time elapsed during download operation: 00:00:00.400
Verifying ...

[그림 3.33] 디버그 모드 종료 및 보드 실행

13. LED 4개 제어하기

[그림 3.34]와 같이 설정 및 Generate Code를 발생시키고 MX_GPIO_Init() 함수에 들어가 보면 GPIO 4개 출력하는 방법이 나와 있다. 같은 포트 경우 OR 연산자(|)를 사용하여 하나 이상을 출력하고 있다.

```
HAL_GPIO_WritePin(GPIOA, LD2_Pin|GPIO_PIN_6|GPIO_PIN_7, GPIO_PIN_RESET);
HAL_GPIO_WritePin(GPIOC, GPIO_PIN_4, GPIO_PIN_RESET);
```

LD2_Pin을 마우스로 선택 한 다음 단축키 F3 누르면 [그림3.35]과 같이 창이 뜬다. 프로그램을 [그림 3.36] 같이 작성하고 하드웨어를 구성하고 디거깅 및 실행을 해본다.

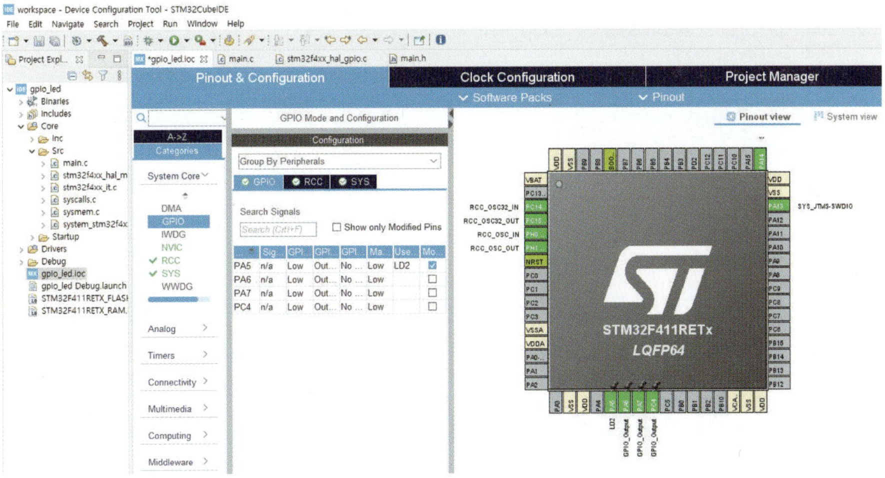

[그림 3.34] LED 4개 설정

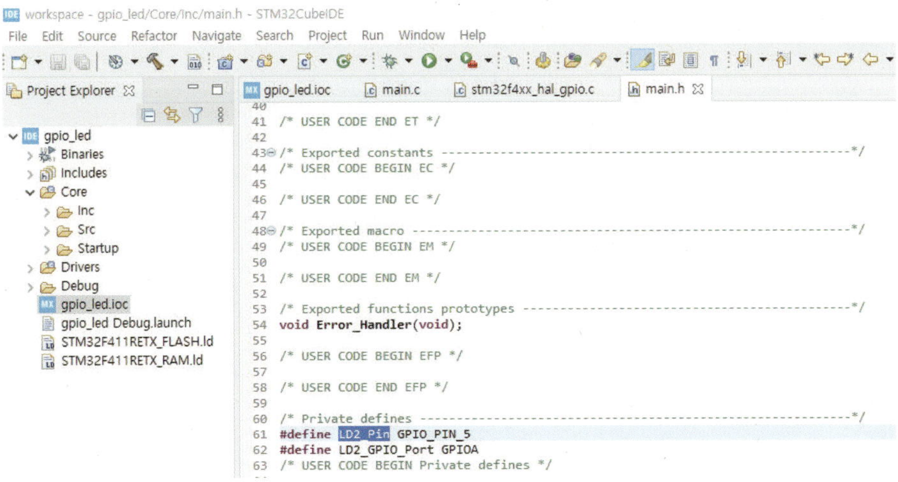

[그림 3.35] GPIO PIN 정의

```
/* Infinite loop */
/* USER CODE BEGIN WHILE */
while (1)
{
   HAL_GPIO_WritePin(GPIOA, LD2_Pin|GPIO_PIN_6|GPIO_PIN_7, GPIO_PIN_SET);
   HAL_GPIO_WritePin(GPIOC, GPIO_PIN_4, GPIO_PIN_SET);
   HAL_Delay(100);
   HAL_GPIO_WritePin(GPIOA, LD2_Pin|GPIO_PIN_6|GPIO_PIN_7, GPIO_PIN_RESET);
   HAL_GPIO_WritePin(GPIOC, GPIO_PIN_4, GPIO_PIN_RESET);
   HAL_Delay(100);

  /* USER CODE END WHILE */

  /* USER CODE BEGIN 3 */
}
/* USER CODE END 3 */
```

[그림 3.36] 프로그램 작성

<GPIO 입력 기능>

1. STM32 프로젝트 시작

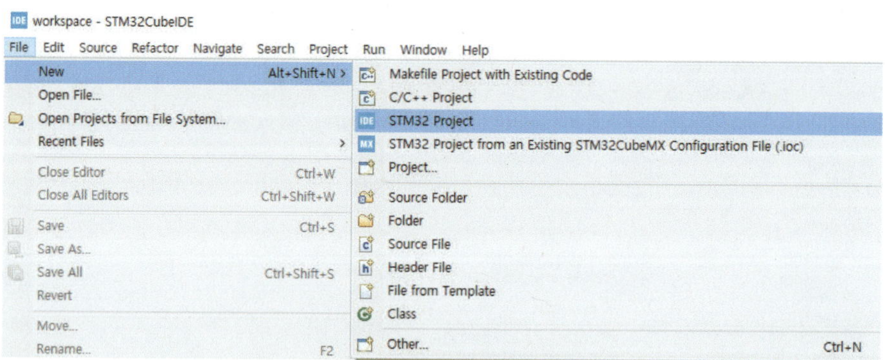

[그림 3b.1] 프로그램 시작

2. 보드 선정: NUCLEO–F411RE

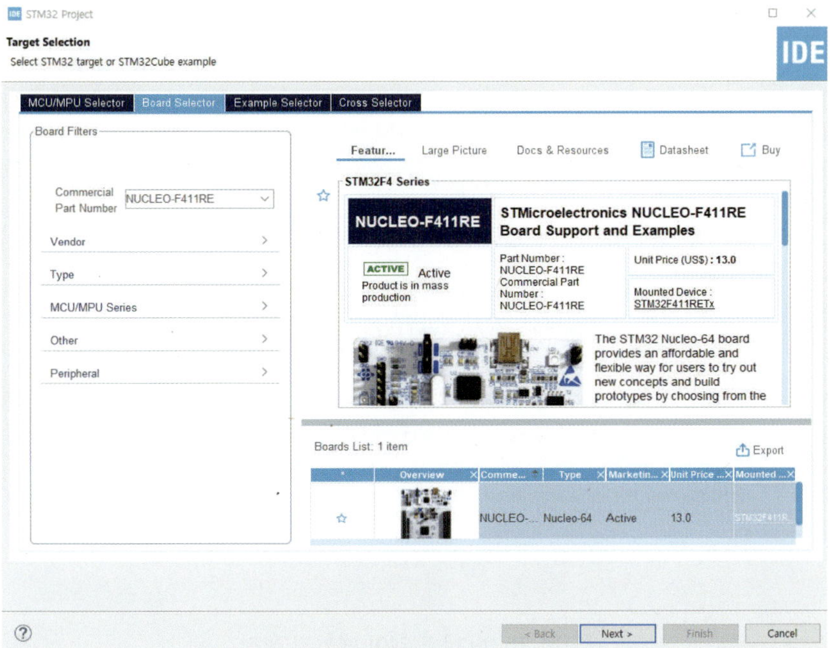

[그림 3b.2] 보드 선정

3. 프로젝트 이름 결정 및 보드 구성 형태로 초기화:

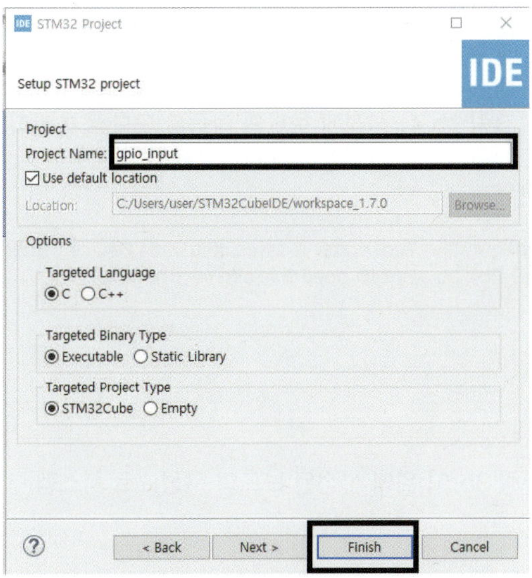

[그림 3b.3] 프로젝트 이름 결정

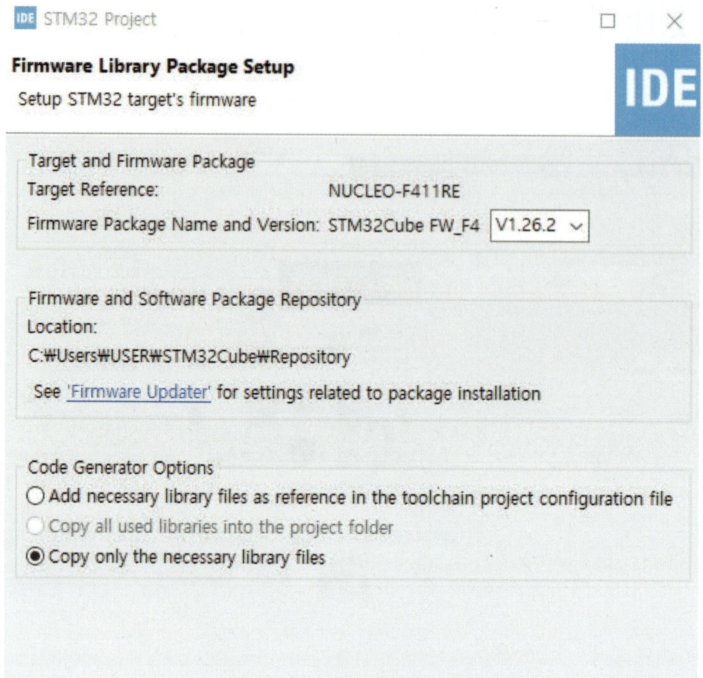

[그림 3b.3-1] 펌웨어 팩키지 설정

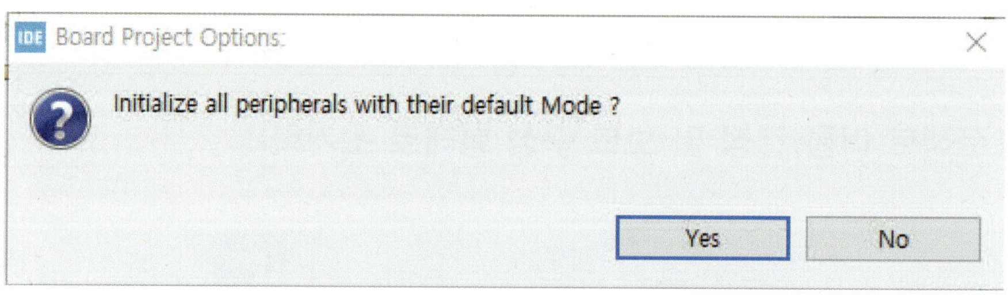

[그림 3b.4] 선정된 보드의 디폴트 모드로 초기화

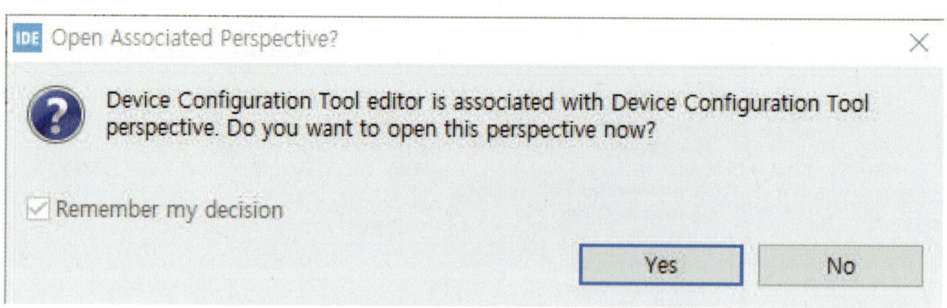

[그림 3b.5] 위의 선정된 보드의 구성으로 시스템 구성

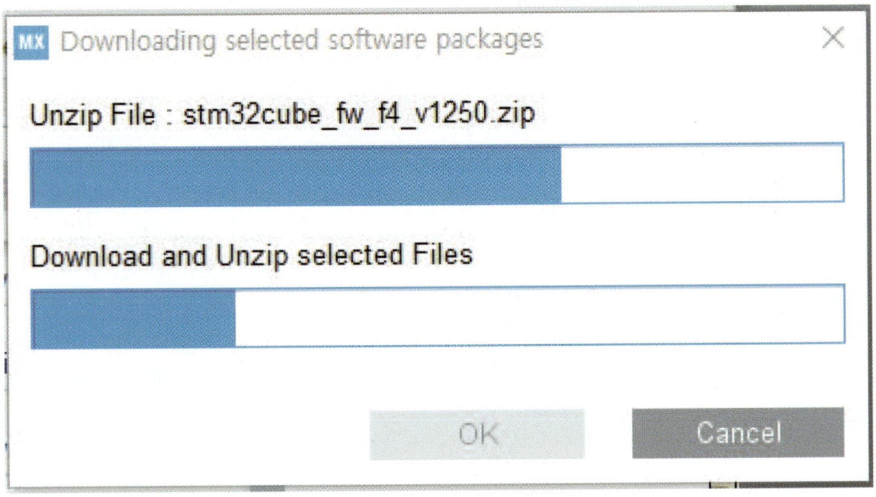

[그림 3b.6] 시스템 구성 진행

4. 실습 보드 확인

실습 보드는 포트 PA5에 LED 1개와 포트 PC13에 push Botton 1개가 연결되어 있다. 본 실습에서는 폴링 방식 스위치의 기능에 대해서 실습한다.

> LED GPIO 출력: LD2 : PA5, Yellow color
> Push Botton 입력: B1:PC13, Blue Tack switch

5. 클럭 설정

[그림 3b.7]~[그림 3b.8]은 클럭을 설정한 경우를 나타낸다. 메인 클럭은 외부 클럭(Crystal/Ceramic Resonator)과 내부 클럭[HSI RC(BYPASS Clock Source)] 둘 중에 하나를 이용하여야 한다. 내부 클럭보다는 외부 클럭이 정확하며, 본 보드에서는 Crystal/Ceramic Resonator는 연결되어 있지 않아 사용하지 않으며 ST-LINK MCO 출력 8MHz를 사용한다.

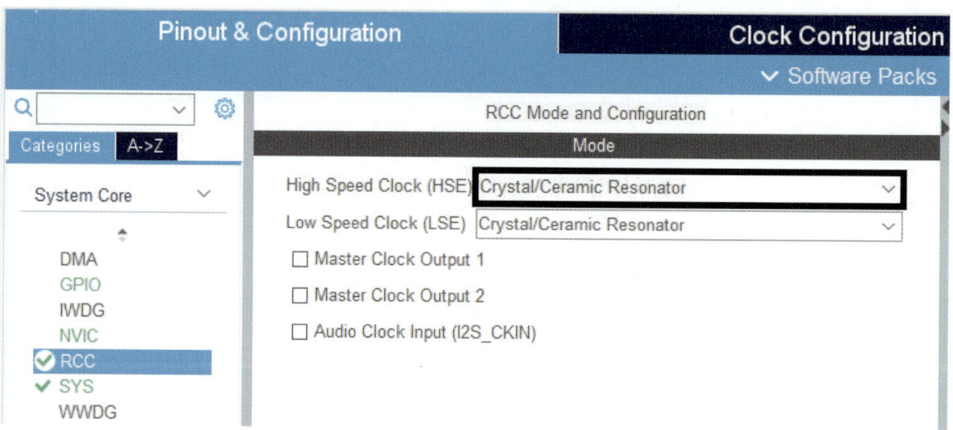

[그림 3b.7] NUCLEO-F411RE 클럭 설정

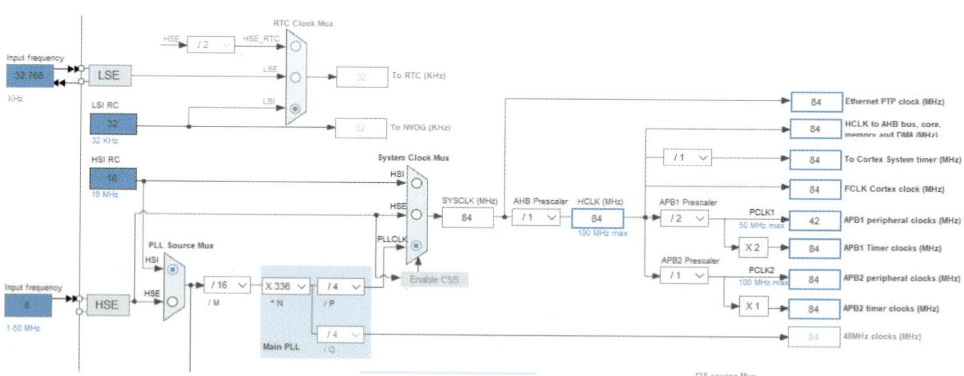

[그림 3b.8] NUCLEO-F411RE 디폴트 클럭설정

[그림 3b.8]은 주변 장치의 클럭이 할당된 것을 알 수 있다. [그림 3b.8]을 보면 APB1에 연결되어 있으면 42MHz, APB2에 연결되어 있으면 84MHz로 동작함을 알 수 있다.

6. Pinout 초기화 설정

보드의 기능을 새로 사용자에 맞도록 하는 경우로 [그림 3b.9]의 경우처럼 핀아웃을 클리어하고 사용자에 맞게 새로 환경 설정을 하는 경우에 해당한다. 이 경우 결과는 [그림 3b.10]과 같이 된다.

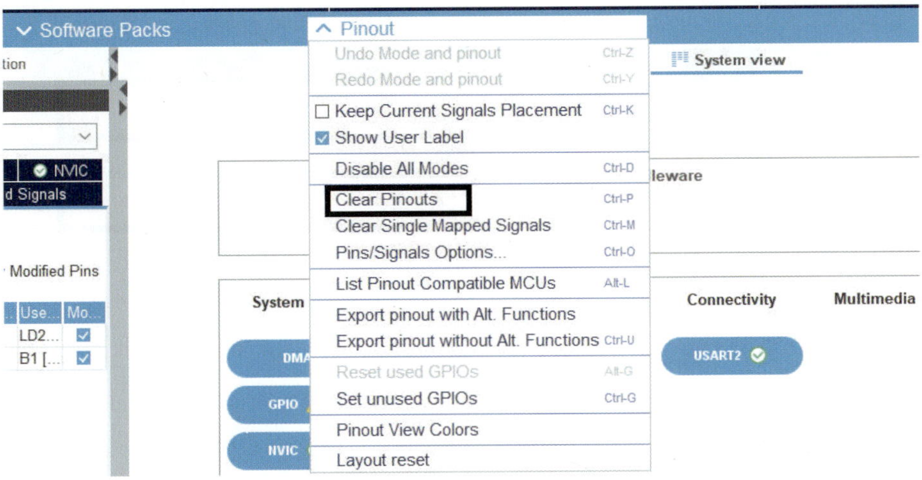

[그림 3b.9] 디바이스 Pinout설정

[그림 3b.10] 디바이스 Pinout 초기화 화면

7. 디바이스 기본 설정

클럭은 [그림 3b.7]와 같이 설정하고 시스템을 아래 [그림3b.11]과 같이 설정한다.

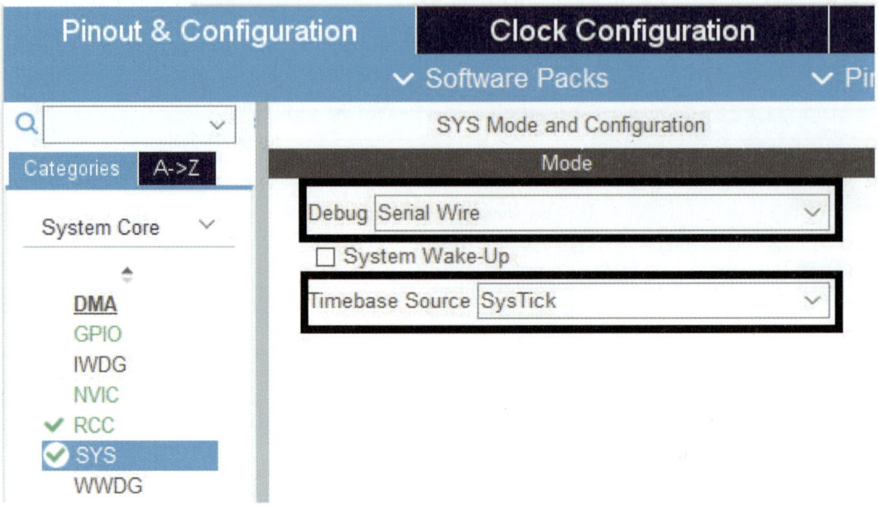

[그림 3b.11] 시리얼 디버그 및 타임베이스 설정

8. GPIO 설정

LED가 PA5에 연결되어 있으므로 GPIO 출력 포트로 설정하면 된다. [그림 3b.12]~[그림 3b.14]은 GPIO_input과 GPIO_Output 설정을 나타내고 있다.

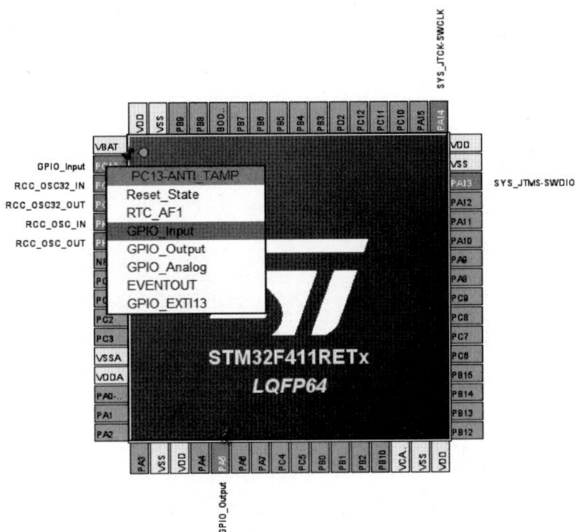

[그림 3b.12] NUCLEO-F411RE PA5 출력, PC13 입력 설정

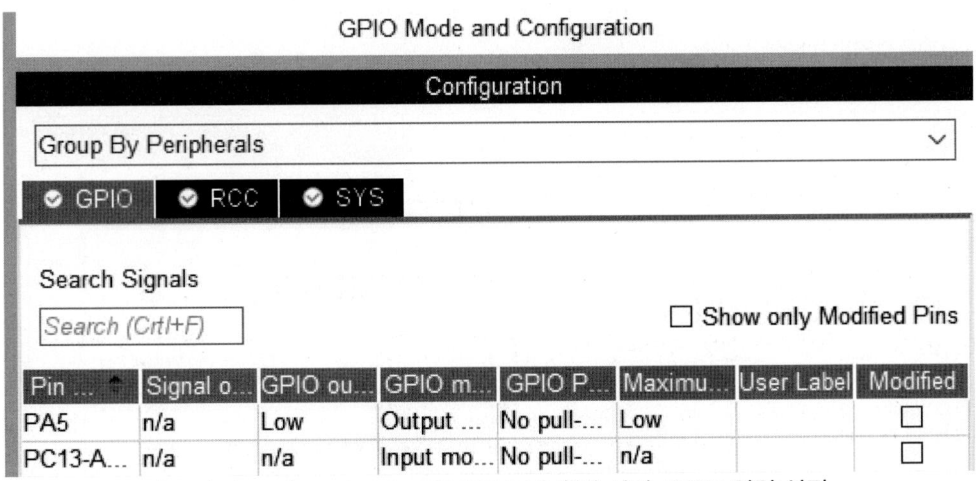

[그림 3b.13] NUCLEO-F411RE PA5 출력 설정, PC13 입력 설정

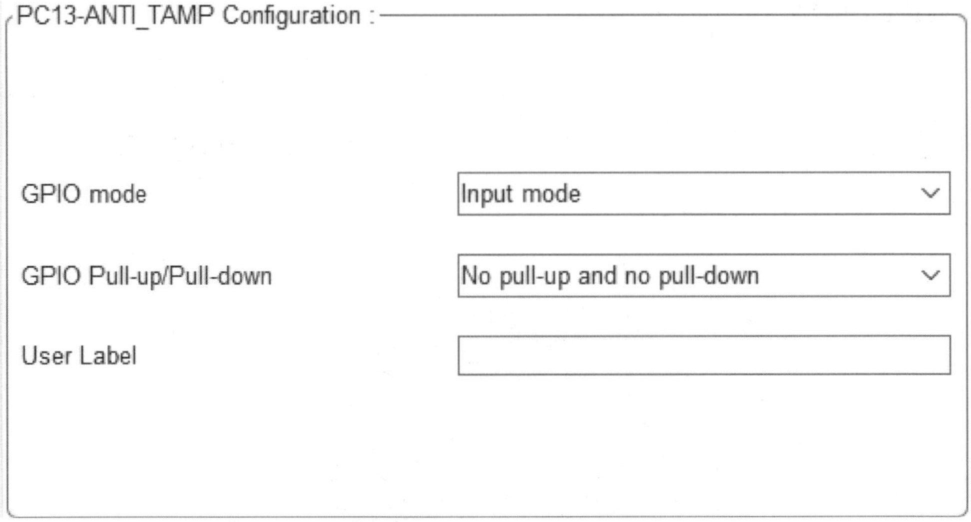

[그림 3b.14] NUCLEO-F411RE PC13 입력 설정

9. Generate Code 생성

[그림 3b.15]와 [그림 3b.16]은 코드를 생성하기 위한 화면이다. 순차적으로 클릭한다.

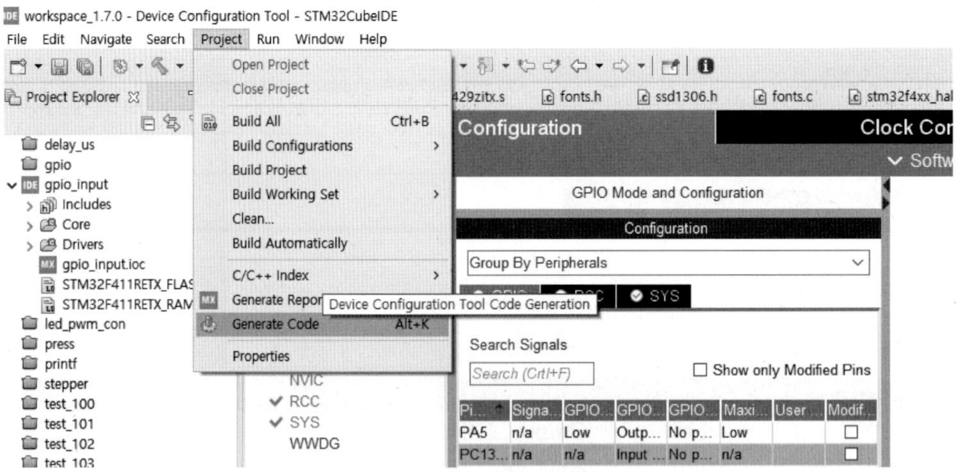

[그림 3b.15] 프로젝트 코드를 생성

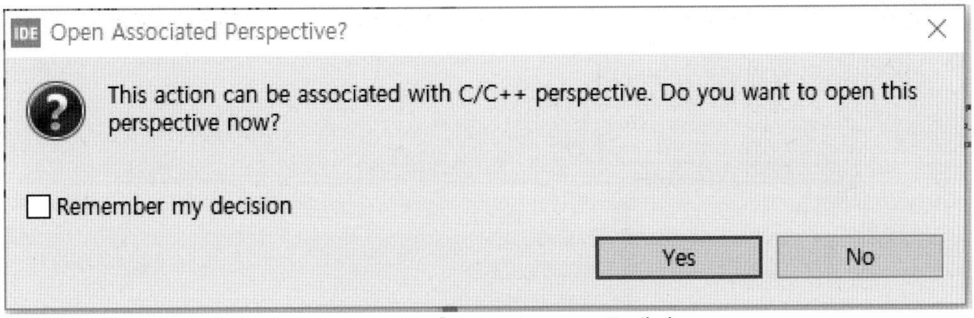
Open Associated Perspective?
This action can be associated with C/C++ perspective. Do you want to open this perspective now?
Remember my decision
Yes
No

[그림 3b.16] 프로젝트 코드를 생성

10. Source Code 작성하기

소스 코드는 [그림 3b.17]의 좌측 프로젝트 폴더에서 Core▷Src▷main.c에 작성한다. Main.c 파일을 위에서부터 아래까지 한 번 보면 함수 2개가 전역변수로 선언되었고, 메인에는 사용할 함수가 선언되었으며 메인 아래에는 함수들이 작성되어 있다. 일반적으로 소스 코드는 while(1) 루프에 작성한다.

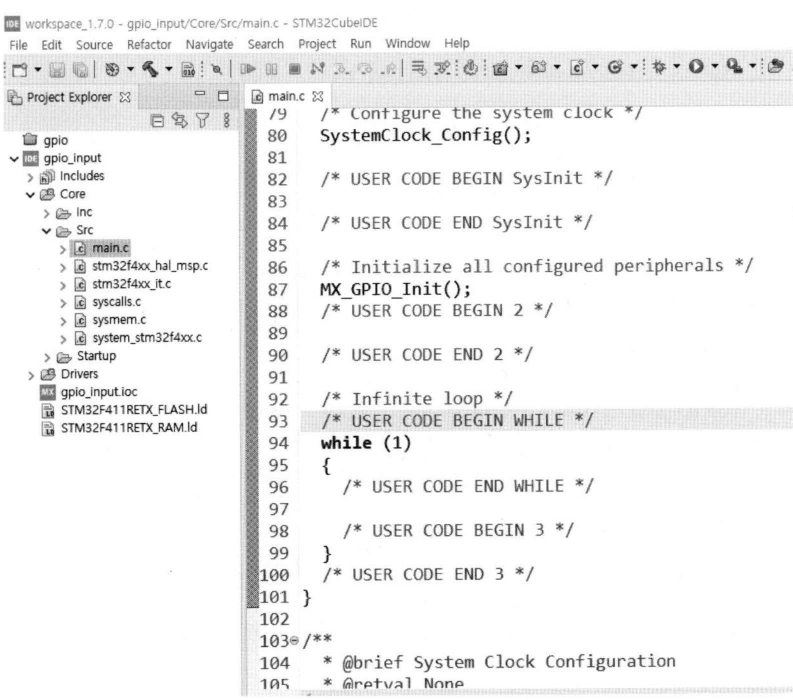
workspace_1.7.0 - gpio_input/Core/Src/main.c - STM32CubeIDE
File Edit Source Refactor Navigate Search Project Run Window Help
Project Explorer
gpio
gpio_input
Includes
Core
Inc
Src
main.c
stm32f4xx_hal_msp.c
stm32f4xx_it.c
syscalls.c
sysmem.c
system_stm32f4xx.c
Startup
Drivers
gpio_input.ioc
STM32F411RETX_FLASH.ld
STM32F411RETX_RAM.ld
main.c
SystemClock_Config();
/* USER CODE BEGIN SysInit */
/* USER CODE END SysInit */
/* Initialize all configured peripherals */
MX_GPIO_Init();
/* USER CODE BEGIN 2 */
/* USER CODE END 2 */
/* Infinite loop */
/* USER CODE BEGIN WHILE */
while (1)
{
/* USER CODE END WHILE */
/* USER CODE BEGIN 3 */
}
/* USER CODE END 3 */
}
/**
* @brief System Clock Configuration

[그림 3b.17] 소스코드 작성 위치

▶ GPIO 명령 사용하기 1: [그림 3b.18]와 같이 마우스로 **MX_GPIO_Init**를 클릭한 다음 단축키 F3를 누른다.

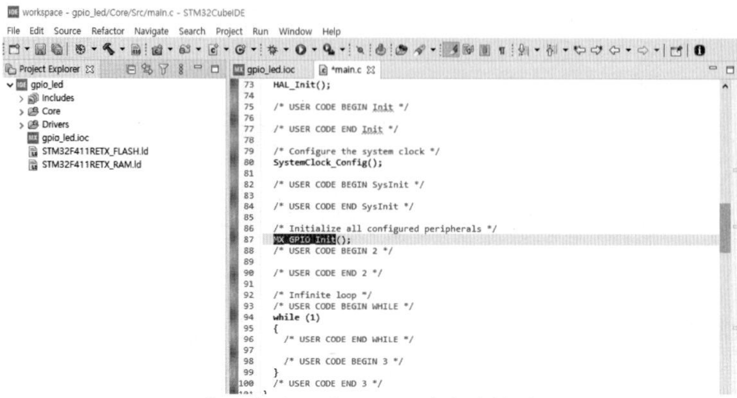

workspace - gpio_led/Core/Src/main.c - STM32CubeIDE
File Edit Source Refactor Navigate Search Project Run Window Help
Project Explorer
gpio_led
Includes
Core
Drivers
gpio_led.ioc
STM32F411RETX_FLASH.ld
STM32F411RETX_RAM.ld
gpio_led.ioc
*main.c
HAL_Init();
/* USER CODE BEGIN Init */
/* USER CODE END Init */
/* Configure the system clock */
SystemClock_Config();
/* USER CODE BEGIN SysInit */
/* USER CODE END SysInit */
/* Initialize all configured peripherals */
MX_GPIO_Init();
/* USER CODE BEGIN 2 */
/* USER CODE END 2 */
/* Infinite loop */
/* USER CODE BEGIN WHILE */
while (1)
{
/* USER CODE END WHILE */
/* USER CODE BEGIN 3 */
}
/* USER CODE END 3 */

[그림 3b.18] GPIO 명령어 찾기

▶ GPIO 명령 사용하기 2: [그림3b.19]과 같이 마우스로 **HAL_GPIO_WritePin**를 클릭한 다음 단축키 F3를 누른다.

```
static void MX_GPIO_Init(void)
{
  GPIO_InitTypeDef GPIO_InitStruct = {0};

  /* GPIO Ports Clock Enable */
  __HAL_RCC_GPIOC_CLK_ENABLE();
  __HAL_RCC_GPIOH_CLK_ENABLE();
  __HAL_RCC_GPIOA_CLK_ENABLE();

  /*Configure GPIO pin Output Level */
  HAL_GPIO_WritePin(GPIOA, GPIO_PIN_5, GPIO_PIN_RESET);

  /*Configure GPIO pin : PC13 */
  GPIO_InitStruct.Pin = GPIO_PIN_13;
  GPIO_InitStruct.Mode = GPIO_MODE_INPUT;
  GPIO_InitStruct.Pull = GPIO_NOPULL;
  HAL_GPIO_Init(GPIOC, &GPIO_InitStruct);

  /*Configure GPIO pin : PA5 */
  GPIO_InitStruct.Pin = GPIO_PIN_5;
  GPIO_InitStruct.Mode = GPIO_MODE_OUTPUT_PP;
  GPIO_InitStruct.Pull = GPIO_NOPULL;
  GPIO_InitStruct.Speed = GPIO_SPEED_FREQ_LOW;
  HAL_GPIO_Init(GPIOA, &GPIO_InitStruct);

}
```

[그림 3b.19] GPIO 명령어 찾기

▶GPIO 명령 사용하기 3: [그림3b.20]는 HAL_GPIO_WritePin 함수를 나타내고 그 아래에는 HAL_GPIO_TogglePin 함수 위에는 HAL_GPIO_ReadPin가 있다.

```
/**
  * @brief  Sets or clears the selected data port bit.
  *
  * @note   This function uses GPIOx_BSRR register to allow atomic read/modify
  *         accesses. In this way, there is no risk of an IRQ occurring between
  *         the read and the modify access.
  *
  * @param  GPIOx where x can be (A..K) to select the GPIO peripheral for STM32F429X device or
  *                      x can be (A..I) to select the GPIO peripheral for STM32F40XX and STM32F427X devices.
  * @param  GPIO_Pin specifies the port bit to be written.
  *          This parameter can be one of GPIO_PIN_x where x can be (0..15).
  * @param  PinState specifies the value to be written to the selected bit.
  *          This parameter can be one of the GPIO_PinState enum values:
  *            @arg GPIO_PIN_RESET: to clear the port pin
  *            @arg GPIO_PIN_SET: to set the port pin
  * @retval None
  */
void HAL_GPIO_WritePin(GPIO_TypeDef* GPIOx, uint16_t GPIO_Pin, GPIO_PinState PinState)
{
  /* Check the parameters */
  assert_param(IS_GPIO_PIN(GPIO_Pin));
  assert_param(IS_GPIO_PIN_ACTION(PinState));

  if(PinState != GPIO_PIN_RESET)
  {
    GPIOx->BSRR = GPIO_Pin;
  }
  else
  {
    GPIOx->BSRR = (uint32_t)GPIO_Pin << 16U;
  }
}

/**
  * @brief  Toggles the specified GPIO pins.
  * @param  GPIOx Where x can be (A..K) to select the GPIO peripheral for STM32F429X device or
  *                      x can be (A..I) to select the GPIO peripheral for STM32F40XX and STM32F427X devices.
  * @param  GPIO_Pin Specifies the pins to be toggled.
  * @retval None
  */
void HAL_GPIO_TogglePin(GPIO_TypeDef* GPIOx, uint16_t GPIO_Pin)
{
```

[그림 3b.20] GPIO 명령어 찾기 1

```
/**
  * @brief  Reads the specified input port pin.
  * @param  GPIOx where x can be (A..K) to select the GPIO peripheral for STM32F429X device or
  *                     x can be (A..I) to select the GPIO peripheral for STM32F40XX and STM32F427X devices.
  * @param  GPIO_Pin specifies the port bit to read.
  *         This parameter can be GPIO_PIN_x where x can be (0..15).
  * @retval The input port pin value.
  */
GPIO_PinState HAL_GPIO_ReadPin(GPIO_TypeDef* GPIOx, uint16_t GPIO_Pin)
{
  GPIO_PinState bitstatus;

  /* Check the parameters */
  assert_param(IS_GPIO_PIN(GPIO_Pin));

  if((GPIOx->IDR & GPIO_Pin) != (uint32_t)GPIO_PIN_RESET)
  {
    bitstatus = GPIO_PIN_SET;
  }
  else
  {
    bitstatus = GPIO_PIN_RESET;
  }
  return bitstatus;
```

[그림 3b.20] GPIO 명령어 찾기 2

▶GPIO 명령 사용하기 4: GPIO 정리

GPIOx :x= (A..K)

GPIO_Pin : GPIO_PIN_x(x= (0..15))

PinState : GPIO_PinState(GPIO_PIN_RESET: 0, GPIO_PIN_SET: 1)

> HAL_GPIO_WritePin(GPIO_TypeDef* GPIOx, uint16_t GPIO_Pin, GPIO_PinState PinState);

> PORT PB0을 0으로 출력하기 위한 설정

ex) HAL_GPIO_WritePin(GPIOB, GPIO_PIN_0, GPIO_PIN_RESET);

> PORT PB0을 1으로 출력하기 위한 설정

ex) HAL_GPIO_WritePin(GPIOB, GPIO_PIN_0, GPIO_PIN_SET);

> HAL_GPIO_TogglePin(GPIO_TypeDef* GPIOx, uint16_t GPIO_Pin);

> PORT PB0을 주기적으로 1과 0으로 출력하기 위한 설정

ex) HAL_GPIO_TogglePin(GPIOB, GPIO_PIN0);

> HAL_GPIO_ReadPin(GPIO_TypeDef* GPIOx, uint16_t GPIO_Pin)

> PORT PC13에 스위치 상태를 체크하는 설정

ex) if(!HAL_GPIO_ReadPin(GPIOC, GPIO_PIN_13)):만약 포트 PC13의 상태값이 Low (0) 수행

▶HAL_Delay() 함수 사용하기

HAL_D를 입력하고 단축키 CTRL+Space bar를 치면 관련 함수들이 나온다[그림 3b.21].해당 함수를 선택하면 된다. HAL_Delay(uint32_t Delay)함수는 SysTick timer를 이용하고 기본 단위가 ms 이다. ex) 100ms , HAL_Delay(100)

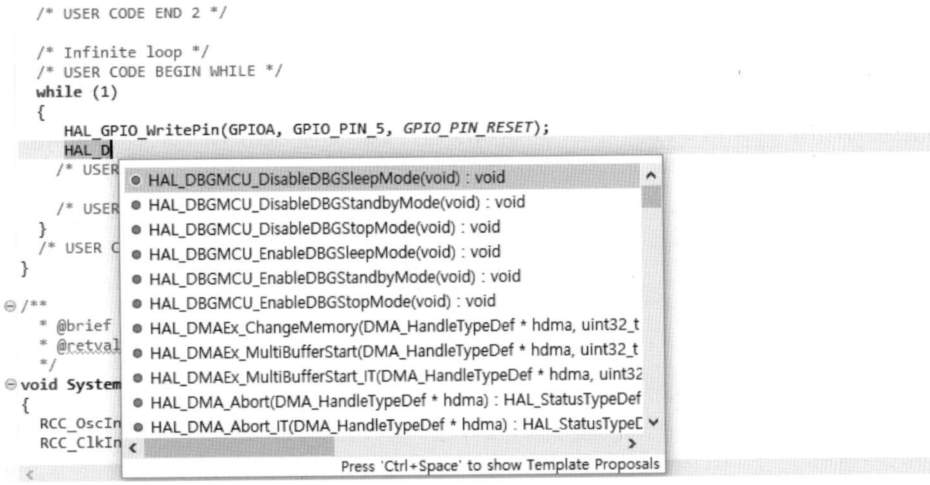
/* USER CODE END 2 */
/* Infinite loop */
/* USER CODE BEGIN WHILE */
while (1)
{
HAL_GPIO_WritePin(GPIOA, GPIO_PIN_5, GPIO_PIN_RESET);
HAL_D
HAL_DBGMCU_DisableDBGSleepMode(void) : void
HAL_DBGMCU_DisableDBGStandbyMode(void) : void
HAL_DBGMCU_DisableDBGStopMode(void) : void
HAL_DBGMCU_EnableDBGSleepMode(void) : void
HAL_DBGMCU_EnableDBGStandbyMode(void) : void
HAL_DBGMCU_EnableDBGStopMode(void) : void
HAL_DMAEx_ChangeMemory(DMA_HandleTypeDef * hdma, uint32_t
HAL_DMAEx_MultiBufferStart(DMA_HandleTypeDef * hdma, uint32_t
HAL_DMAEx_MultiBufferStart_IT(DMA_HandleTypeDef * hdma, uint32
HAL_DMA_Abort(DMA_HandleTypeDef * hdma) : HAL_StatusTypeDef
HAL_DMA_Abort_IT(DMA_HandleTypeDef * hdma) : HAL_StatusTypeD
Press 'Ctrl+Space' to show Template Proposals

[그림 3b.21] GPIO 명령어 찾기

▶ Main 함수에 프로그램 작성하기

```
/* Infinite loop */
/* USER CODE BEGIN WHILE */
while (1)
{
    if(!HAL_GPIO_ReadPin(GPIOC, GPIO_PIN_13)){
        HAL_GPIO_TogglePin(GPIOA, GPIO_PIN_5);
        HAL_Delay(100);
    }

  /* USER CODE END WHILE */

  /* USER CODE BEGIN 3 */
}
/* USER CODE END 3 */
}
```

[그림 3b.22] 메인 함수 작성

11. 디버그 및 다운로드하기

메인 함수에 프로그램이 작성되었다면 다음으로 디버그는 [그림3b.23]의 [아이콘]를 클릭하거나 단축키 F11을 누르면 디버깅과 프로그램 다운로드가 완료된다[그림 3b.24]~[그림 3b.25].

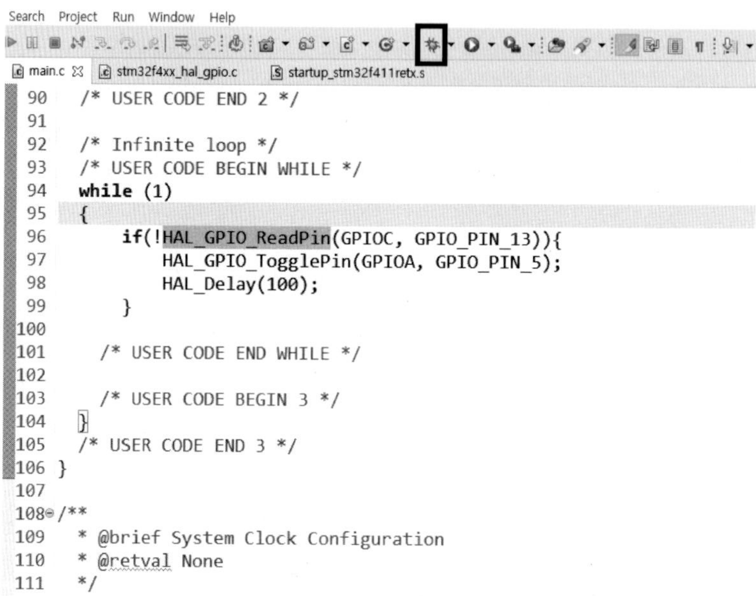
Search Project Run Window Help
main.c
stm32f4xx_hal_gpio.c
startup_stm32f411retx.s
/* USER CODE END 2 */
/* Infinite loop */
/* USER CODE BEGIN WHILE */
while (1)
{
if(!HAL_GPIO_ReadPin(GPIOC, GPIO_PIN_13)){
HAL_GPIO_TogglePin(GPIOA, GPIO_PIN_5);
HAL_Delay(100);
}
/* USER CODE END WHILE */
/* USER CODE BEGIN 3 */
}
/* USER CODE END 3 */
}
/**
* @brief System Clock Configuration
* @retval None
*/

[그림 3b.23] 프로그램 디버그

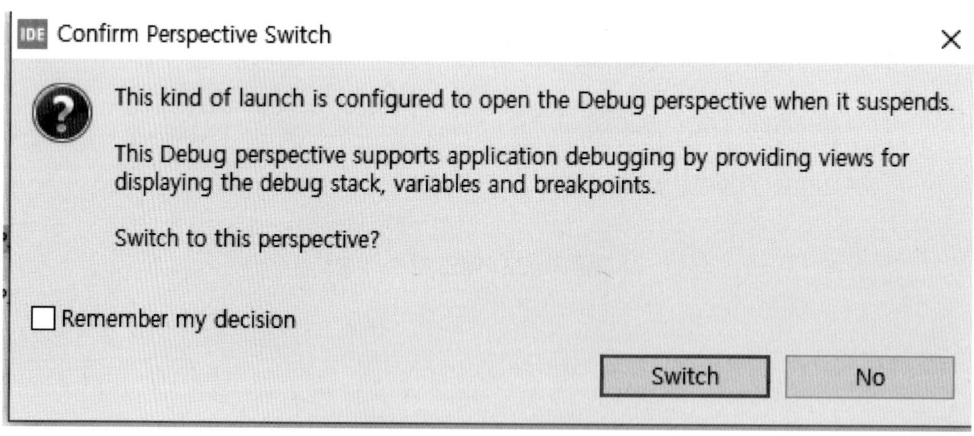
Confirm Perspective Switch
This kind of launch is configured to open the Debug perspective when it suspends.
This Debug perspective supports application debugging by providing views for displaying the debug stack, variables and breakpoints.
Switch to this perspective?
Remember my decision
Switch
No

[그림 3b.24] 프로그램 디버그

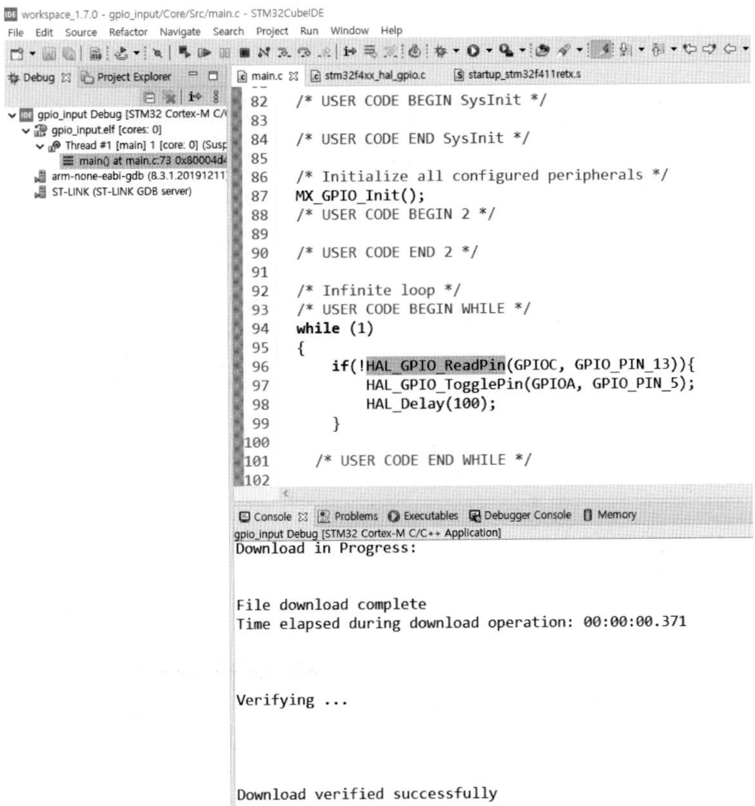
workspace_1.7.0 - gpio_input/Core/Src/main.c - STM32CubeIDE
File Edit Source Refactor Navigate Search Project Run Window Help
Debug
Project Explorer
main.c
stm32f4xx_hal_gpio.c
startup_stm32f411retx.s
gpio_input Debug [STM32 Cortex-M C/
gpio_input.elf [cores: 0]
Thread #1 [main] 1 [core: 0] (Susp
main() at main.c:73 0x80004d
arm-none-eabi-gdb (8.3.1.20191211
ST-LINK (ST-LINK GDB server)
/* USER CODE BEGIN SysInit */
/* USER CODE END SysInit */
/* Initialize all configured peripherals */
MX_GPIO_Init();
/* USER CODE BEGIN 2 */
/* USER CODE END 2 */
/* Infinite loop */
/* USER CODE BEGIN WHILE */
while (1)
{
if(!HAL_GPIO_ReadPin(GPIOC, GPIO_PIN_13)){
HAL_GPIO_TogglePin(GPIOA, GPIO_PIN_5);
HAL_Delay(100);
}
/* USER CODE END WHILE */
Console
Problems
Executables
Debugger Console
Memory
gpio_input Debug [STM32 Cortex-M C/C++ Application]
Download in Progress:
File download complete
Time elapsed during download operation: 00:00:00.371
Verifying ...
Download verified successfully

[그림 3b.25] Console 창 내용

12. 보드 프로그램 실행하기

[그림 3b.26] 상단 중앙에 아이콘 ■ 클릭하면 디버그 모드가 종료되어 보드의 프로그램이 실행된다.

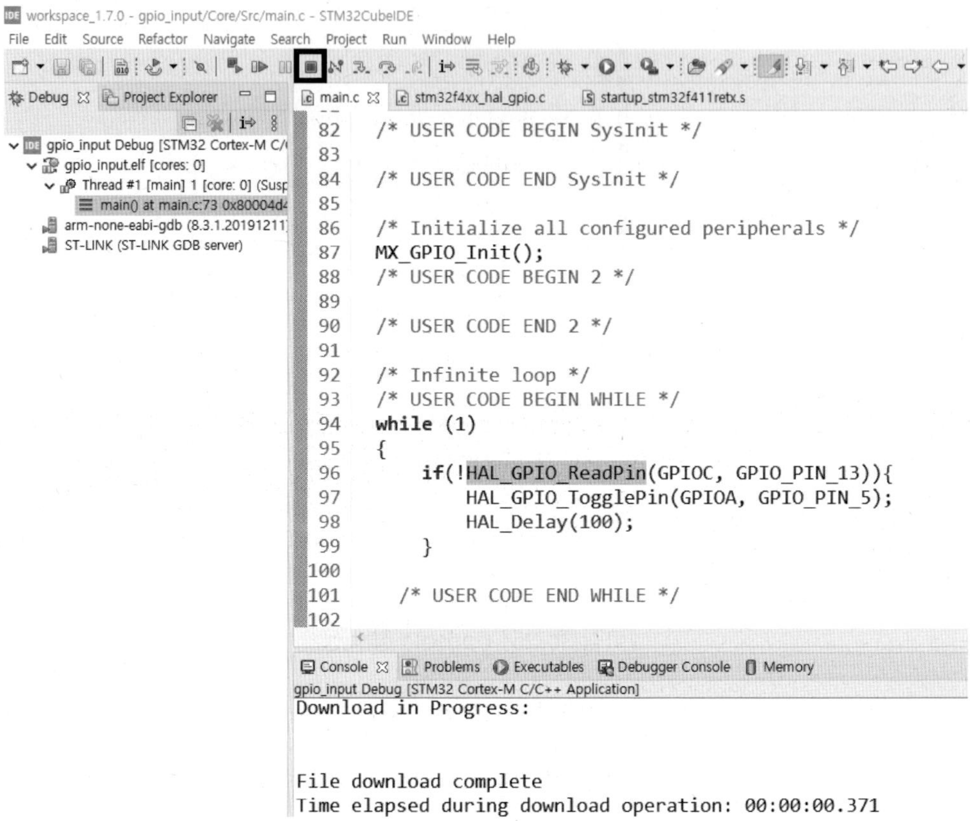

[그림 3b.26] 디버그 모드 종료 및 보드 실행

CHAPTER 4

EXTI(외부 인터럽트) 활용하기

1. EXTI(외부 인터럽트/이벤트) 라인 맵핑
2. EXTI(외부 인터럽트/이벤트) 하드웨어 구성
3. 프로젝트시작 및 설정
4. 실습 보드 구성
5. 클럭 설정
6. Pinout 초기화 설정
7. 디바이스 기본 설정
8. GPIO 인터럽트 설정
9. Generate Code 생성
10. Source Code 작성하기
11. 디버그 및 다운로드하기
12. 보드 프로그램 실행하기

CHAPTER 04

Cortex-M4 기초 활용(ARM Core STM32)

EXTI(외부 인터럽트) 활용하기

1. EXTI(외부 인터럽트/이벤트) 라인 맵핑

[그림 4.1]을 보면 23개 라인이 인터럽트 소스를 구성하고 있음을 알 수 있다. [그림 4.2]는 EXTI0~EXTI4까지는 개별 어드레스를 가지고 있다. 또한, EXTI9_5와 EXTI15_10도 하나의 어드레스가 지정되어 있음을 알 수 있다. 따라서 EXTI9_5와 EXTI15_10은 어느 GPIO에서 인터럽트 소스가 발생했는지는 해당 어드레스에 해당하는 함수에서 구분하여야 한다. EXTI0는 PA0~PK0 중 하나를 선택해서 인터럽트 신호로 사용한다.

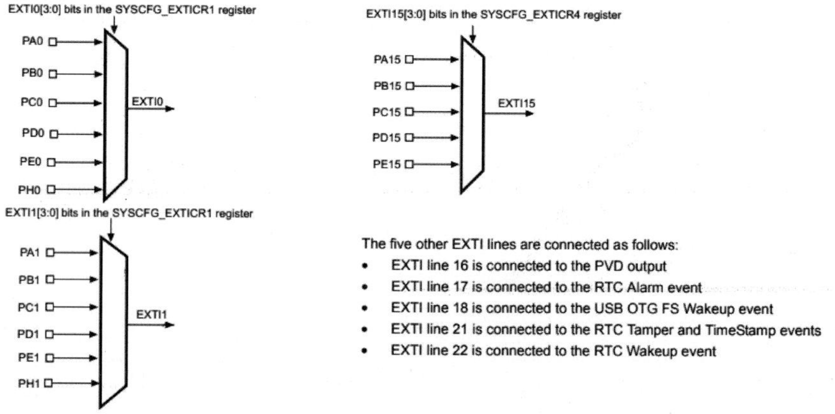

[그림 4.1] EXIT GPIO Mapping

Position	Priority	Type of priority	Acronym	Description	Address
6	13	settable	EXTI0	EXTI Line0 interrupt	0x0000 0058
7	14	settable	EXTI1	EXTI Line1 interrupt	0x0000 005C
8	15	settable	EXTI2	EXTI Line2 interrupt	0x0000 0060
9	16	settable	EXTI3	EXTI Line3 interrupt	0x0000 0064
10	17	settable	EXTI4	EXTI Line4 interrupt	0x0000 0068
23	30	settable	EXTI9_5	EXTI Line[9:5] interrupts	0x0000 009C
40	47	settable	EXTI15_10	EXTI Line[15:10] interrupts	0x0000 00E0

[그림 4.2] EXIT Vector Table

2. EXTI(외부 인터럽트/이벤트) 하드웨어 구성

실습 보드의 Port를 이용하여 LED 및 스위치를 [그림 4.3]과 같이 할당하여 관련된 기본 실습을 한다.

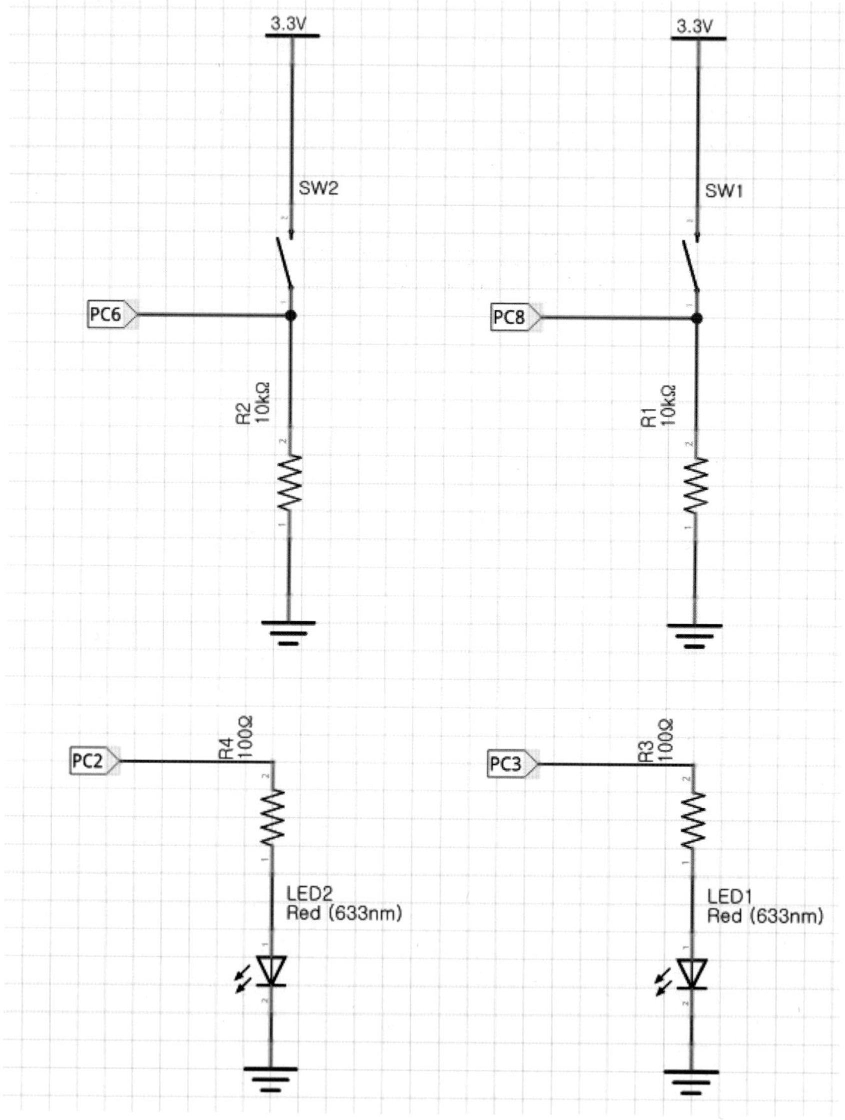

[그림 4.3] LED 및 스위치 구성

3. 프로젝트 시작 및 설정

3.1. STM32 프로젝트 시작

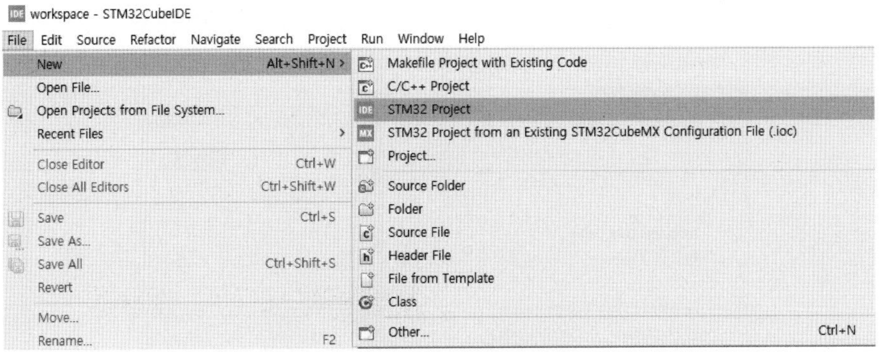

[그림 4.4] 프로그램 시작

3.2. 보드 선정: NUCLEO-F411RE

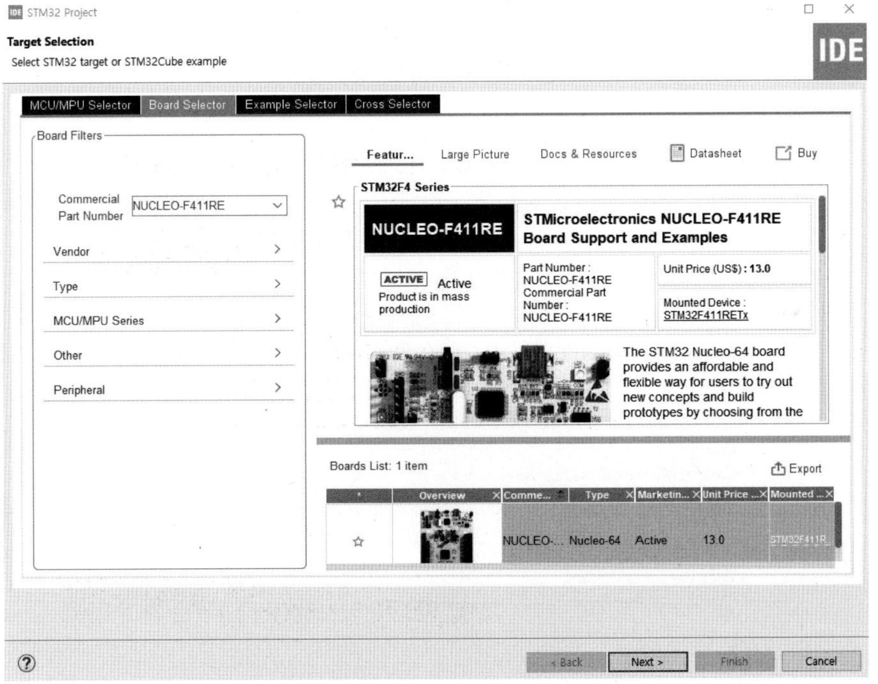

[그림 4.5] 보드 선정

3.3. 프로젝트 이름 결정 및 보드 구성 형태로 초기화:

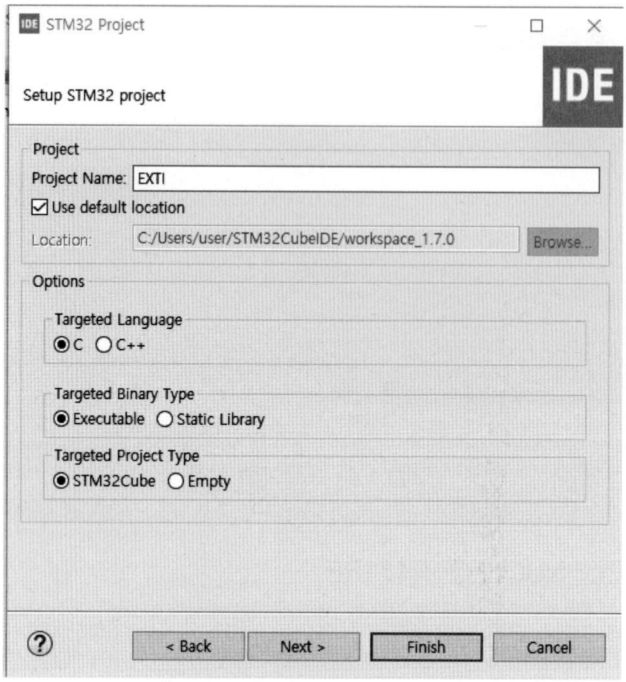

[그림 4.6] 프로젝트 이름 결정

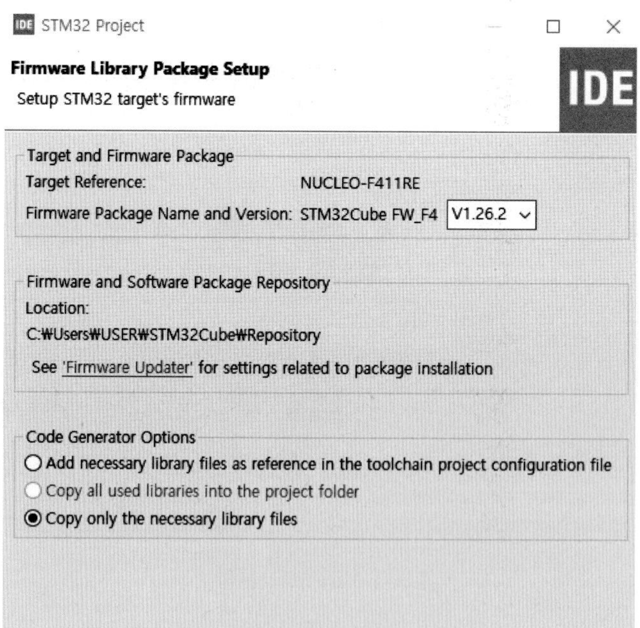

[그림 4.7] 펌웨어 패키지 설정

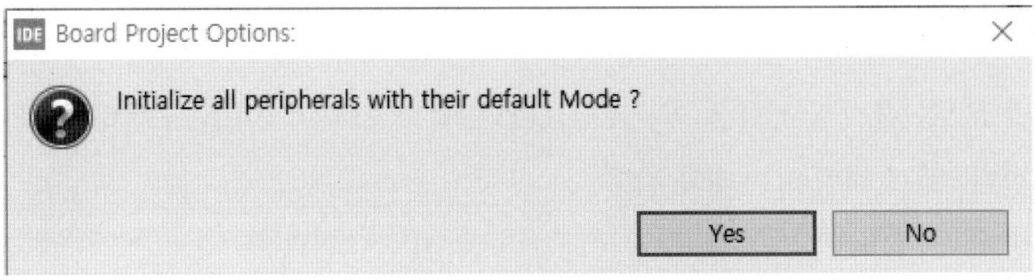
Board Project Options:
Initialize all peripherals with their default Mode ?
Yes
No

[그림 4.8] 선정된 보드의 디폴트 모드로 초기화

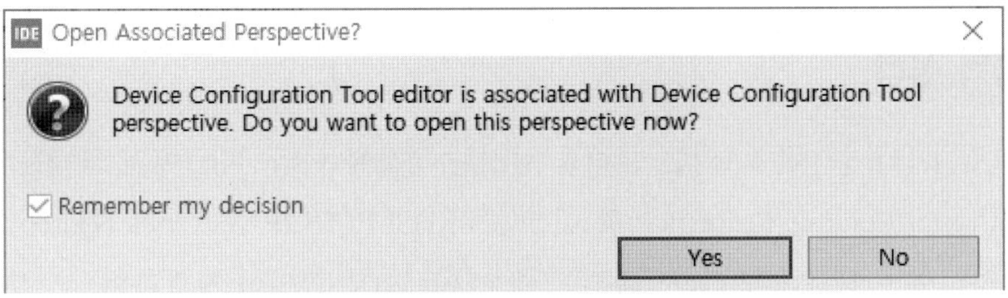
Open Associated Perspective?
Device Configuration Tool editor is associated with Device Configuration Tool perspective. Do you want to open this perspective now?
Remember my decision
Yes
No

[그림 4.9] 위의 선정된 보드의 구성으로 시스템 구성

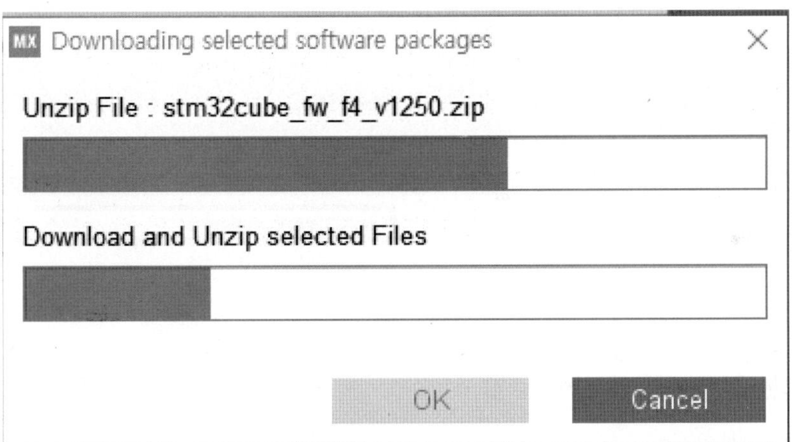
Downloading selected software packages
Unzip File : stm32cube_fw_f4_v1250.zip
Download and Unzip selected Files
OK
Cancel

[그림 4.10] 시스템 구성 진행

4. 실습 보드 구성

실습 보드는 [그림 4.3]을 기준으로 LED 2개 및 스위치 2개를 구성하여 관련 실습을 한다. 포트 배정은 아래와 같다.

> LED GPIO 출력: PC2, 3
> Push Botton 입력: PC6, 8

5. 클럭 설정

[그림 4.11]~[그림 4.12]는 클럭을 설정한 경우를 나타낸다. 메인 클럭은 외부 클럭(Crystal/Ceramic Resonator)과 내부 클럭(HSI RC(BYPASS Clock Source)) 둘 중에 하나를 이용하여야 한다. 내부 클럭보다는 외부 클럭이 정확하며, 본 보드에서는 Crystal/Ceramic Resonator는 연결되어 있지 않아 사용하지 않으며 ST-LINK MCO 출력 8MHz를 사용한다.

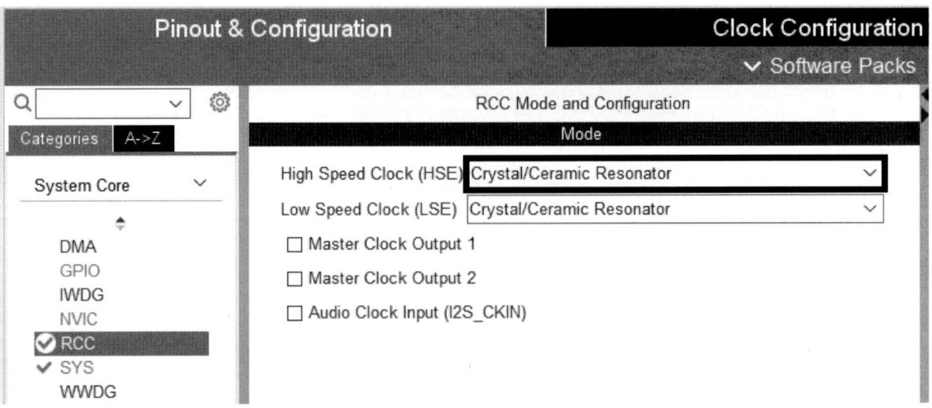

[그림 4.11] NUCLEO-F411RE 클럭 설정

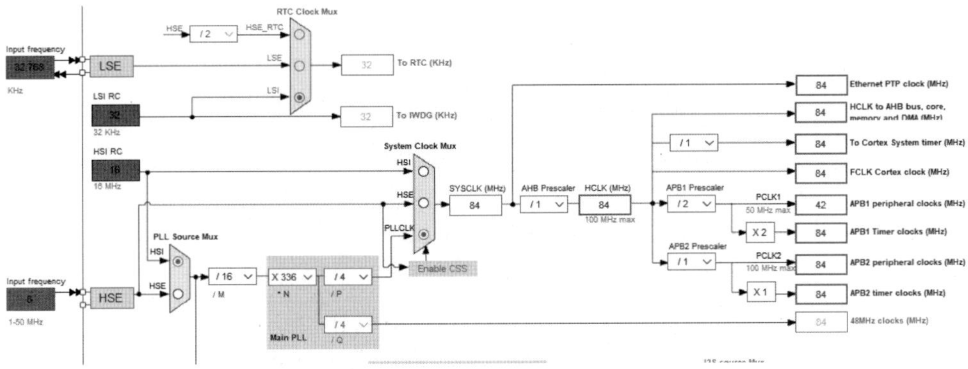

[그림 4.12] NUCLEO-F411RE 디폴트 클럭 설정

[그림 4.12]는 주변 장치의 클럭이 할당된 것을 알 수 있고 APB1에 연결되어 있으면 42MHz, APB2에 연결되어 있으면 84MHz로 동작함을 알 수 있다.

6. Pinout 초기화 설정

보드의 기능을 새로 사용자에 맞도록 하는 경우로 [그림 4.13]의 경우처럼 핀아웃을 클리어하고 사용자에 맞게 새로 환경 설정을 하는 경우에 해당한다. 이 경우 결과는 [그림 4.14]와 같이 된다.

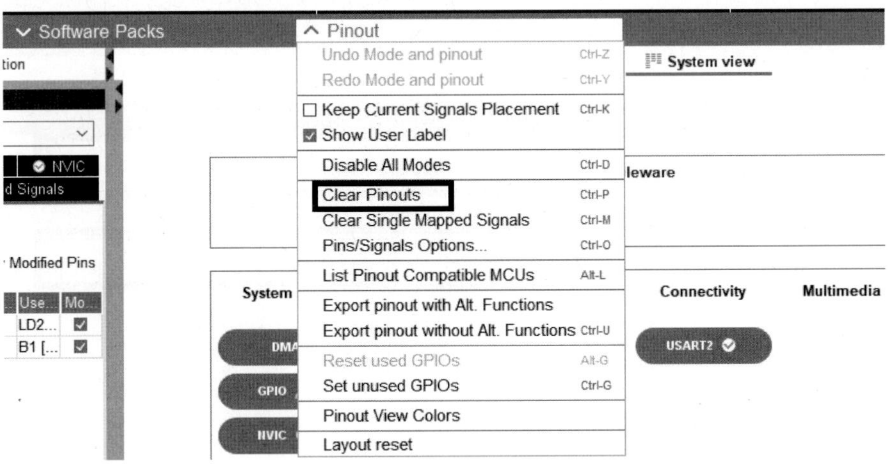

[그림 4.13] 디바이스 Pinout 설정

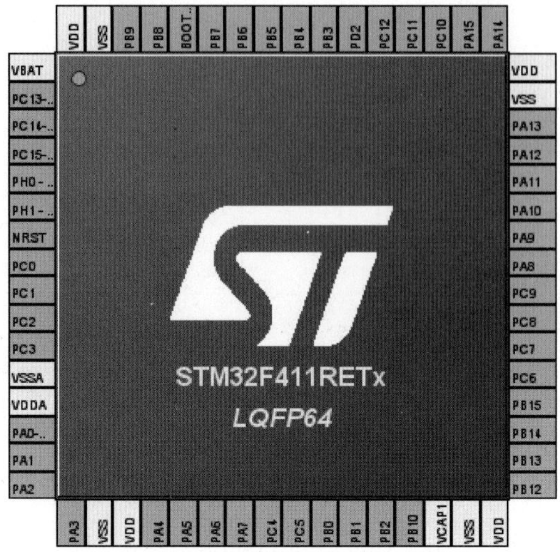

[그림 4.14] 디바이스 Pinout 초기화 화면

7. 디바이스 기본 설정

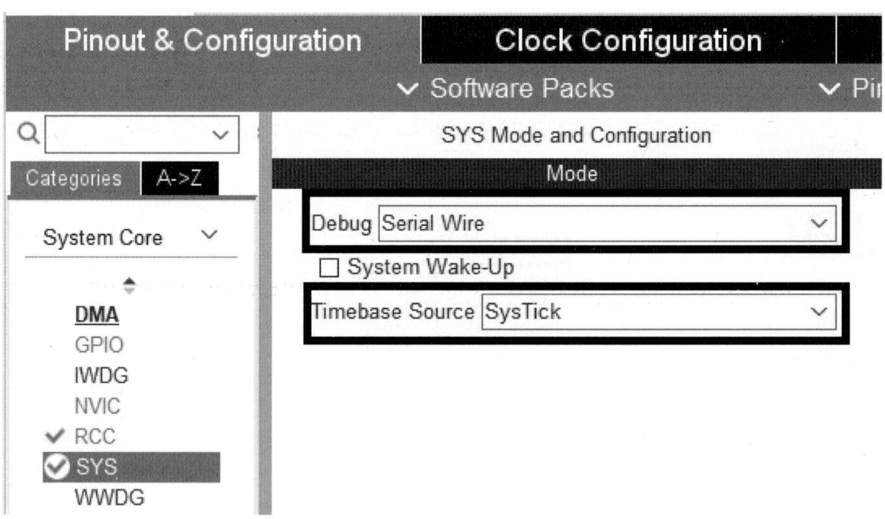

[그림 4.15] 시리얼 디버그 및 타임베이스 설정

8. GPIO 인터럽트 설정

LED가 PC0, PC1에 연결되어 있으므로 GPIO 출력 포트로 설정하면 된다. [그림 4.3]과 같이 SW1[PC8]과 SW2[PC6]를 인터럽트로 설정하고 [그림 4.16], [그림 4.17]에서 GPIO mode를 설정한다.

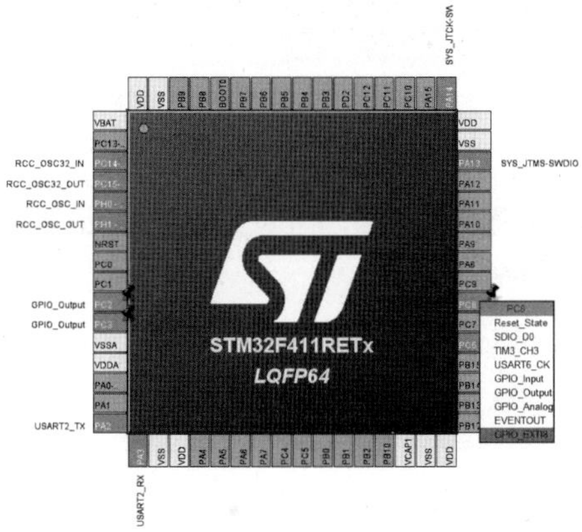

[그림 4.16] GPIO 인터럽트 설정

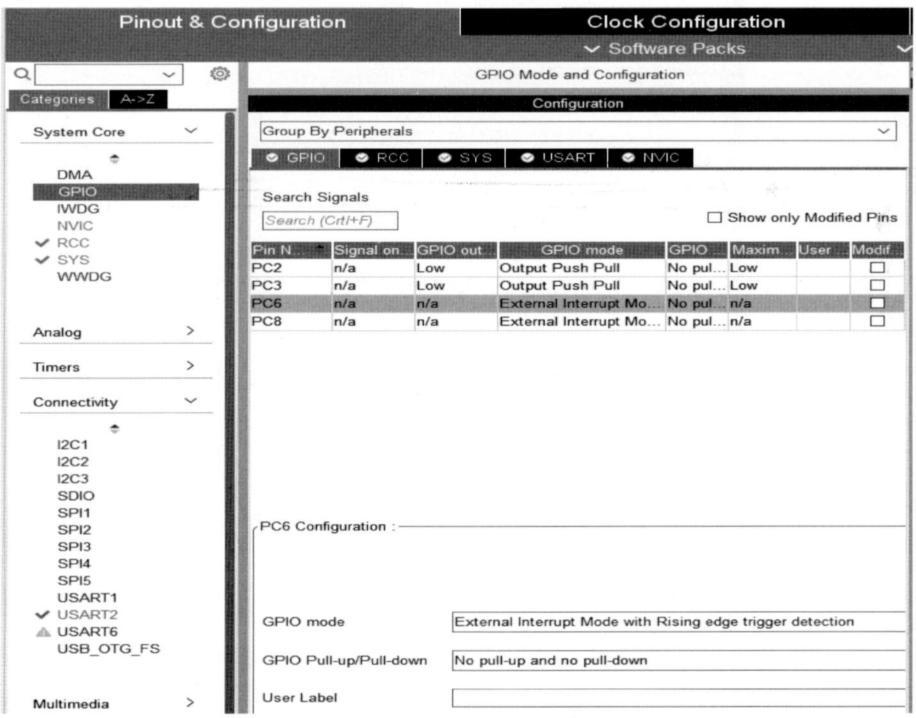

[그림 4.17] GPIO 인터럽트 설정

System Core ▷ NVIC에서 [그림 4.18]과 같이 Enable 체크하여 인터럽트 처리가 생성될 수 있도록 한다.

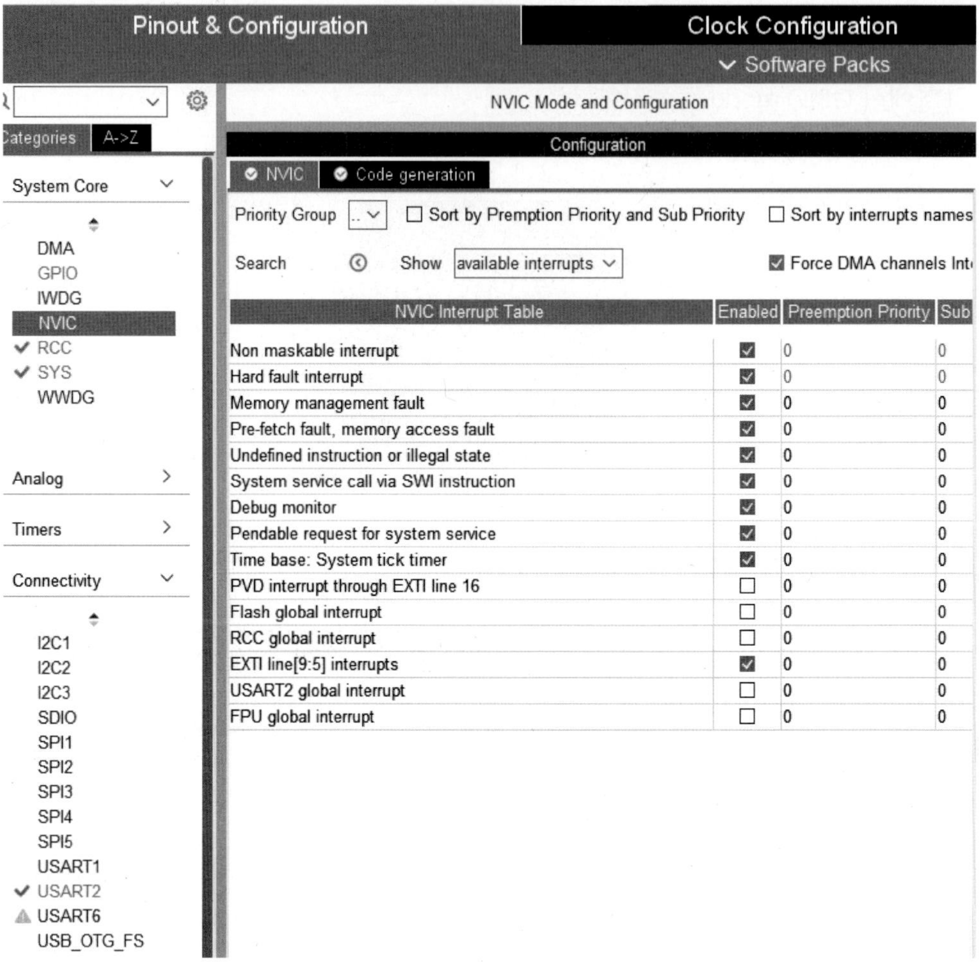

[그림 4.18] NVIC 인터럽트 Enabled 확인

9. Generate Code 생성

[그림 4.19]와 [그림 4.20]은 코드를 생성하기 위한 화면이다.  클릭한다.

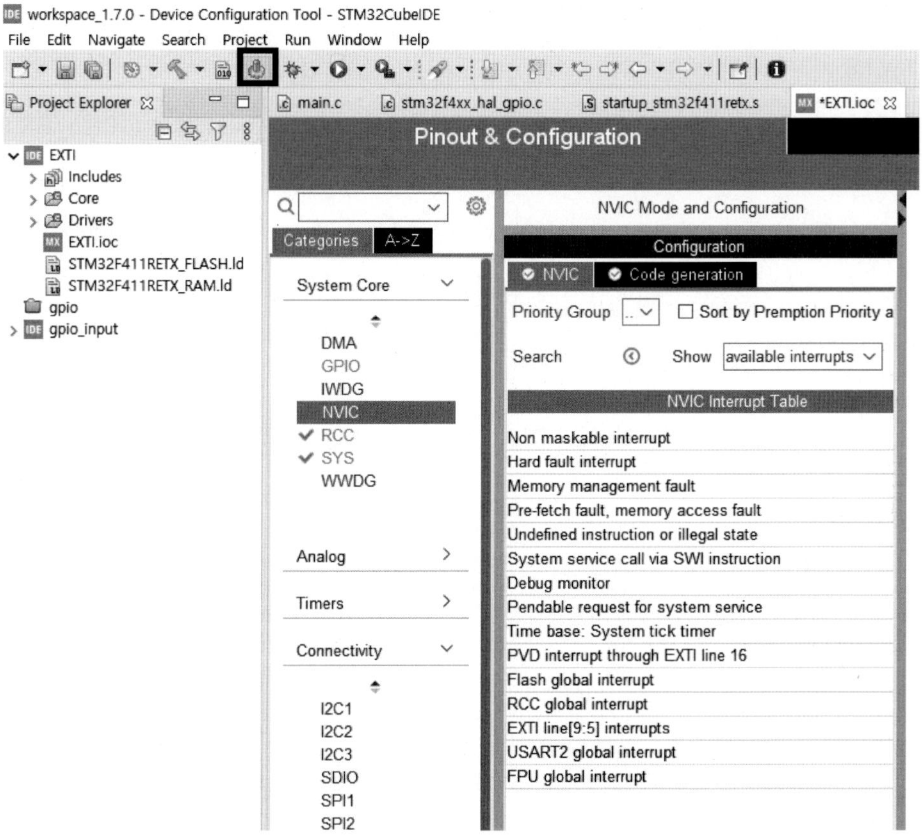

[그림 4.19] 프로젝트 코드를 생성

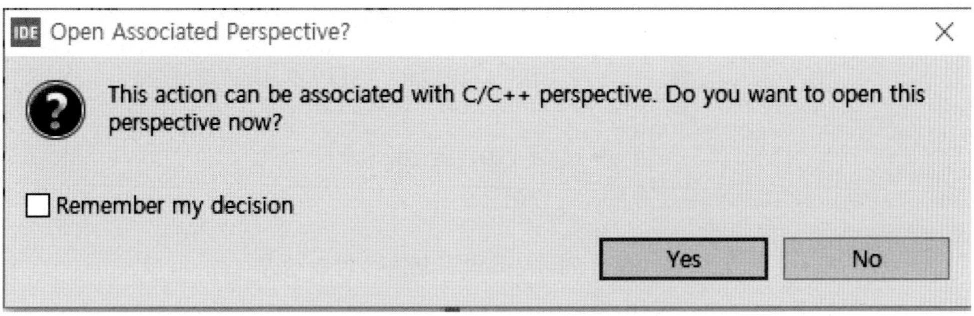

[그림 4.20] 프로젝트 코드를 생성

10. Source Code 작성하기

[그림 4.21]의 좌측 프로젝트 폴더에서 Core▷Src▷stm32f4xx_it.c을 클릭하여 들어가면 외부 인터럽트 처리 핸들러 함수들이 있다. 만약 GPIO PF3 인터럽트 신호가 발생하면 EXT9_5_IRQHandler(void) 함수가 실행되고 또한 그 내부에 있는 HAL_GPIO_EXTI_IRQHandler(GPIO_PIN_6),HAL_GPIO_EXTI_IRQHandler(GPIO_PIN_8) 함수가 실행됨을 알 수 있다. 핸들러 내부 함수들의 이름이 모두 같음을 짐작할 수 있다. 이 내부 함수들은 모두 HAL_GPIO_EXTI_Callback() 함수를 호출한다. 여기에 인터럽트 프로그램을 작성하면 된다.

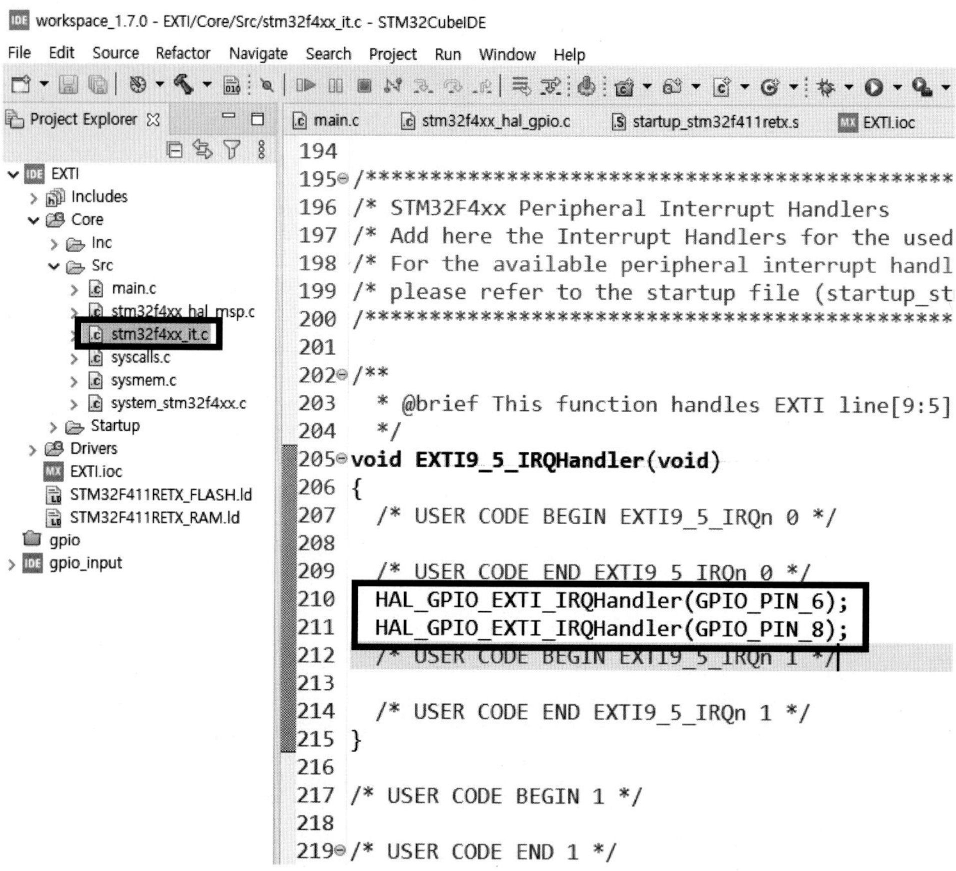
workspace_1.7.0 - EXTI/Core/Src/stm32f4xx_it.c - STM32CubeIDE
File Edit Source Refactor Navigate Search Project Run Window Help
Project Explorer
main.c
stm32f4xx_hal_gpio.c
startup_stm32f411retx.s
EXTI.ioc
EXTI
Includes
Core
Inc
Src
main.c
stm32f4xx_hal_msp.c
stm32f4xx_it.c
syscalls.c
sysmem.c
system_stm32f4xx.c
Startup
Drivers
EXTI.ioc
STM32F411RETX_FLASH.ld
STM32F411RETX_RAM.ld
gpio
gpio_input
194
195 /**
196 /* STM32F4xx Peripheral Interrupt Handlers
197 /* Add here the Interrupt Handlers for the used
198 /* For the available peripheral interrupt handl
199 /* please refer to the startup file (startup_st
200 /**
201
202 /**
203 * @brief This function handles EXTI line[9:5]
204 */
205 void EXTI9_5_IRQHandler(void)
206 {
207 /* USER CODE BEGIN EXTI9_5_IRQn 0 */
208
209 /* USER CODE END EXTI9_5_IRQn 0 */
210 HAL_GPIO_EXTI_IRQHandler(GPIO_PIN_6);
211 HAL_GPIO_EXTI_IRQHandler(GPIO_PIN_8);
212 /* USER CODE BEGIN EXTI9_5_IRQn 1 */
213
214 /* USER CODE END EXTI9_5_IRQn 1 */
215 }
216
217 /* USER CODE BEGIN 1 */
218
219 /* USER CODE END 1 */

[그림 4.21] 해당 인터럽트 핸들러 함수 생성

▶ 프로그램 작성을 위해서 Callback 함수를 찾는다. [그림 4.22]와 같이 마우스로 HAL_GPIO_EXTI_IRQHandler(GPIO_PIN_6);를 클릭한 다음 단축키 F3을 누른다.

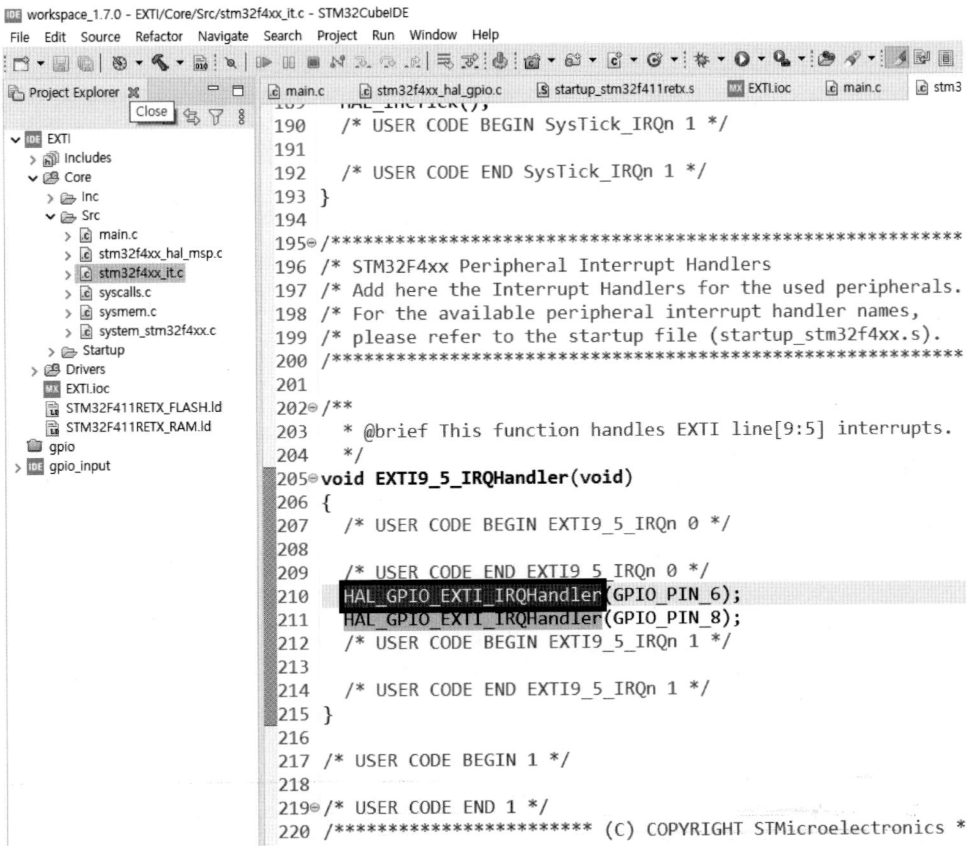

workspace_1.7.0 - EXTI/Core/Src/stm32f4xx_it.c - STM32CubeIDE
File Edit Source Refactor Navigate Search Project Run Window Help
Project Explorer
Close
EXTI
Includes
Core
Inc
Src
main.c
stm32f4xx_hal_msp.c
stm32f4xx_it.c
syscalls.c
sysmem.c
system_stm32f4xx.c
Startup
Drivers
EXTI.ioc
STM32F411RETX_FLASH.ld
STM32F411RETX_RAM.ld
gpio
gpio_input
main.c
stm32f4xx_hal_gpio.c
startup_stm32f411retx.s
EXTI.ioc
main.c
stm3
190 /* USER CODE BEGIN SysTick_IRQn 1 */
191
192 /* USER CODE END SysTick_IRQn 1 */
193 }
194
195 /**
196 /* STM32F4xx Peripheral Interrupt Handlers
197 /* Add here the Interrupt Handlers for the used peripherals.
198 /* For the available peripheral interrupt handler names,
199 /* please refer to the startup file (startup_stm32f4xx.s).
200 /**
201
202 /**
203 * @brief This function handles EXTI line[9:5] interrupts.
204 */
205 void EXTI9_5_IRQHandler(void)
206 {
207 /* USER CODE BEGIN EXTI9_5_IRQn 0 */
208
209 /* USER CODE END EXTI9_5_IRQn 0 */
210 HAL_GPIO_EXTI_IRQHandler(GPIO_PIN_6);
211 HAL_GPIO_EXTI_IRQHandler(GPIO_PIN_8);
212 /* USER CODE BEGIN EXTI9_5_IRQn 1 */
213
214 /* USER CODE END EXTI9_5_IRQn 1 */
215 }
216
217 /* USER CODE BEGIN 1 */
218
219 /* USER CODE END 1 */
220 /************************** (C) COPYRIGHT STMicroelectronics *

[그림 4.22] Callback 함수로 이동

▶ weak void HAL_GPIO_EXTI_Callback 함수 전체를 복사[그림 4.23]하여 main.c에 복사하여 __weak를 제거한다.

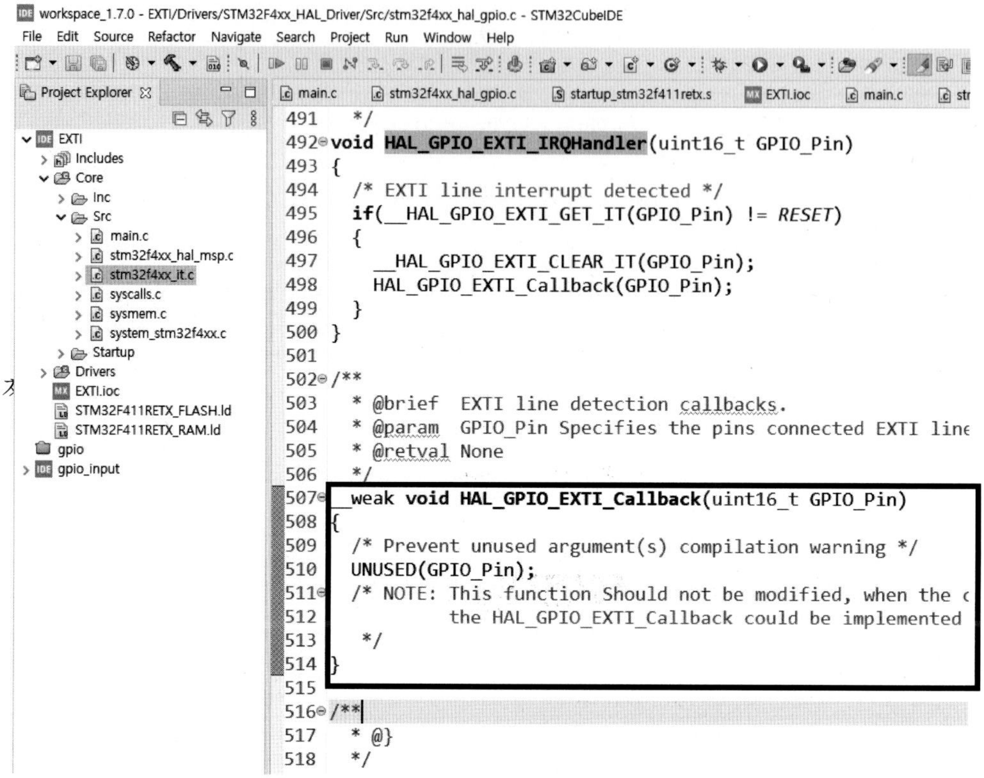

[그림 4.23] __weak void HAL_GPIO_EXTI_Callback 함수 복사

▶ 복사한 Callback 함수에 프로그램 작성한다[그림 4.24]. 스위치를 누르면 LED가 토글되는 프로그램이다. GPIO 포트번호만 필요하다(인터럽트 라인 맵핑 참조).

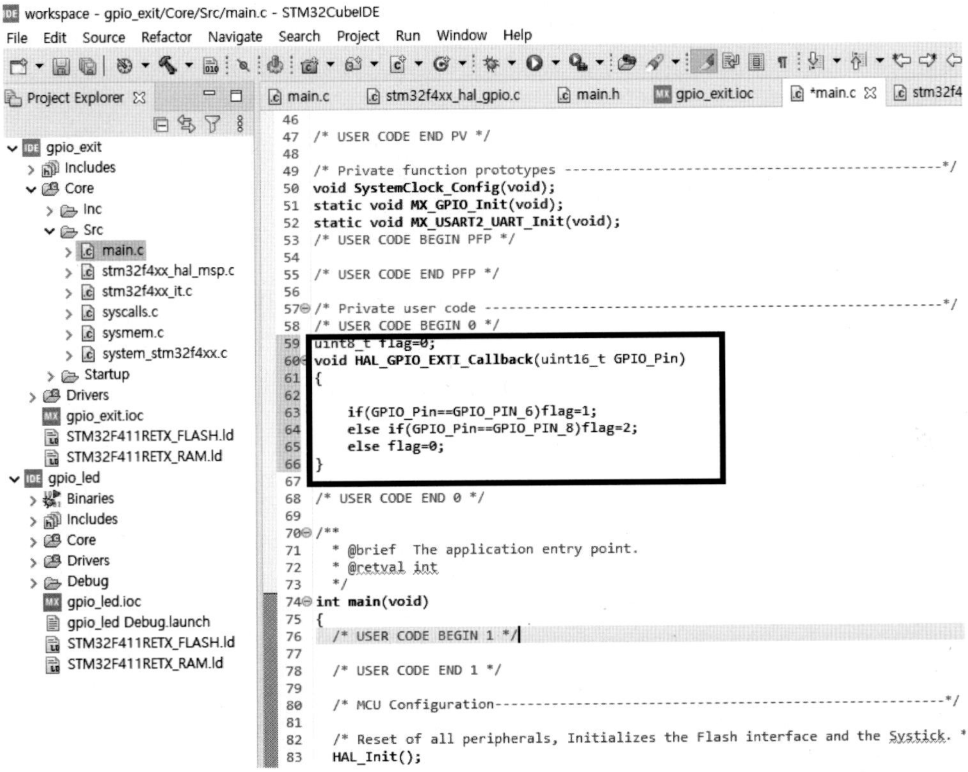

[그림 4.24] void HAL_GPIO_EXTI_Callback 작성

▶ 메인 프로그램 작성

```
/* Infinite loop */
/* USER CODE BEGIN WHILE */
while (1)
{
  if(flag==1){
      HAL_GPIO_TogglePin(GPIOC, GPIO_PIN_2);
      HAL_Delay(100);
  }
  else if(flag==2){
          HAL_GPIO_TogglePin(GPIOC, GPIO_PIN_3);
          HAL_Delay(100);
      }
  else{
      HAL_GPIO_WritePin(GPIOC, GPIO_PIN_2 | GPIO_PIN_3, 0);
  }
  /* USER CODE END WHILE */

  /* USER CODE BEGIN 3 */
}
/* USER CODE END 3 */
}
```

[그림 4.25] 메인 프로그램 작성

11. 디버그 및 다운로드하기

메인 함수에 프로그램이 작성되었다면 다음으로 디버그는 [그림 4.26]의 를 클릭하거나 단축키 F11을 누르면 디버깅과 프로그램 다운로드가 완료된다[그림 4.27]~[그림 4.28].

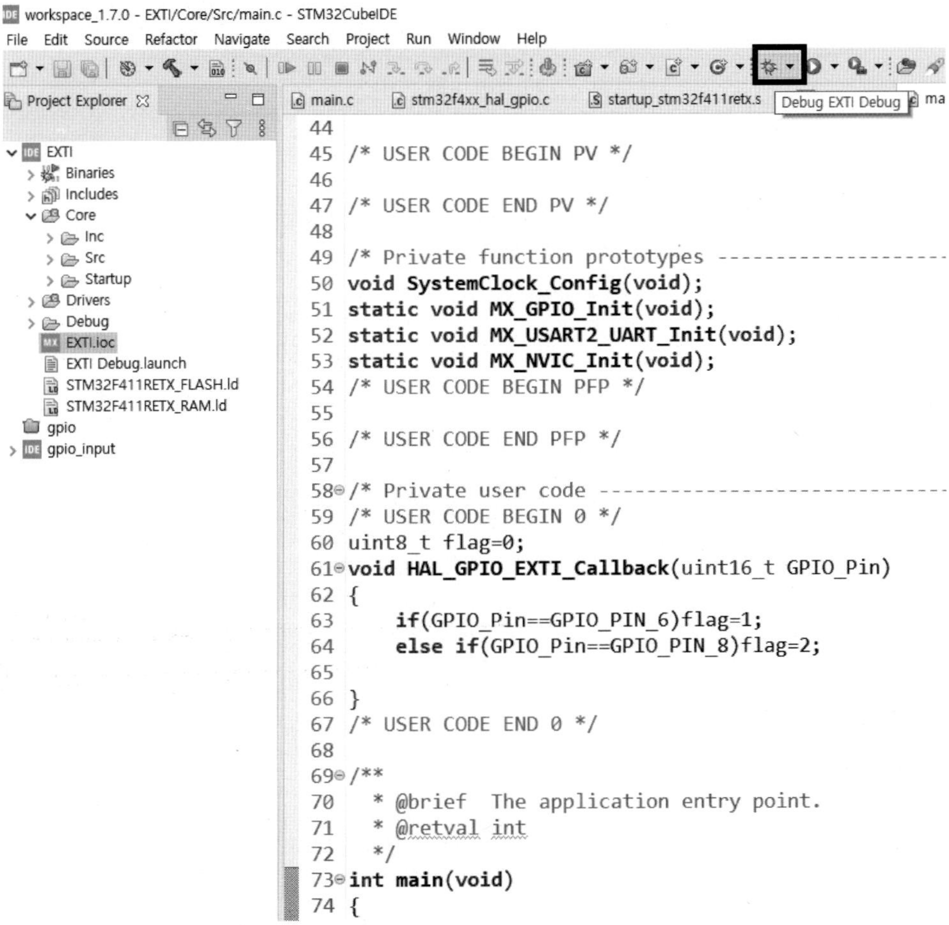
workspace_1.7.0 - EXTI/Core/Src/main.c - STM32CubeIDE
File Edit Source Refactor Navigate Search Project Run Window Help
Project Explorer
main.c
stm32f4xx_hal_gpio.c
startup_stm32f411retx.s
Debug EXTI Debug
EXTI
Binaries
Includes
Core
Inc
Src
Startup
Drivers
Debug
EXTI.ioc
EXTI Debug.launch
STM32F411RETX_FLASH.ld
STM32F411RETX_RAM.ld
gpio
gpio_input
44
45 /* USER CODE BEGIN PV */
46
47 /* USER CODE END PV */
48
49 /* Private function prototypes ---------------------
50 void SystemClock_Config(void);
51 static void MX_GPIO_Init(void);
52 static void MX_USART2_UART_Init(void);
53 static void MX_NVIC_Init(void);
54 /* USER CODE BEGIN PFP */
55
56 /* USER CODE END PFP */
57
58 /* Private user code ---------------------------
59 /* USER CODE BEGIN 0 */
60 uint8_t flag=0;
61 void HAL_GPIO_EXTI_Callback(uint16_t GPIO_Pin)
62 {
63 if(GPIO_Pin==GPIO_PIN_6)flag=1;
64 else if(GPIO_Pin==GPIO_PIN_8)flag=2;
65
66 }
67 /* USER CODE END 0 */
68
69 /**
70 * @brief The application entry point.
71 * @retval int
72 */
73 int main(void)
74 {

[그림 4.26] 프로그램 디버그

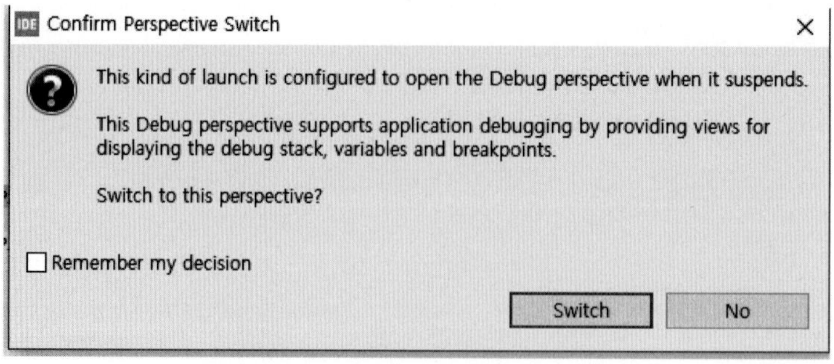
Confirm Perspective Switch
This kind of launch is configured to open the Debug perspective when it suspends.
This Debug perspective supports application debugging by providing views for displaying the debug stack, variables and breakpoints.
Switch to this perspective?
Remember my decision
Switch
No

[그림 4.27] 프로그램 디버그

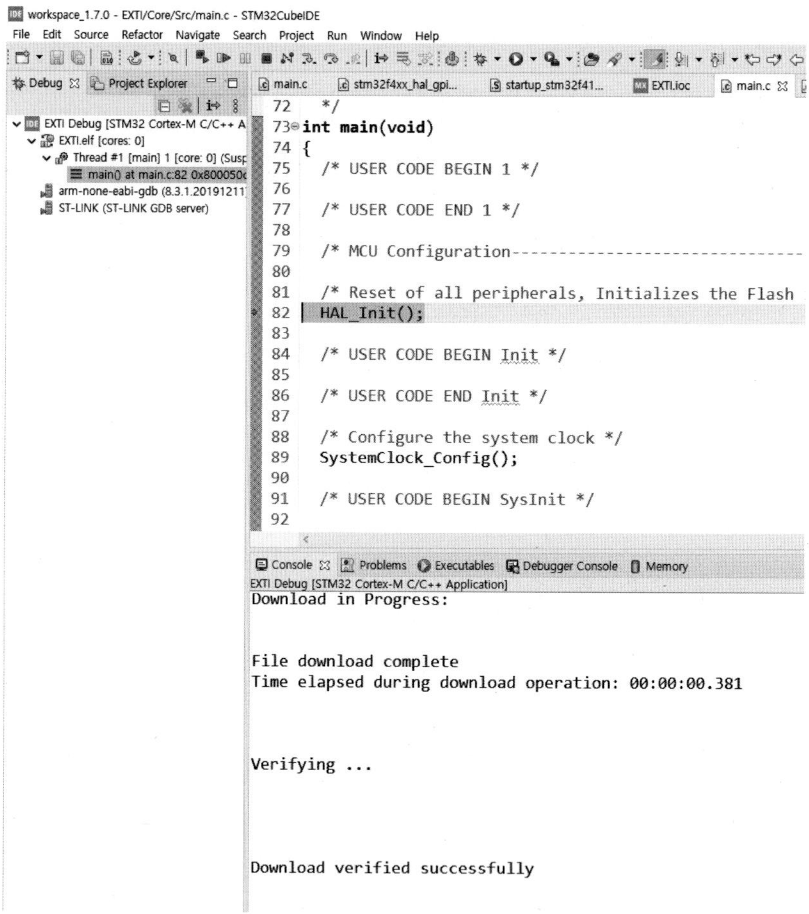

[그림 4.28] Console 창 내용

12. 보드 프로그램 실행하기

[그림 4.29] 상단 중앙에 아이콘 ■ 클릭하면 디버그 모드가 종료되어 보드의 프로그램이 실행된다. 스위치를 누르면 LED가 토글된다.

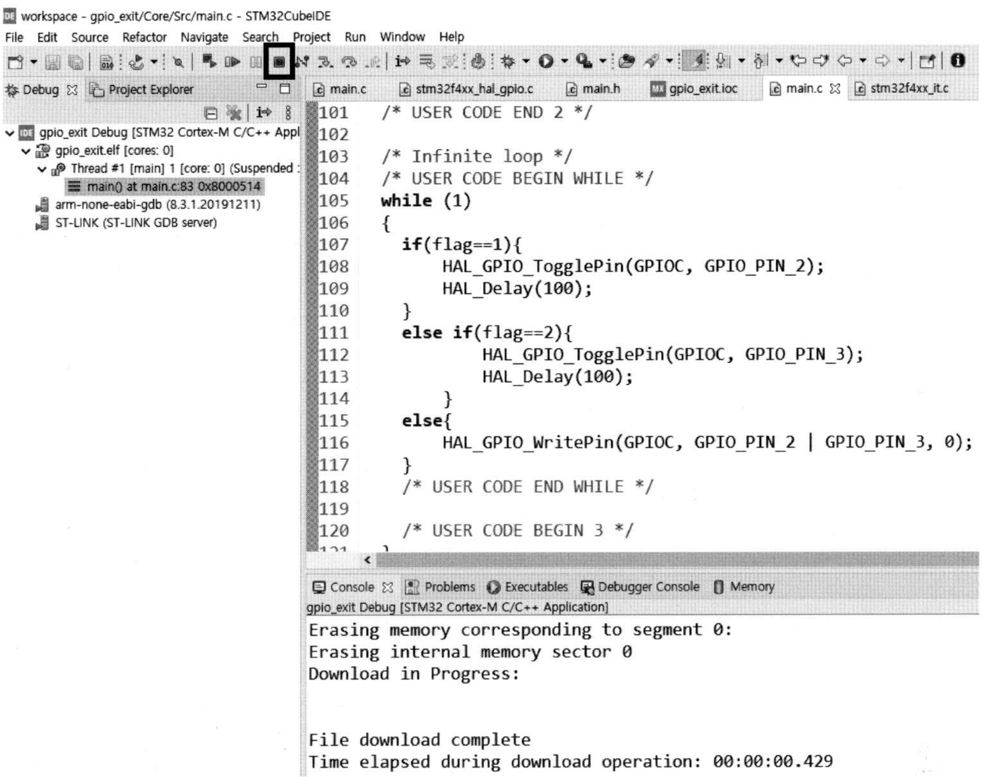
workspace - gpio_exit/Core/Src/main.c - STM32CubeIDE
File Edit Source Refactor Navigate Search Project Run Window Help
Debug Project Explorer
gpio_exit Debug [STM32 Cortex-M C/C++ Appl
gpio_exit.elf [cores: 0]
Thread #1 [main] 1 [core: 0] (Suspended
main() at main.c:83 0x8000514
arm-none-eabi-gdb (8.3.1.20191211)
ST-LINK (ST-LINK GDB server)
main.c stm32f4xx_hal_gpio.c main.h gpio_exit.ioc main.c stm32f4xx_it.c
/* USER CODE END 2 */
/* Infinite loop */
/* USER CODE BEGIN WHILE */
while (1)
{
if(flag==1){
HAL_GPIO_TogglePin(GPIOC, GPIO_PIN_2);
HAL_Delay(100);
}
else if(flag==2){
HAL_GPIO_TogglePin(GPIOC, GPIO_PIN_3);
HAL_Delay(100);
}
else{
HAL_GPIO_WritePin(GPIOC, GPIO_PIN_2 | GPIO_PIN_3, 0);
}
/* USER CODE END WHILE */
/* USER CODE BEGIN 3 */
Console Problems Executables Debugger Console Memory
gpio_exit Debug [STM32 Cortex-M C/C++ Application]
Erasing memory corresponding to segment 0:
Erasing internal memory sector 0
Download in Progress:
File download complete
Time elapsed during download operation: 00:00:00.429

[그림 4.29] 디버그 모드 종료 및 보드실행

CHAPTER 5

USART 활용하기 (Polling, Interrupt, DMA)

1. STM32 USART 구성
2. USART 하드웨어 설정
3. Polling 방식 (printf())
4. STM32IDE console 활용 Debug 사용하기
5. Interrupt 방식
6. DMA 방식

CHAPTER 05

Cortex-M4 기초 활용(ARM Core STM32)

USART 활용하기 (Polling, Interrupt, DMA)

<USART>

직렬 통신의 개념보다는 쉽게 활용할 수 있는 내용 위주로 구성을 하였으며 폴링 방식, 인터럽트 방식 및 DMA 방식에 대하여 기술한다.

1. STM32 USART 구성

[그림 5.1]을 보면 3개의 특징을 나타내고 있고 USART 방식은 3개(1,2,6)를 지원하고 있다. 실습에는 보드의 디폴트로 설정되어 있는 USART2을 이용한다.

USART feature comparison

USART name	Standard features	Modem (RTS/CTS)	LIN	SPI master	irDA	Smartcard (ISO 7816)	Max. baud rate in Mbit/s (oversampling by 16)	Max. baud rate in Mbit/s (oversampling by 8)	APB mapping
USART1	X	X	X	X	X	X	6.25	12.5	APB2 (max. 100 MHz)
USART2	X	X	X	X	X	X	3.12	6.25	APB1 (max. 50 MHz)
USART6	X	N.A	X	X	X	X	6.25	12.5	APB2 (max. 100 MHz)

[그림 5.1] USART 특징 비교

2. USART 하드웨어 설정

USART3은 ST-LINK 및 통신 포트로 활용할 수 있으므로 본 실습에는 기본 통신 포트로 활용하고 별도의 하드웨어가 없어도 된다.

Pin name	Function
PA2	TX
PA3	RX

[그림 5.2] USART3 핀 설정

3. Polling 방식 (printf())

3.1. USART 설정

System Core에서 RCC 및 SYS는 이전과 같이 설정한다. USART2의 TX RX 단자는 [PA2, PA3]를 사용할 수가 있도록 [그림 5.3]같이 파라메터를 설정한다. [그림 5.4]와 [그림 5.5]은 핀 설정 및 폴링 방식으로 설정하였음을 알 수 있다.

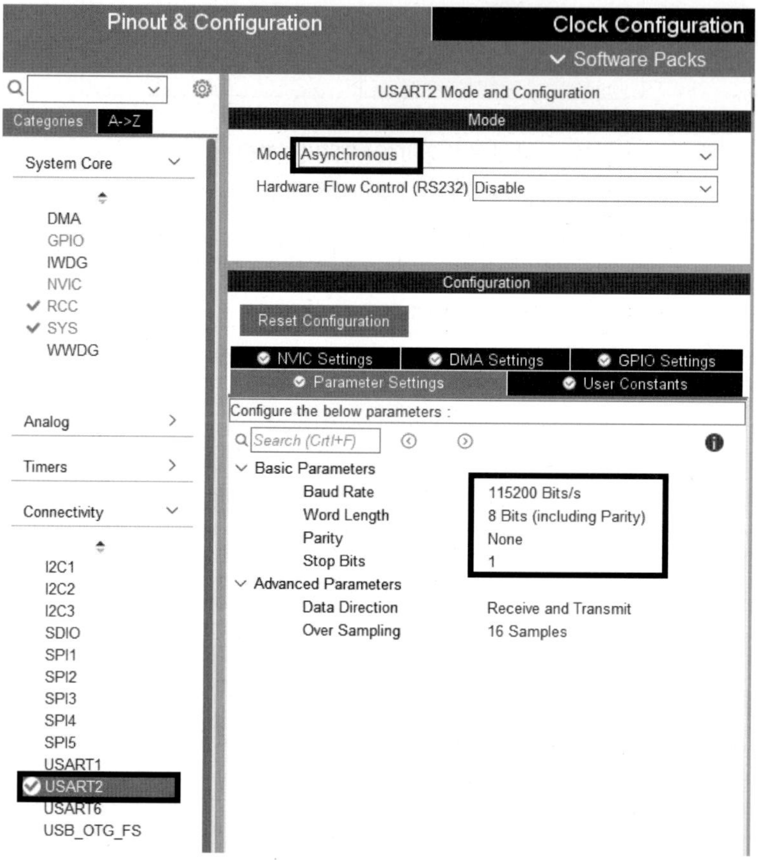

[그림 5.3] USART2 설정

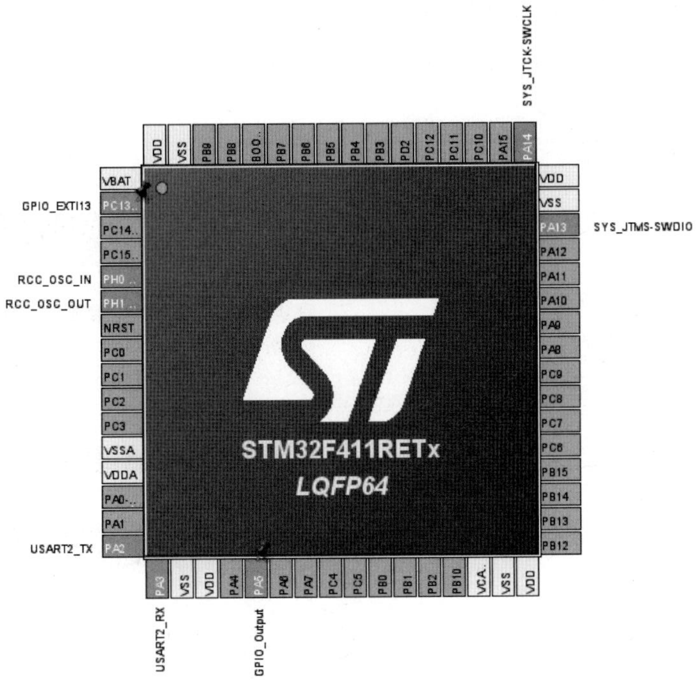

[그림 5.4] USART2 PINOUT 설정

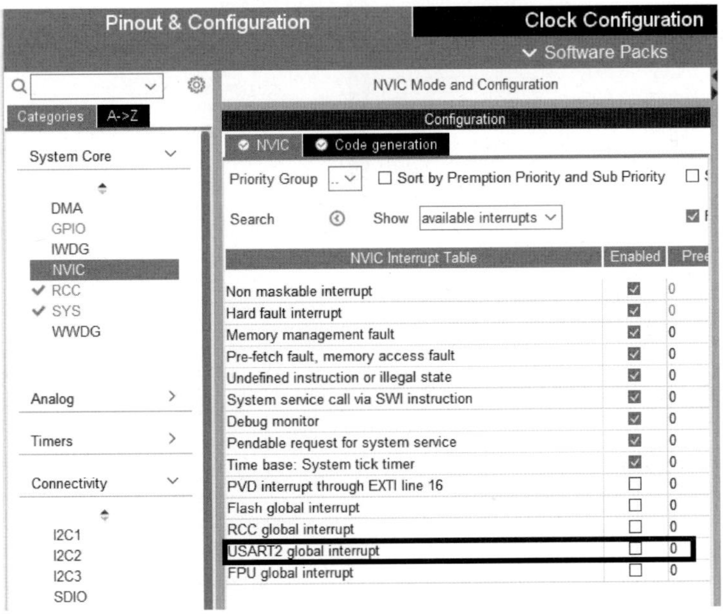

[그림5.5] 인터럽트 미설정(폴링 방식)

3.2. Generate Code 생성

[그림 5.6]과 [그림 5.7]은 코드를 생성하기 위한 화면이다. 순차적으로 클릭한다.

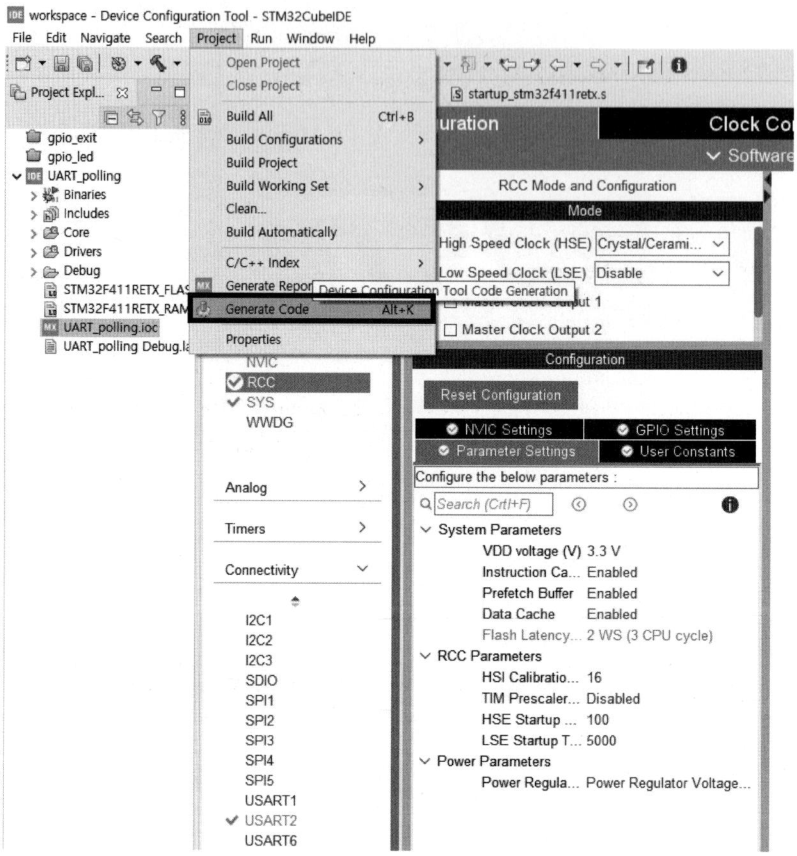

[그림 5.6] 프로그램 코드를 생성

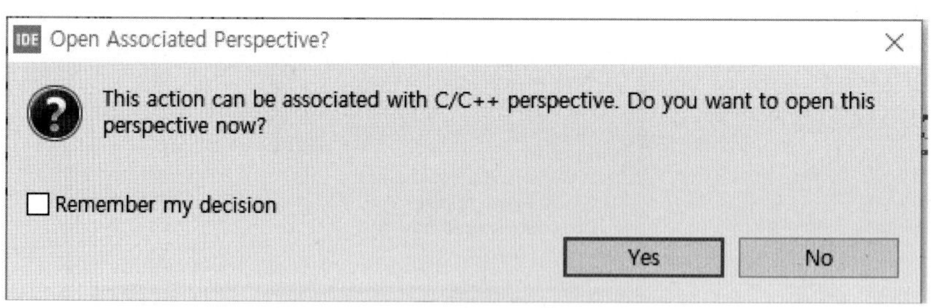

[그림 5.7] 프로젝트 코드를 생성

3.3. Source Code 작성하기

Polling 방식을 사용하기 위해서는 전송과 수신을 위한 관련 함수를 이용하면 된다. 관련 함수는 HAL_UART_Transmit, HAL_UART_Receive 가 있다.

▶ 데이터 전송 함수

HAL_UART_Transmit(UART_HandleTypeDef *huart, uint8_t *pData, uint16_t Size, uint32_t Timeout)

> 사용 방법: uart2에 특정 바이트의 데이터를 100밀리 이내 보내고자 할 때

```
UART_HandleTypeDef huart2 ;
uint8_t tx_data[4]={1234};
uint8_t tx_data1= ' 1 ' ;

HAL_UART_Transmit(&huart2,tx_data,4,100); //4바이트 전송
HAL_UART_Transmit(&huart2,&tx_data1,1,100); //1바이트 전송
```

▶ 데이터 수신 함수

HAL_UART_Receive(UART_HandleTypeDef *huart, uint8_t *pData, uint16_t Size, uint32_t Timeout)

> 사용 방법: uart2를 통해서 특정 바이트의 데이터를 100밀리 이내 받고자 할 때

```
UART_HandleTypeDef huart2 ;
uint8_t rx_data[4];
uint8_t rx_data1 ' ;

HAL_UART_Receive(&huart2,rx_data,4,100); //4바이트 전송
HAL_UART_Transmit(&huart2,&rx_data1,1,100); //1바이트 전송
```

▶ **HAL_UART_Transmit 대신 printf** 사용 경우 아래 함수 메인에 추가할 것

```
#include " stdio.h "
int __io_putchar(int ch)
{
HAL_UART_Transmit(&UartHandle, (uint8_t *)&ch, 1, 0xFFFF);
return ch;
}
```

▶ 다음과 같은 프로토콜로 LED 2개를 on/off 해보자.
>LED 연결 :PC2 , PC3
>통신 방식(Polling), 통신 채널: USART2

stx	LED	On/off	etx
0x02	1	0	0x03
0x02	2	1	0x03

LED:1,2
On/Off: 1:on, 0:off

[그림 5.8] LED 제어 프로토콜

만약 "2113"을 전송하면 1번 LED "On" 의미이고 "2103"을 전송하면 1번 LED "Off"를 의미한다.

```
  uint8_t rx_data[4];
  while (1)
  {
      if(HAL_UART_Receive(&huart2, rx_data, 4, 100)==HAL_OK){
          if(rx_data[0]=='2' && rx_data[1]=='1' && rx_data[3]=='3'){
              if(rx_data[2]=='1') HAL_GPIO_WritePin(GPIOC, GPIO_PIN_2, 1);
              else HAL_GPIO_WritePin(GPIOC, GPIO_PIN_2, 0);
          }
          else if(rx_data[0]=='2' && rx_data[1]=='0' && rx_data[3]=='3'){
              if(rx_data[2]=='1') HAL_GPIO_WritePin(GPIOC, GPIO_PIN_3, 1);
              else HAL_GPIO_WritePin(GPIOC, GPIO_PIN_3, 0);
          }
          HAL_UART_Transmit(&huart2, rx_data, sizeof(rx_data), 100);
      }
    /* USER CODE END WHILE */

    /* USER CODE BEGIN 3 */
  }
  /* USER CODE END 3 */
}
```

[그림 5.9] 메인 프로그램 내용

3.4. 디버그 및 다운로드하기

메인 함수에 프로그램이 작성되었다면 다음으로 디버그는 [그림 5.10]의 를 클릭하거나 단축키 F11을 누르면 디버깅과 프로그램 다운로드가 완료된다.

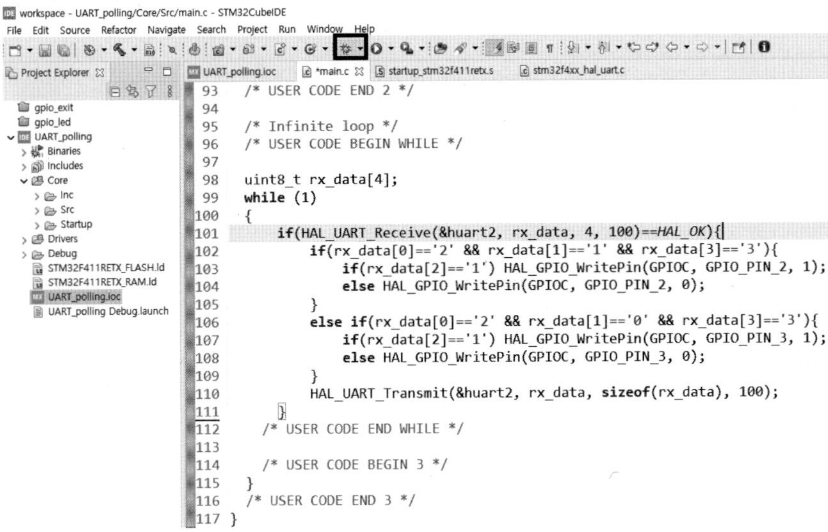
workspace - UART_polling/Core/Src/main.c - STM32CubeIDE
File Edit Source Refactor Navigate Search Project Run Window Help
Project Explorer
gpio_exit
gpio_led
UART_polling
Binaries
Includes
Core
Inc
Src
Startup
Drivers
Debug
STM32F411RETX_FLASH.ld
STM32F411RETX_RAM.ld
UART_polling.ioc
UART_polling Debug.launch
UART_polling.ioc
*main.c
startup_stm32f411retx.s
stm32f4xx_hal_uart.c
/* USER CODE END 2 */
/* Infinite loop */
/* USER CODE BEGIN WHILE */
uint8_t rx_data[4];
while (1)
{
if(HAL_UART_Receive(&huart2, rx_data, 4, 100)==HAL_OK){
if(rx_data[0]=='2' && rx_data[1]=='1' && rx_data[3]=='3'){
if(rx_data[2]=='1') HAL_GPIO_WritePin(GPIOC, GPIO_PIN_2, 1);
else HAL_GPIO_WritePin(GPIOC, GPIO_PIN_2, 0);
}
else if(rx_data[0]=='2' && rx_data[1]=='0' && rx_data[3]=='3'){
if(rx_data[2]=='1') HAL_GPIO_WritePin(GPIOC, GPIO_PIN_3, 1);
else HAL_GPIO_WritePin(GPIOC, GPIO_PIN_3, 0);
}
HAL_UART_Transmit(&huart2, rx_data, sizeof(rx_data), 100);
}
/* USER CODE END WHILE */
/* USER CODE BEGIN 3 */
}
/* USER CODE END 3 */
}

[그림 5.10] 프로그램 디버그

3.5. 보드 프로그램 실행하기

[그림 5.11]상단 중앙에 아이콘 클릭하면 디버그 모드가 종료되어 보드의 프로그램이 실행된다.

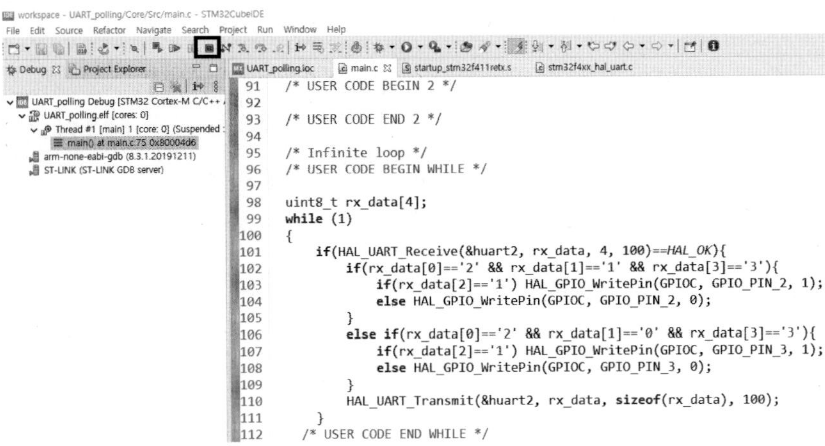
workspace - UART_polling/Core/Src/main.c - STM32CubeIDE
File Edit Source Refactor Navigate Search Project Run Window Help
Debug
Project Explorer
UART_polling Debug [STM32 Cortex-M C/C++
UART_polling.elf [cores: 0]
Thread #1 [main] 1 [core: 0] (Suspended
main() at main.c:75 0x80004d6
arm-none-eabi-gdb (8.3.1.20191211)
ST-LINK (ST-LINK GDB server)
UART_polling.ioc
main.c
startup_stm32f411retx.s
stm32f4xx_hal_uart.c
/* USER CODE BEGIN 2 */
/* USER CODE END 2 */
/* Infinite loop */
/* USER CODE BEGIN WHILE */
uint8_t rx_data[4];
while (1)
{
if(HAL_UART_Receive(&huart2, rx_data, 4, 100)==HAL_OK){
if(rx_data[0]=='2' && rx_data[1]=='1' && rx_data[3]=='3'){
if(rx_data[2]=='1') HAL_GPIO_WritePin(GPIOC, GPIO_PIN_2, 1);
else HAL_GPIO_WritePin(GPIOC, GPIO_PIN_2, 0);
}
else if(rx_data[0]=='2' && rx_data[1]=='0' && rx_data[3]=='3'){
if(rx_data[2]=='1') HAL_GPIO_WritePin(GPIOC, GPIO_PIN_3, 1);
else HAL_GPIO_WritePin(GPIOC, GPIO_PIN_3, 0);
}
HAL_UART_Transmit(&huart2, rx_data, sizeof(rx_data), 100);
}
/* USER CODE END WHILE */

[그림 5.11] 디버그 모드 프로그램 실행

3.6. 프로그램 실행확인

시리얼 통신 프로그램을 확인하기 쉬운 방법은 Tera term 이나 Hercules, docklight 통신 터미널 [그림 5.12]을 이용하면 된다. 인터넷에서 hercules terminal 검색하면 다운로드할 수 있다.

관련 사이트: https://www.hw-group.com/software/hercules-setup-utility

https://tera-term.softonic.kr/

https://docklight.de/downloads/

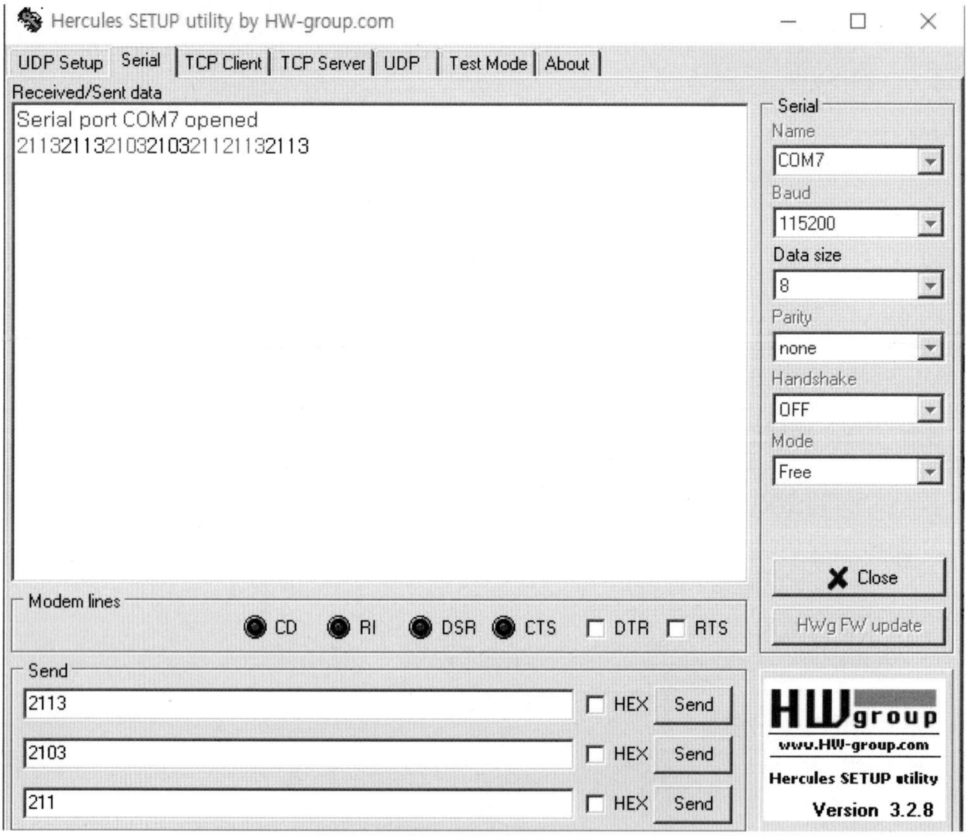

[그림 5.12] 시리얼 통신 터미널 실행

3.7. printf() 함수 사용하기

printf() 함수를 사용하기 위해서는 [그림 5.13]와 관련 함수과 헤더파일을 추가하면 된다.

```
/* USER CODE BEGIN Includes */
#include "stdio.h"
/* USER CODE END Includes */

/* USER CODE BEGIN 0 */
int __io_putchar(int ch)
{
HAL_UART_Transmit(&huart2, (uint8_t *)&ch, 1, 0xFFFF);
 return ch;
}
/* USER CODE END 0 */
 uint8_t count=0;
 while (1)
 {
    printf("count=%d\r\n",count++);
    if(count>256)count=0;
    HAL_Delay(100);
   /* USER CODE END WHILE */

   /* USER CODE BEGIN 3 */
 }
 /* USER CODE END 3 */
}
```

[그림 5.13] printf() 함수 사용 방법

〉 printf() 함수에 부동소수점 사용하기:

이를 위해서는 Project –〉 Properties –〉 C/C++ Build –〉 Settings –〉 MCU Settings에서 [그림 5.14]와 같이 체크를 하면 된다.

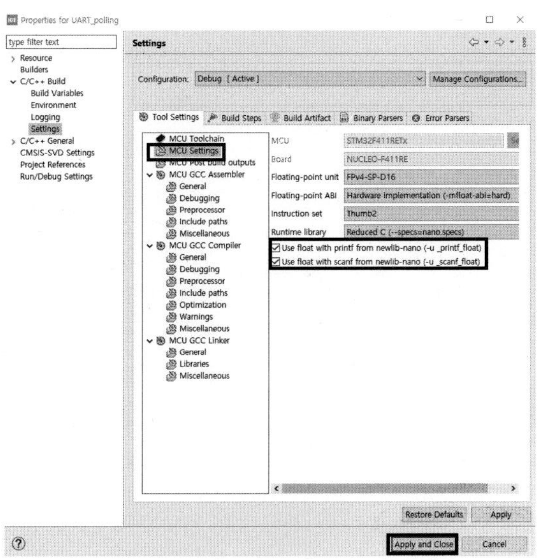

[그림 5.14] printf() float 사용 방법

[그림 5.15]와 같이 코딩 및 디버그한 다음 출력을 확인한다[그림 5.22].

```
uint8_t count=0;
float div;
while (1)
{
    div=(float)count/10;
    printf("count=%d, div=%f\r\n",count++,div);
    if(count>256)count=0;
    HAL_Delay(100);
  /* USER CODE END WHILE */

  /* USER CODE BEGIN 3 */
}
/* USER CODE END 3 */
}
```

[그림 5.15] float형 코딩

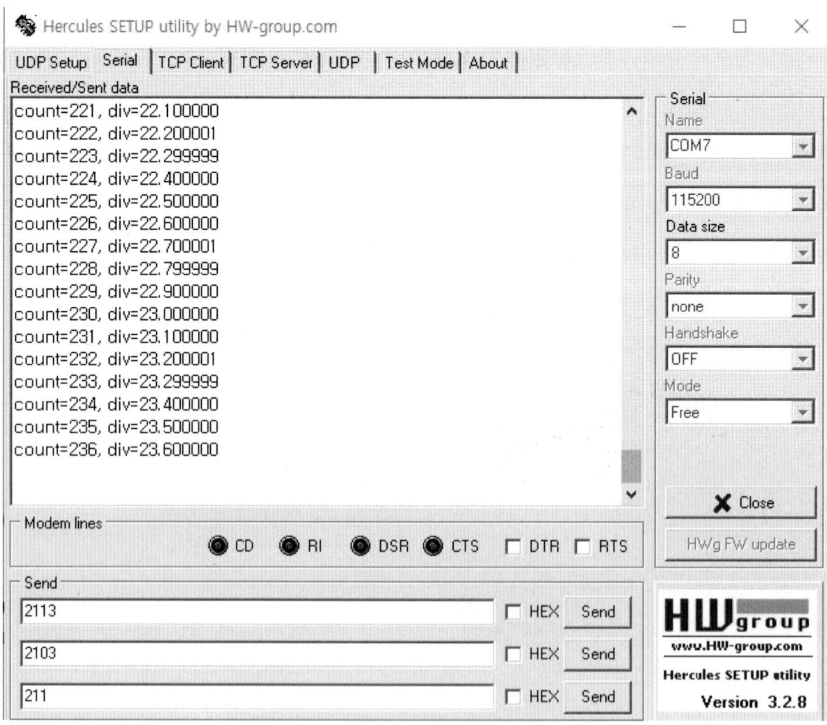

[그림 5.16] float형 출력화면

4. STM32IDE console 활용 Debug 사용하기

STM32IDE에는 프로그램 디버그 용도로 사용하는 콘솔이 있다. 이를 이용하면 별도의 출력 장치가 없어도 출력을 확인할 수 있다. 설정은 프로그램을 작성하고 디버그한 다음 아래 그림 순서대로 따라한다[그림5.17]~[그림 5.18].

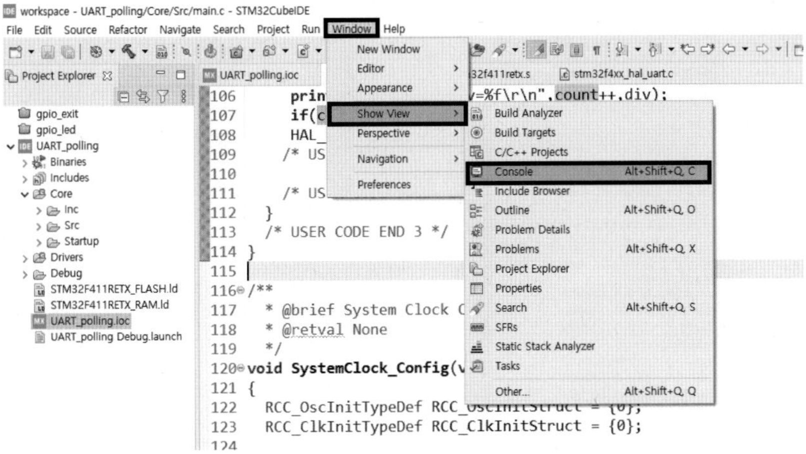

[그림 5.17] STM32IDE console 설정

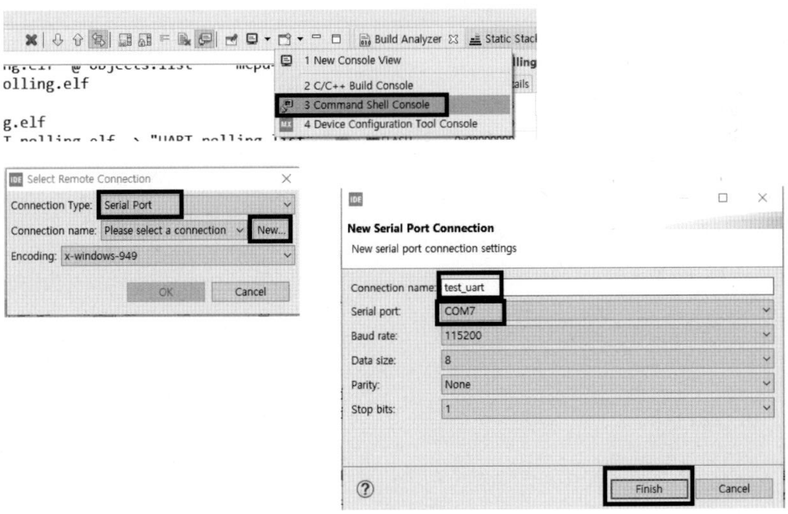

[그림 5.18] STM32IDE console 설정

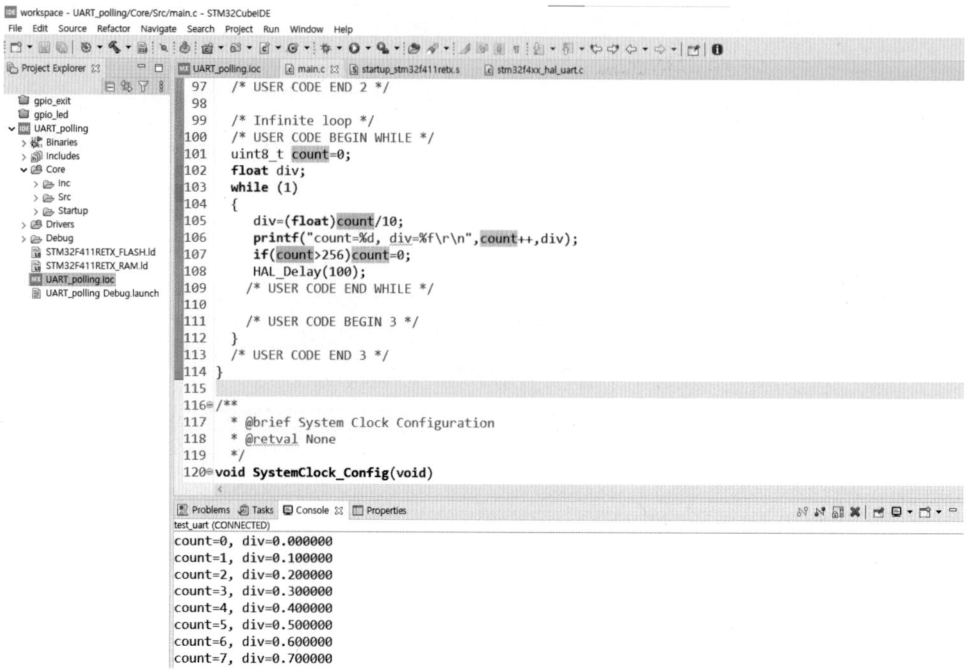
workspace - UART_polling/Core/Src/main.c - STM32CubeIDE
File Edit Source Refactor Navigate Search Project Run Window Help
Project Explorer
gpio_exit
gpio_led
UART_polling
Binaries
Includes
Core
Inc
Src
Startup
Drivers
Debug
STM32F411RETX_FLASH.ld
STM32F411RETX_RAM.ld
UART_polling.ioc
UART_polling Debug.launch
UART_polling.ioc
main.c
startup_stm32f411retx.s
stm32f4xx_hal_uart.c
/* USER CODE END 2 */
/* Infinite loop */
/* USER CODE BEGIN WHILE */
uint8_t count=0;
float div;
while (1)
{
div=(float)count/10;
printf("count=%d, div=%f\r\n",count++,div);
if(count>256)count=0;
HAL_Delay(100);
/* USER CODE END WHILE */
/* USER CODE BEGIN 3 */
}
/* USER CODE END 3 */
}
/**
* @brief System Clock Configuration
* @retval None
*/
void SystemClock_Config(void)
Problems Tasks Console Properties
test_uart (CONNECTED)
count=0, div=0.000000
count=1, div=0.100000
count=2, div=0.200000
count=3, div=0.300000
count=4, div=0.400000
count=5, div=0.500000
count=6, div=0.600000
count=7, div=0.700000

[그림 5.19] STM32IDE console 활용

5. Interrupt 방식

5.1. USART 설정

System Core에서 RCC 및 SYS는 이전과 같이 설정한다. USART2의 TX RX 단자는 [PA2, PA3]를 사용할 수가 있도록 [그림 5.20]같이 파라메터를 설정한다. [그림 5.21]과 [그림 5.22]는 핀 설정 및 Interrupt 방식으로 설정하였음을 알 수 있다.

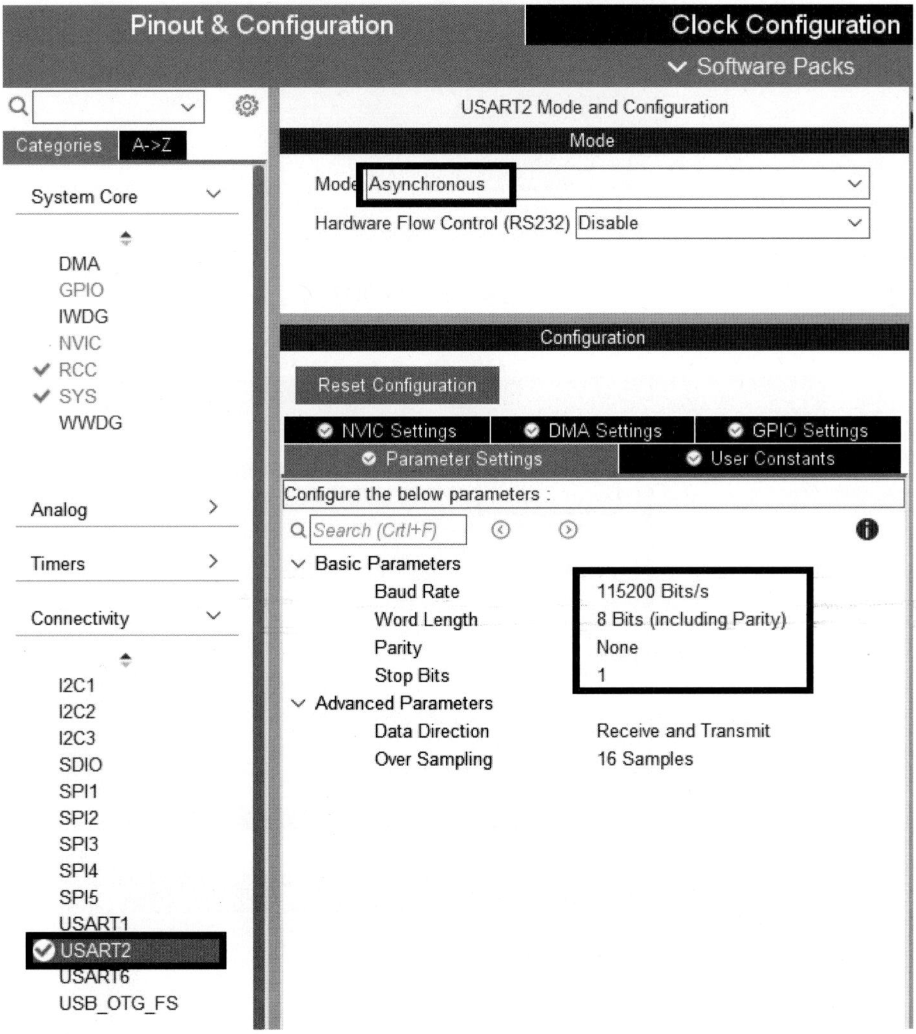

[그림 5.20] USART2 설정

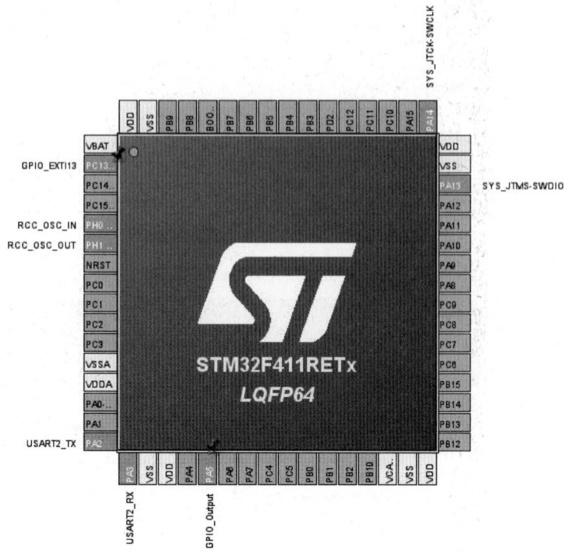

[그림 5.21] USART2 PINOUT 설정

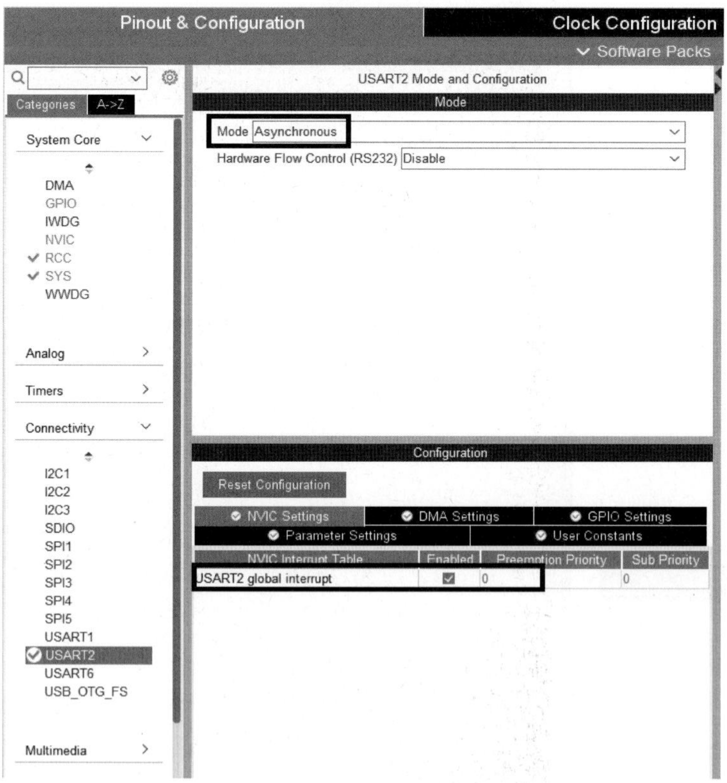

[그림 5.22] 인터럽트 설정

5.2. Generate Code 생성

[그림 5.23]와 [그림 5.24]는 코드를 생성하기 위한 화면이다. 순차적으로 클릭한다.

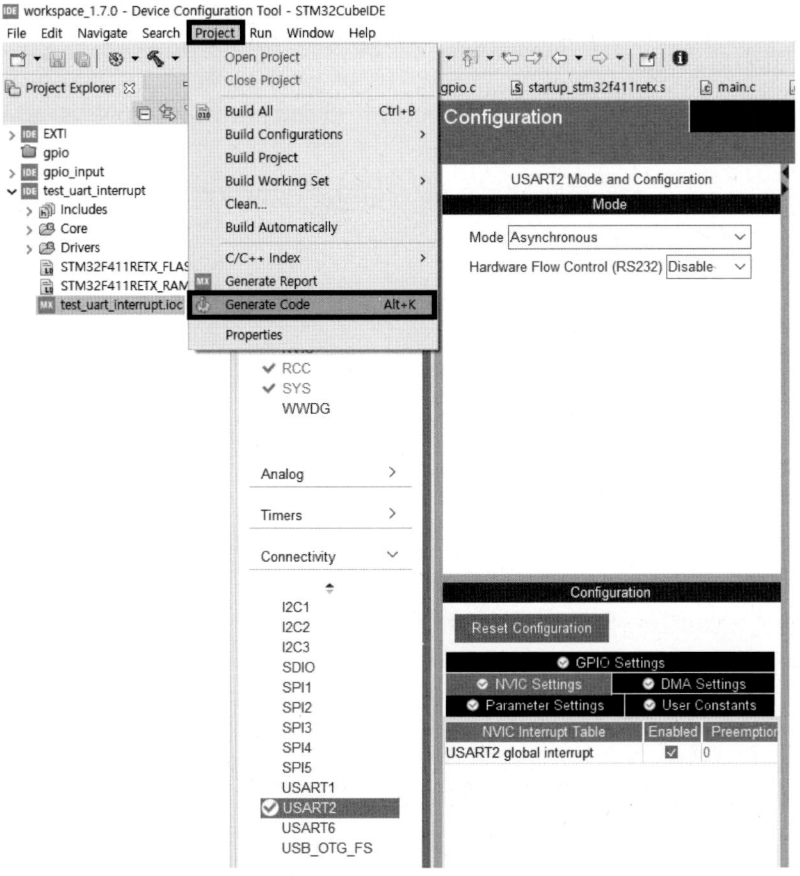

[그림 5.23] 프로그램 코드를 생성

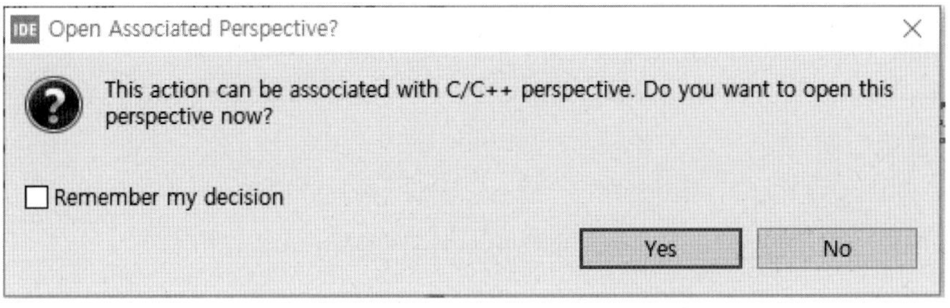

[그림 5.24] 프로젝트 코드를 생성

5.3. Source Code 작성하기

Interrupt 방식을 사용하기 위해서는 전송과 수신 및 인터럽트 Callback 함수를 이용하면 된다.

▶ 데이터 전송 함수

```
HAL_UART_Transmit_IT(UART_HandleTypeDef *huart, uint8_t *pData, uint16_t
Size)
> 사용 방법: uart2에 특정 바이트의 데이터를 보내고자 할 때
UART_HandleTypeDef huart2 ;
uint8_t tx_data[4]={1234};
uint8_t tx_data1=’1’;
HAL_UART_Transmit_IT(&huart2,tx_data,4); //4바이트 전송
HAL_UART_Transmit_IT(&huart2,&tx_data1,1);  //1바이트 전송
```

▶ 데이터 수신 함수

```
HAL_UART_Receive_IT(UART_HandleTypeDef *huart, uint8_t *pData, uint16_t
Size)
사용 방법: uart2를 통해서 특정 바이트의 데이터를 받고자 할 때
UART_HandleTypeDef huart2 ;
uint8_t rx_data[4];
uint8_t rx_data1’;
HAL_UART_Receive_IT(&huart2,rx_data,4); //4바이트 전송
HAL_UART_Transmit(&huart2,&rx_data1,1);  //1바이트 전송
```

▶ 인터럽트 콜백 함수

```
void HAL_UART_RxCpltCallback(UART_HandleTypeDef *huart)
{
수신 프로그램
}
void HAL_UART_TxCpltCallback(UART_HandleTypeDef *huart)
{
송신 프로그램
}
```

▶ 다음과 같은 프로토콜로 LED 2개를 on/off 해보자.

> LED 연결 :PC2, PC3

> 통신 방식(Interrupt), 통신 채널: USART2

stx	LED	On/off	etx
0x02	1	0	0x03
0x02	2	1	0x03

LED:1,2
On/Off: 1:on, 0:off

[그림 5.25] LED 제어 프로토콜

만약 “2113”을 전송하면 1번 LED “On”의미이고 “2103”을 전송하면 1번 LED “Off”를 의미한다.

```
/* USER CODE BEGIN 0 */
uint8_t rx_data[4];
void HAL_UART_RxCpltCallback(UART_HandleTypeDef *huart)
{
    if(huart->Instance==USART2){
        if(rx_data[0]=='2' && rx_data[3]=='3' && rx_data[1]=='1'){
                if(rx_data[2]=='1') HAL_GPIO_WritePin(GPIOC, GPIO_PIN_2, 1);
                else if(rx_data[2]=='0')HAL_GPIO_WritePin(GPIOC, GPIO_PIN_2, 0);
            }
        else if(rx_data[0]=='2' && rx_data[3]=='3' && rx_data[1]=='2'){
                if(rx_data[2]=='1') HAL_GPIO_WritePin(GPIOC, GPIO_PIN_3, 1);
                else if(rx_data[2]=='0')HAL_GPIO_WritePin(GPIOC, GPIO_PIN_3, 0);
            }
        else if(rx_data[0]=='2' && rx_data[3]=='3' && rx_data[1]=='0'){
            HAL_GPIO_WritePin(GPIOC, GPIO_PIN_2, 0);
            HAL_GPIO_WritePin(GPIOC, GPIO_PIN_3, 0);
            }
    HAL_UART_Transmit_IT(&huart2, rx_data, 4); //Echo mode
    HAL_UART_Receive_IT(&huart2, rx_data, 4);  //재 인터럽트요청
    }
}
/* USER CODE END 0 */
```

```
  /* USER CODE BEGIN 2 */
  HAL_UART_Receive_IT(&huart2, rx_data, 4);
  /* USER CODE END 2 */

  /* Infinite loop */
  /* USER CODE BEGIN WHILE */
  while (1)
  {

    /* USER CODE END WHILE */

    /* USER CODE BEGIN 3 */
  }
  /* USER CODE END 3 */
}
```

[그림 5.26] 메인 프로그램 내용

5.4. 디버그 및 다운로드하기

메인 함수에 프로그램이 작성되었다면 다음으로 디버그는 [그림 5.27]의 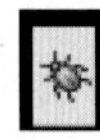를 클릭하거나 단축키 F11을 누르면 디버깅과 프로그램 다운로드가 완료된다.

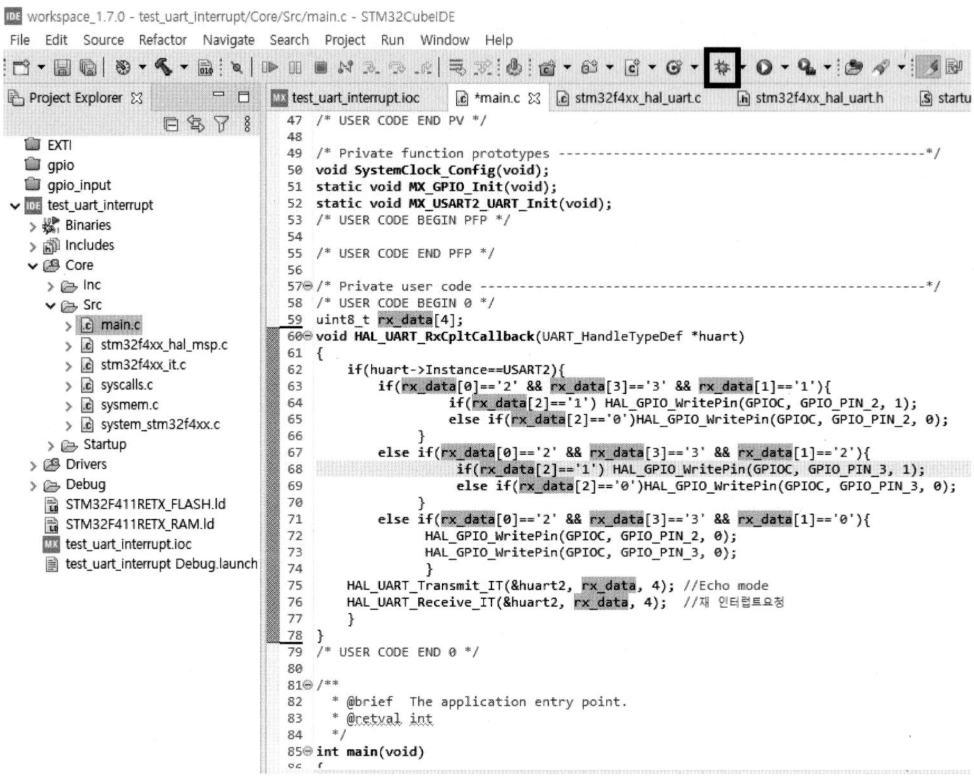

[그림 5.27] 프로그램 디버그

5.5. 보드 프로그램 실행하기

[그림 5.28] 상단 중앙에 아이콘 ■ 클릭하면 디버그 모드가 종료되어 보드의 프로그램이 실행된다.

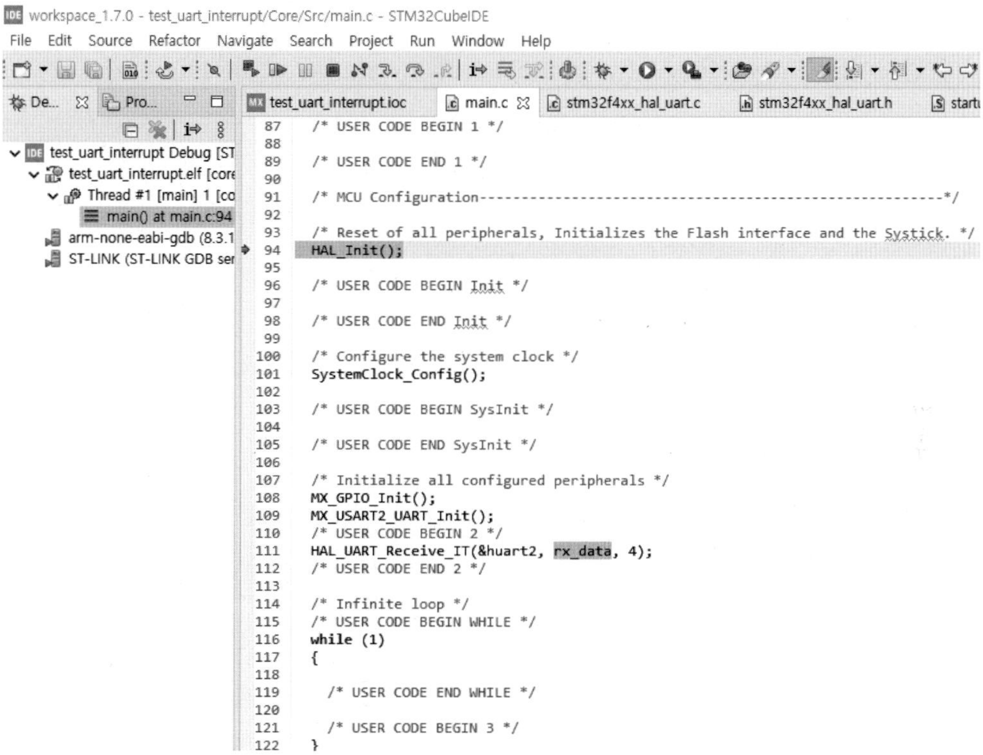
workspace_1.7.0 - test_uart_interrupt/Core/Src/main.c - STM32CubeIDE
File Edit Source Refactor Navigate Search Project Run Window Help
De... Pro...
test_uart_interrupt.ioc
main.c
stm32f4xx_hal_uart.c
stm32f4xx_hal_uart.h
test_uart_interrupt Debug [ST
test_uart_interrupt.elf [core
Thread #1 [main] 1 [co
main() at main.c:94
arm-none-eabi-gdb (8.3.1
ST-LINK (ST-LINK GDB ser
/* USER CODE BEGIN 1 */
/* USER CODE END 1 */
/* MCU Configuration--*/
/* Reset of all peripherals, Initializes the Flash interface and the Systick. */
HAL_Init();
/* USER CODE BEGIN Init */
/* USER CODE END Init */
/* Configure the system clock */
SystemClock_Config();
/* USER CODE BEGIN SysInit */
/* USER CODE END SysInit */
/* Initialize all configured peripherals */
MX_GPIO_Init();
MX_USART2_UART_Init();
/* USER CODE BEGIN 2 */
HAL_UART_Receive_IT(&huart2, rx_data, 4);
/* USER CODE END 2 */
/* Infinite loop */
/* USER CODE BEGIN WHILE */
while (1)
{
/* USER CODE END WHILE */
/* USER CODE BEGIN 3 */
}

[그림 5.28] 디버그 모드 프로그램 실행

5.6. 프로그램 실행 확인

시리얼 통신 프로그램을 확인하기 쉬운 방법은 Tera term 이나 Hercules,docklight 통신 터미널 [그림 5.29]를 이용하면 된다. 인터넷에서 hercules terminal 검색하면 다운로드할 수 있다.

관련사이트 : https://www.hw-group.com/software/hercules-setup-utility
https://tera-term.softonic.kr/
https://docklight.de/downloads/

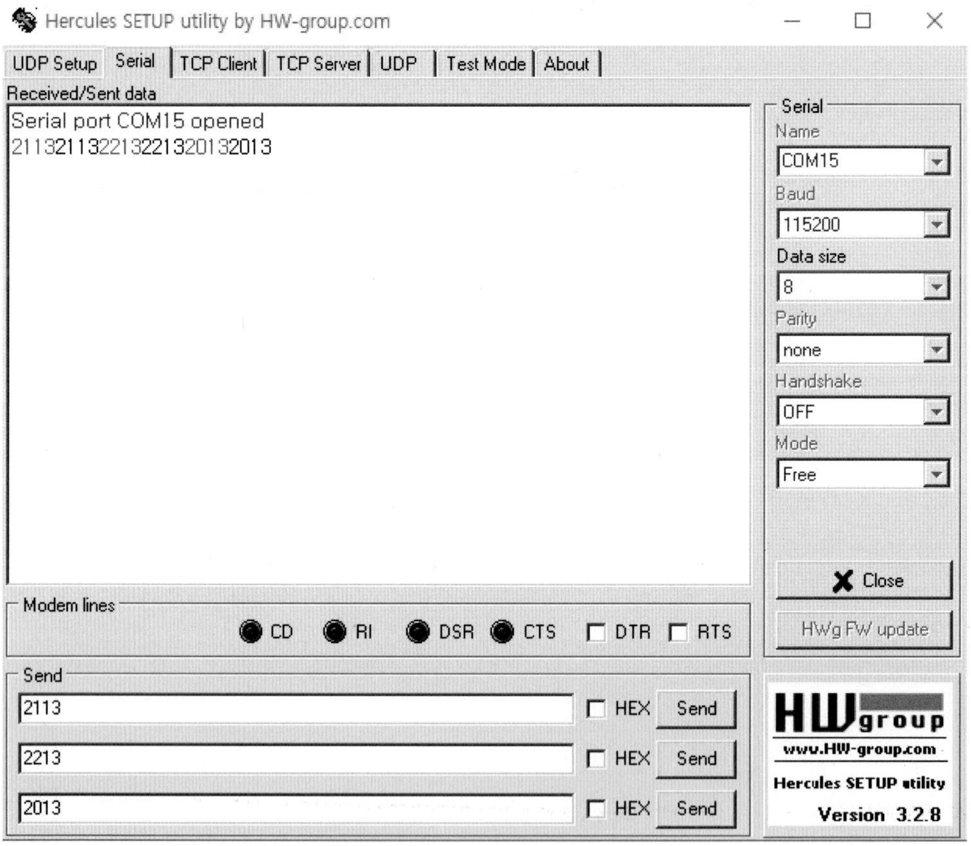

[그림 5.29] 시리얼 통신 터미널 실행

6. DMA 방식

6.1. USART 설정

System Core에서 RCC 및 SYS는 이전과 같이 설정한다. USART2의 TX RX 단자는 [PA2, PA3]를 사용할 수 있도록 [그림 5.30]같이 파라메터를 설정한다. [그림 5.31]과 [그림 5.32]는 핀 설정 및 DMA 방식으로 설정하였음을 알 수 있다.

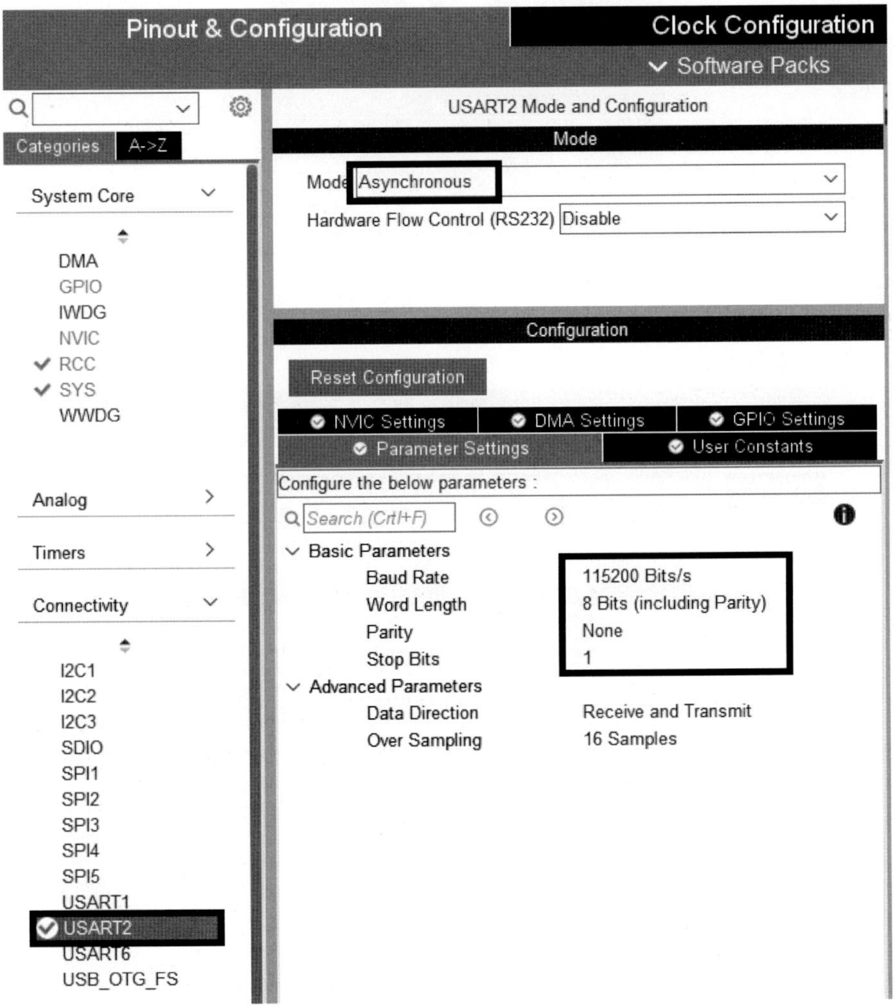

[그림 5.30] USART2 설정

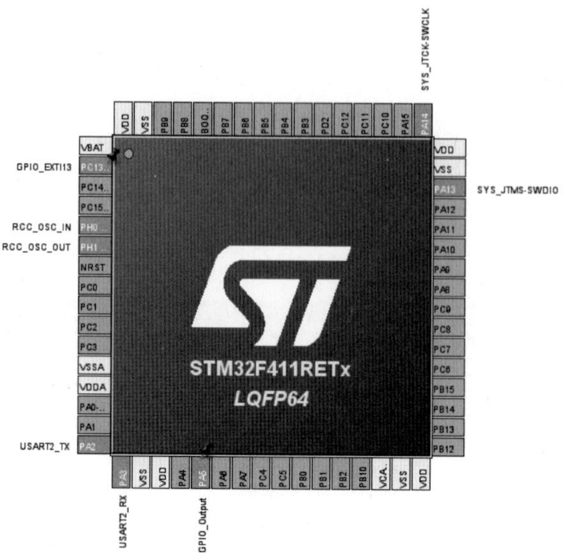

[그림 5.31] USART2 PINOUT 설정

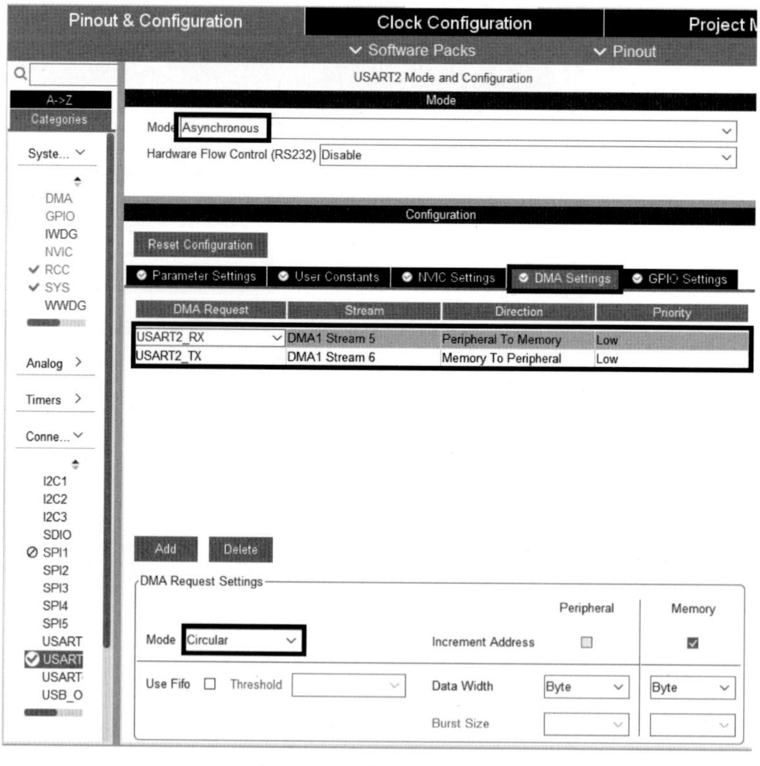

[그림 5.32] DMA 설정

6.2. Generate Code 생성

[그림 5.33]과 [그림 5.34]는 코드를 생성하기 위한 화면이다. 순차적으로 클릭한다.

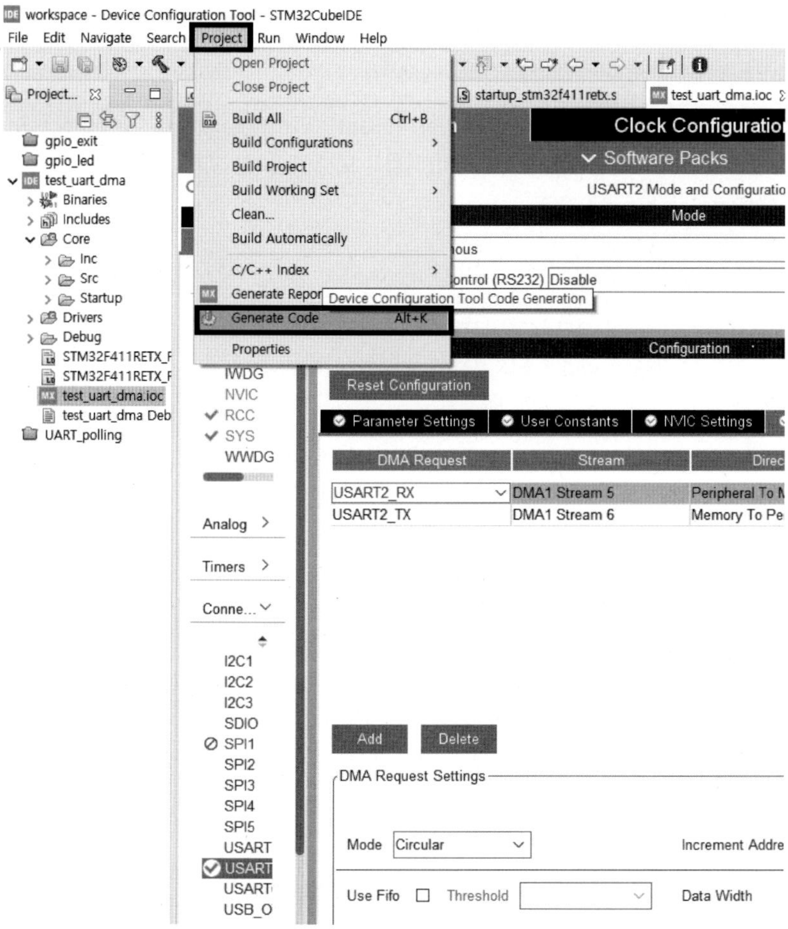

[그림 5.33] 프로그램 코드를 생성

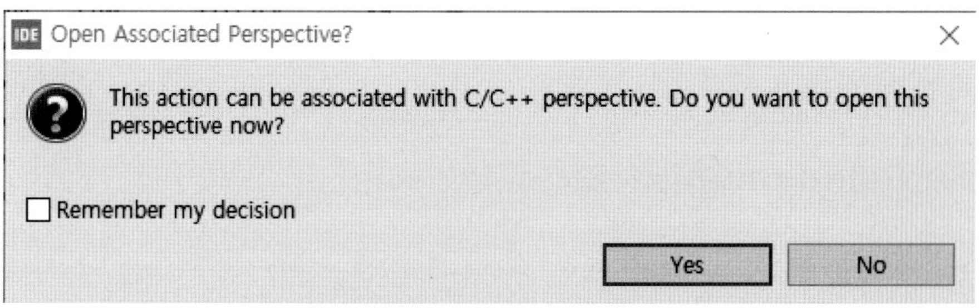

[그림 5.34] 프로젝트 코드를 생성

6.3. Source Code 작성하기

Interrupt 방식을 사용하기 위해서는 전송과 수신 및 DMA Callback 함수를 이용하면 된다.

▶ 데이터 전송 함수

```
HAL_UART_Transmit_DMA(UART_HandleTypeDef *huart, uint8_t *pData, uint16_t Size)
> 사용 방법: uart2에 특정 바이트의 데이터를  보내고자 할 때
UART_HandleTypeDef huart2 ;
uint8_t tx_data[4]={1234};
uint8_t tx_data1= ' 1 ' ;
HAL_UART_Transmit_DMA(&huart2,tx_data,4); //4바이트 전송
HAL_UART_Transmit_DMA(&huart2,&tx_data1,1);  //1바이트 전송
```

▶ 데이터 수신 함수

```
HAL_UART_Receive_DMA(UART_HandleTypeDef *huart, uint8_t *pData, uint16_t Size)
사용 방법: uart2를 통해서 특정 바이트의 데이터를 받고자 할 때
UART_HandleTypeDef huart2 ;
uint8_t rx_data[4];
uint8_t rx_data1 ' ;
HAL_UART_Receive_DMA(&huart2,rx_data,4); //4바이트 전송
HAL_UART_Receive_DMA(&huart2,&rx_data1,1); //1바이트 전송
```

▶ 인터럽트 콜백 함수

```
void HAL_UART_RxCpltCallback(UART_HandleTypeDef *huart)
{
수신이 완료되면 동작
}
void HAL_UART_TxCpltCallback(UART_HandleTypeDef *huart)
{
송신이 완료되면 동작
}
HAL_UART_RxHalfCpltCallback(UART_HandleTypeDef *huart)
{
설정된 데이터의 절반이 도착하면  동작
}
```

▶ 다음과 같은 프로토콜로 LED 2개를 on/off 해보자.

> LED 연결: PC2 , PC3
> 통신 방식(DMA), 통신채널: USART2

stx	LED	On/off	etx
0x02	1	0	0x03
0x02	2	1	0x03

LED:1,2
On/Off: 1:on, 0:off

[그림 5.35a] LED 제어 프로토콜

만약 "2113"을 전송하면 1번 LED "On" 의미이고 "2103"을 전송하면 1번 LED "Off"를 의미한다.

```
uint8_t rx_data[4];
uint8_t data_arrived=0;
void HAL_UART_RxCpltCallback(UART_HandleTypeDef *huart)
{
    data_arrived=1;

    HAL_UART_Transmit(&huart2, (uint8_t *)rx_data, 4,100);
    HAL_UART_Receive_DMA(&huart2, (uint8_t *)rx_data, 4);
}
void HAL_UART_RxHalfCpltCallback(UART_HandleTypeDef *huart){

    HAL_GPIO_TogglePin(GPIOA, GPIO_PIN_5);

}
```

```
HAL_UART_Receive_DMA(&huart2, (uint8_t *)rx_data, 4);
while(!data_arrived);
/* USER CODE END 2 */

/* Infinite loop */
/* USER CODE BEGIN WHILE */
while (1)
{
  if(data_arrived==1){
      if(rx_data[0]=='2' && rx_data[3]=='3' && rx_data[1]=='1'){
          if(rx_data[2]=='1') HAL_GPIO_WritePin(GPIOC,GPIO_PIN_2, 1);
          else HAL_GPIO_WritePin(GPIOC,GPIO_PIN_2, 0);
      }
      else if(rx_data[0]=='2' && rx_data[3]=='3' && rx_data[1]=='2'){
          if(rx_data[2]=='1') HAL_GPIO_WritePin(GPIOC,GPIO_PIN_3, 1);
          else HAL_GPIO_WritePin(GPIOC,GPIO_PIN_3, 0);
      }
  data_arrived=0;
  }
  HAL_Delay(100);
```

[그림 5.35b] 메인 프로그램 내용

6.4. 디버그 및 다운로드하기

메인 함수에 프로그램이 작성되었다면 다음으로 디버그는 [그림 5.36]의 를 클릭하거나 단축키 F11을 누르면 디버깅과 프로그램 다운로드가 완료된다.

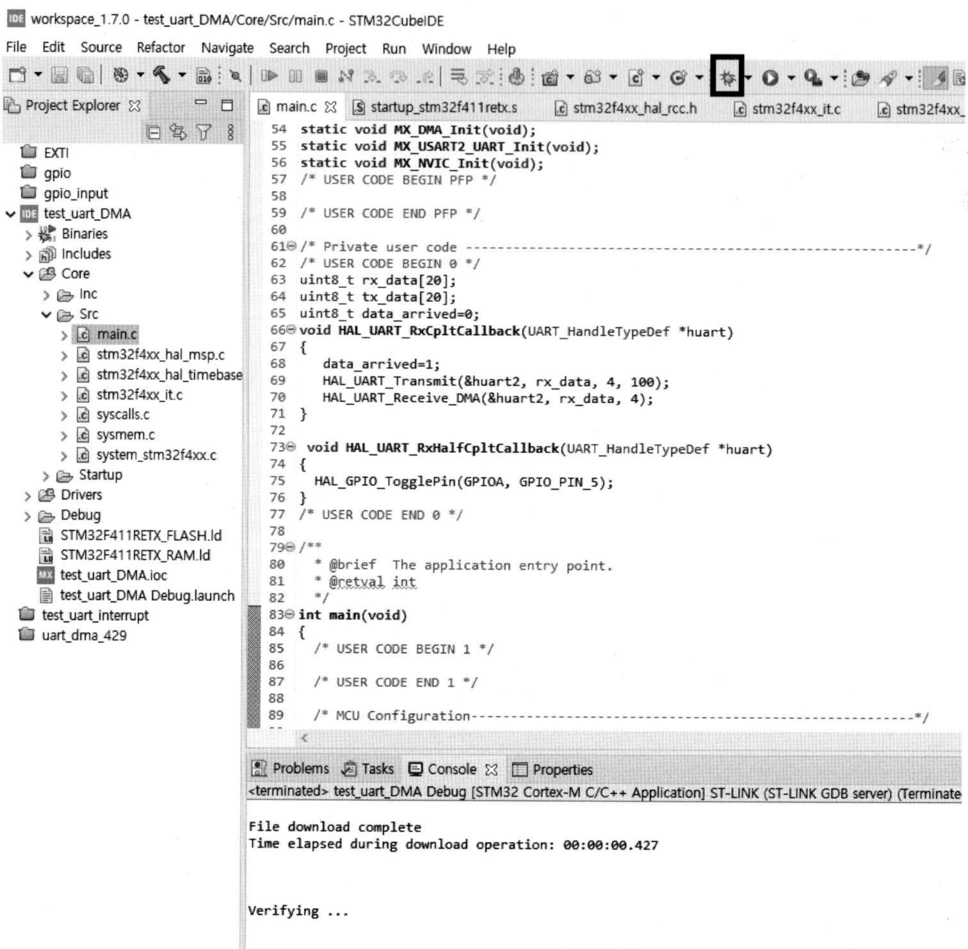
workspace_1.7.0 - test_uart_DMA/Core/Src/main.c - STM32CubeIDE
File Edit Source Refactor Navigate Search Project Run Window Help
Project Explorer
main.c
startup_stm32f411retx.s
stm32f4xx_hal_rcc.h
stm32f4xx_it.c
EXTI
gpio
gpio_input
test_uart_DMA
Binaries
Includes
Core
Inc
Src
main.c
stm32f4xx_hal_msp.c
stm32f4xx_hal_timebase
stm32f4xx_it.c
syscalls.c
sysmem.c
system_stm32f4xx.c
Startup
Drivers
Debug
STM32F411RETX_FLASH.ld
STM32F411RETX_RAM.ld
test_uart_DMA.ioc
test_uart_DMA Debug.launch
test_uart_interrupt
uart_dma_429
static void MX_DMA_Init(void);
static void MX_USART2_UART_Init(void);
static void MX_NVIC_Init(void);
/* USER CODE BEGIN PFP */
/* USER CODE END PFP */
/* Private user code ---*/
/* USER CODE BEGIN 0 */
uint8_t rx_data[20];
uint8_t tx_data[20];
uint8_t data_arrived=0;
void HAL_UART_RxCpltCallback(UART_HandleTypeDef *huart)
{
data_arrived=1;
HAL_UART_Transmit(&huart2, rx_data, 4, 100);
HAL_UART_Receive_DMA(&huart2, rx_data, 4);
}
void HAL_UART_RxHalfCpltCallback(UART_HandleTypeDef *huart)
{
HAL_GPIO_TogglePin(GPIOA, GPIO_PIN_5);
}
/* USER CODE END 0 */
/**
* @brief The application entry point.
* @retval int
*/
int main(void)
{
/* USER CODE BEGIN 1 */
/* USER CODE END 1 */
/* MCU Configuration--*/
Problems Tasks Console Properties
<terminated> test_uart_DMA Debug [STM32 Cortex-M C/C++ Application] ST-LINK (ST-LINK GDB server) (Terminate
File download complete
Time elapsed during download operation: 00:00:00.427
Verifying ...

[그림 5.36] 프로그램 디버그

6.5. 보드 프로그램 실행하기

[그림 5.37] 상단 중앙에 아이콘 클릭하면 디버그 모드가 종료되어 보드의 프로그램이 실행된다.

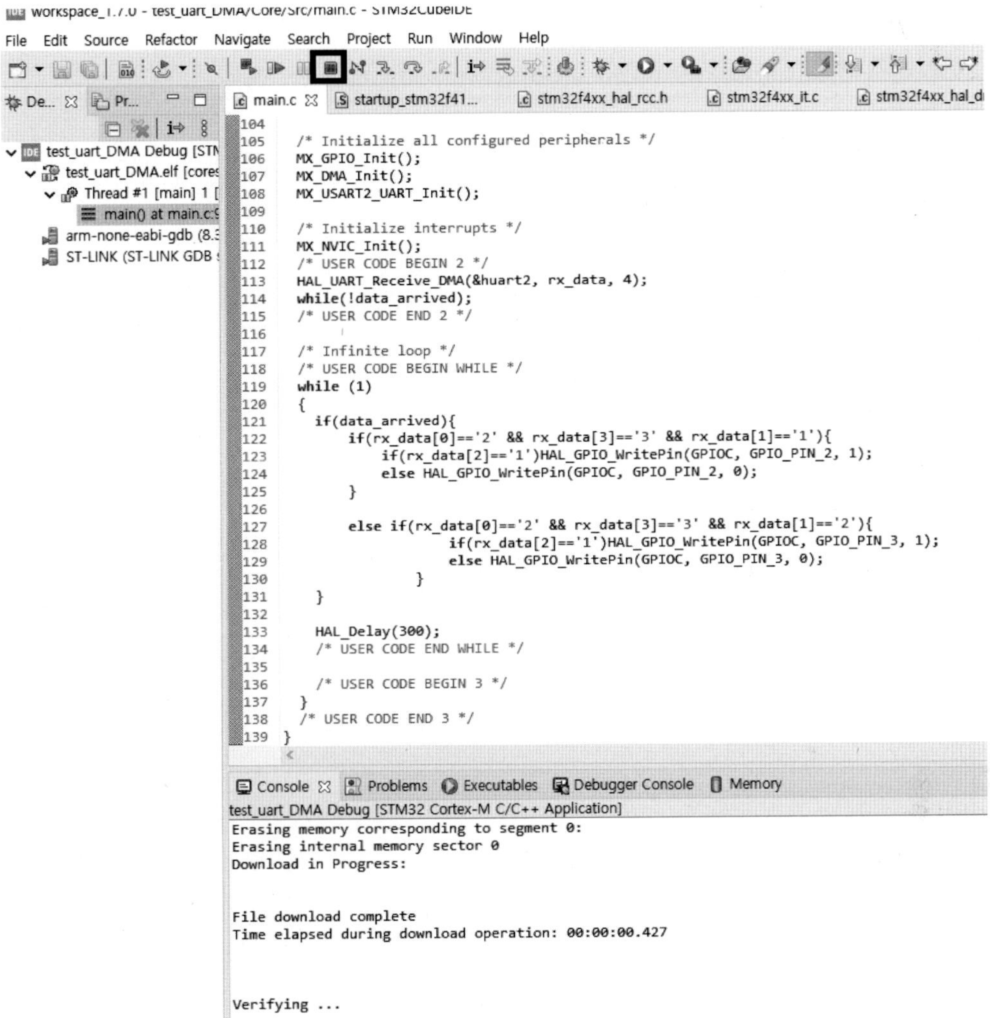

File Edit Source Refactor Navigate Search Project Run Window Help
main.c
startup_stm32f41...
stm32f4xx_hal_rcc.h
stm32f4xx_it.c
test_uart_DMA Debug
test_uart_DMA.elf
Thread #1 [main] 1
main() at main.c
arm-none-eabi-gdb
ST-LINK (ST-LINK GDB
/* Initialize all configured peripherals */
MX_GPIO_Init();
MX_DMA_Init();
MX_USART2_UART_Init();
/* Initialize interrupts */
MX_NVIC_Init();
/* USER CODE BEGIN 2 */
HAL_UART_Receive_DMA(&huart2, rx_data, 4);
while(!data_arrived);
/* USER CODE END 2 */
/* Infinite loop */
/* USER CODE BEGIN WHILE */
while (1)
{
if(data_arrived){
if(rx_data[0]=='2' && rx_data[3]=='3' && rx_data[1]=='1'){
if(rx_data[2]=='1')HAL_GPIO_WritePin(GPIOC, GPIO_PIN_2, 1);
else HAL_GPIO_WritePin(GPIOC, GPIO_PIN_2, 0);
}
else if(rx_data[0]=='2' && rx_data[3]=='3' && rx_data[1]=='2'){
if(rx_data[2]=='1')HAL_GPIO_WritePin(GPIOC, GPIO_PIN_3, 1);
else HAL_GPIO_WritePin(GPIOC, GPIO_PIN_3, 0);
}
}
HAL_Delay(300);
/* USER CODE END WHILE */
/* USER CODE BEGIN 3 */
}
/* USER CODE END 3 */
}
Console Problems Executables Debugger Console Memory
test_uart_DMA Debug [STM32 Cortex-M C/C++ Application]
Erasing memory corresponding to segment 0:
Erasing internal memory sector 0
Download in Progress:
File download complete
Time elapsed during download operation: 00:00:00.427
Verifying ...

[그림 5.37] 디버그 모드 프로그램 실행

6.6. 프로그램 실행 확인

시리얼 통신 프로그램을 확인하기 쉬운 방법은 Tera term 이나 Hercules,docklight 통신 터미널 [그림 5.38]를 이용하면 된다. 인터넷에서 hercules terminal 검색하면 다운로드할 수 있다.

관련사이트: https://www.hw-group.com/software/hercules-setup-utility
https://tera-term.softonic.kr/
https://docklight.de/downloads/

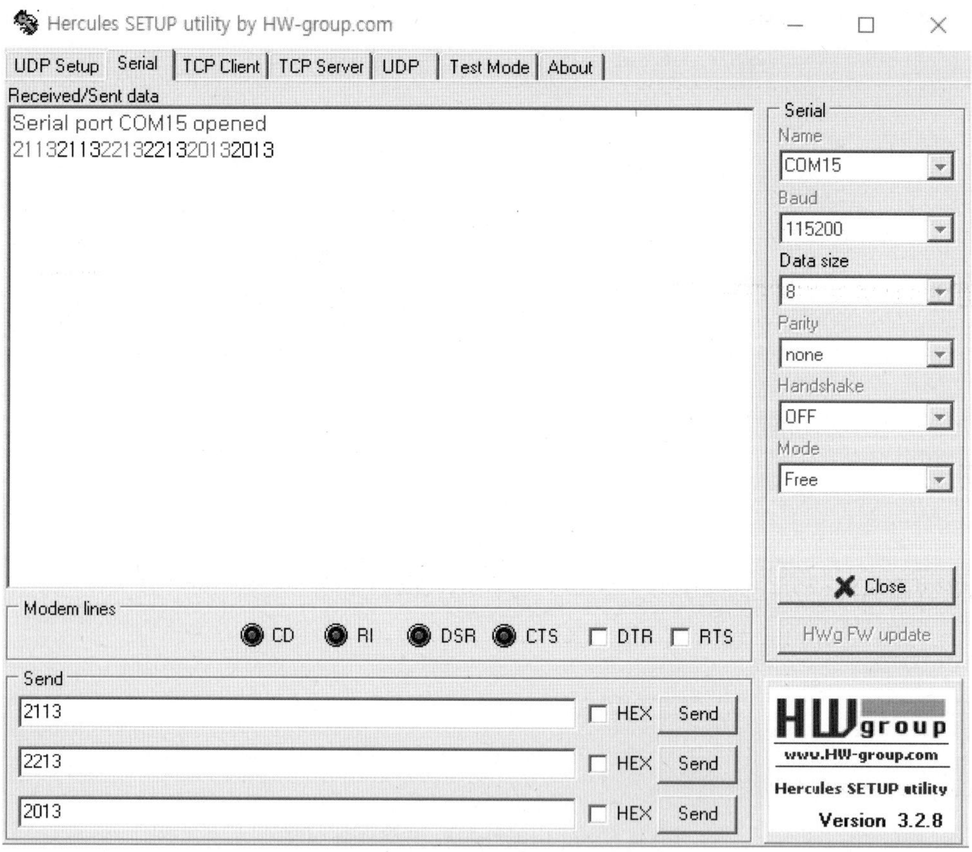

[그림 5.38] 시리얼 통신 터미널 실행

CHAPTER 6

ADC 활용하기 (Polling, Interrupt, DMA)

CHAPTER 06

Cortex-M4 기초 활용(ARM Core STM32)

ADC 활용하기 (Polling, Interrupt, DMA)

1. 개요

ADC는 Analog to Digital Converter의 약자이다. 마이크로컨트롤러들은 디지털 구성체이기 때문에 이진 및 이진화된 신호만 이해할 수 있다. 그러므로 센서의 데이터를 알고자 하면 센서의 물리적 변화를 전압이나 전류로 변환시켜 ADC로 읽으면 된다. ADC의 필요성이 여기에 있다고 볼 수 있다.

STM32F4 시리즈도 종류에 따라 차이는 있지만 2~3개의 ADC 모듈이 있고 single, continuous, scan, discontinuous 모드 및 18개의 채널을 가지고 있다. 2개의 채널은 온도 센서 및 ADC Voltage Reference 용도로 내부에 연결이 되어 있다. ADC를 이용하려면 ADC 모듈과 채널(PIN)을 선택하고 프로그램을 작성하면 된다.

Channel No.	ADC module 1	ADC module 2	ADC module 3
0	PA0	PA0	PA0
1	PA1	PA1	PA1
2	PA2	PA2	PA2
3	PA3	PA3	PA3
4	PA4	PA4	PF6
5	PA5	PA5	PF7
6	PA6	PA6	PF8
7	PA7	PA7	PF9
8	PB0	PB0	PF10
9	PB1	PB1	PF3
10	PC0	PC0	PC0
11	PC1	PC1	PC1
12	PC2	PC2	PC2
13	PC3	PC3	PC3
14	PC4	PC4	PF4
15	PC5	PC5	PF5

[그림 6.1] STM32F4의 ADC 모듈 및 채널

2. ADC 하드웨어 설정

[그림 6.2]와 같이 포트 PA0, PA1, PA4에 PointA ~C를 연결하여 해당 포인트의 전압을 표시하는 과정을 진행한다. 또한, 결과를 시리얼 통신 포트로 표시한다. 시리얼 통신은 5장에서와 같이 USART2를 이용한다.

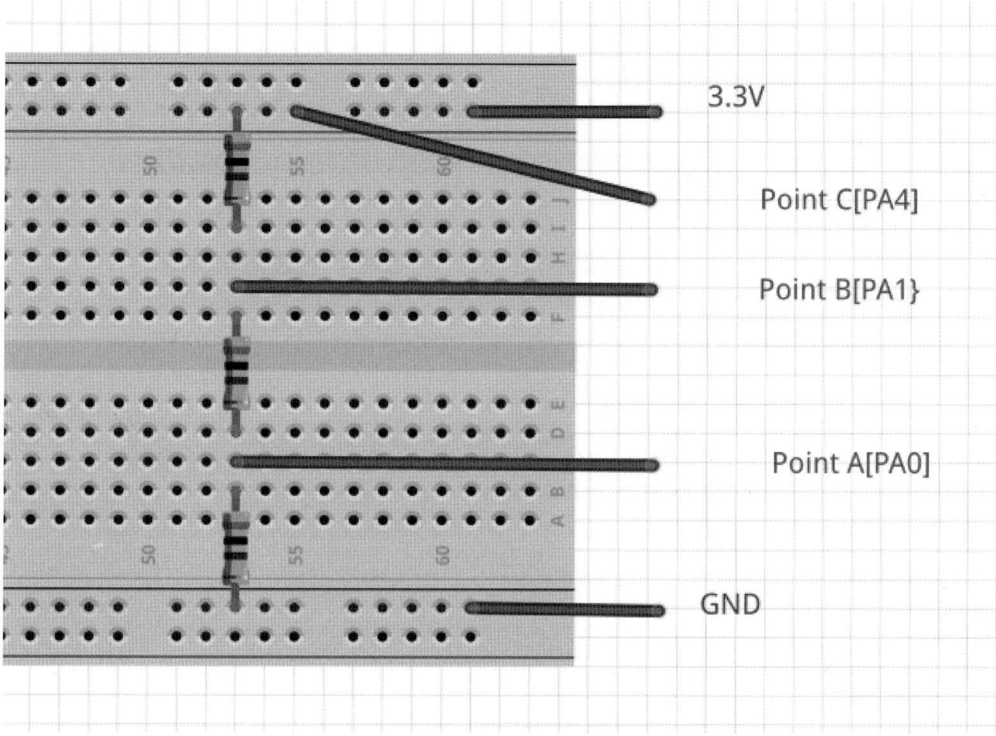

[그림 6.2] ADC PORT 결선 회로

3. Polling 방식 싱글

3.1. ADC 1개 사용 경우 설정

System Core에서 RCC 및 SYS 및 TX RX 단자는 이전과 같이 설정한다. [그림 6.3]은 Channel0을 폴링 방식 선언을 나타내고 있다.

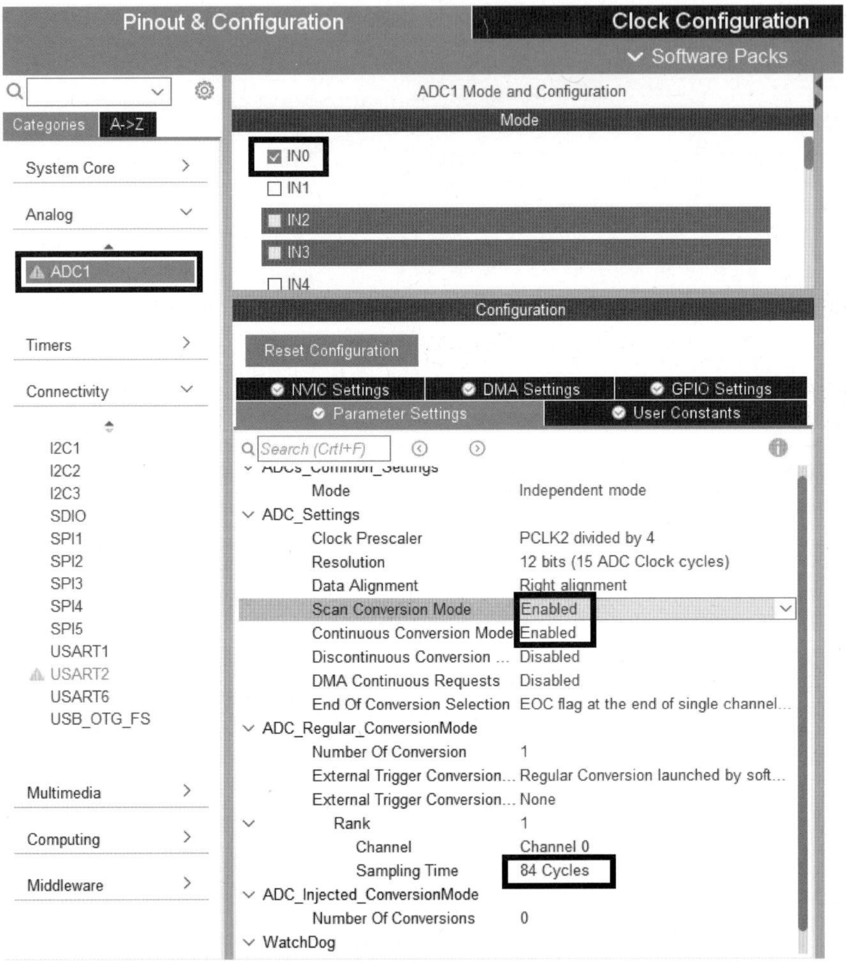

[그림 6.3] ADC1 IN0 설정

3.2. Generate Code 생성

[그림 6.4]는 코드를 생성하기 위한 화면이다. 그림과 같이 클릭해서 프로그램을 생성한다.

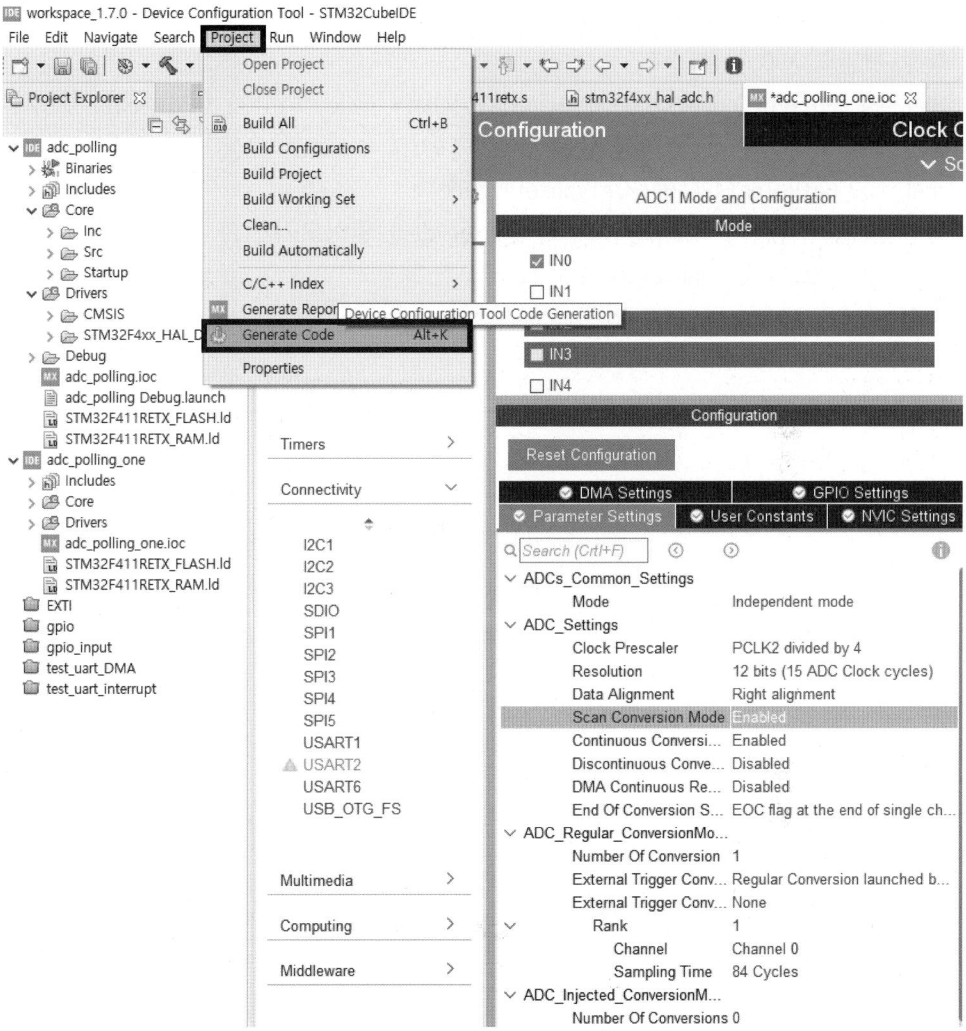

[그림 6.4] 프로그램 코드를 생성

3.3. Source Code 작성하기

ADC를 Polling 방식으로 사용하려면 관련된 함수를 알아야 한다. 함수를 살펴보자.

>HAL_ADC_Start(ADC_HandleTypeDef* hadc):
특정 채널의 ADC 시작 함수
>HAL_ADC_PollForConversion(ADC_HandleTypeDef* hadc, uint32_t Timeout):
특정 채널의 ADC 변환 완료를 위한 관련 함수
>HAL_ADC_GetValue(ADC_HandleTypeDef* hadc):
특정 채널의 데이터 레지스터로부터 변환된 값을 읽어오기
>HAL_ADC_Stop(ADC_HandleTypeDef* hadc):
특정 채널의 변환을 STOP 과 Disable 함수

[그림 6.5]는 메인 코드 내용이다. 디버그 및 다운로드한 다음 실행해 보기 바란다. [그림 6.2] 회로를 기반으로 Point A에서 C까지를 각 포인터를 측정해 보기 바란다.

```
/* USER CODE BEGIN 0 */
int __io_putchar(int ch)
{
HAL_UART_Transmit(&huart2, (uint8_t *)&ch, 1, 0xFFFF);
return ch;
}
/* USER CODE END 0 */
```

```
  /* USER CODE BEGIN WHILE */
  uint32_t adc0;
  float volt;
  while (1)
  {
      HAL_ADC_Start(&hadc1);
      HAL_ADC_PollForConversion(&hadc1, 100);
      adc0=HAL_ADC_GetValue(&hadc1);
      HAL_ADC_Stop(&hadc1);

      volt=(float)adc0*3.3/4096.0;
      printf("volt=%f, adc0=%ld\r\n",volt,adc0);
      HAL_Delay(100);
    /* USER CODE END WHILE */
    /* USER CODE BEGIN 3 */
  }
  /* USER CODE END 3 */
}
```

[그림 6.5] 메인 코드 작성

4. Polling 방식 멀티

4.1. ADC 3개 사용 경우 설정

System Core에서 RCC 및 SYS 및 TX RX 단자는 이전과 같이 설정한다. [그림 6.2]와 같이 구성한다. [그림 6.6]은 Channel0,1,4 3개를 폴링 방식 선언을 나타내고 있다.

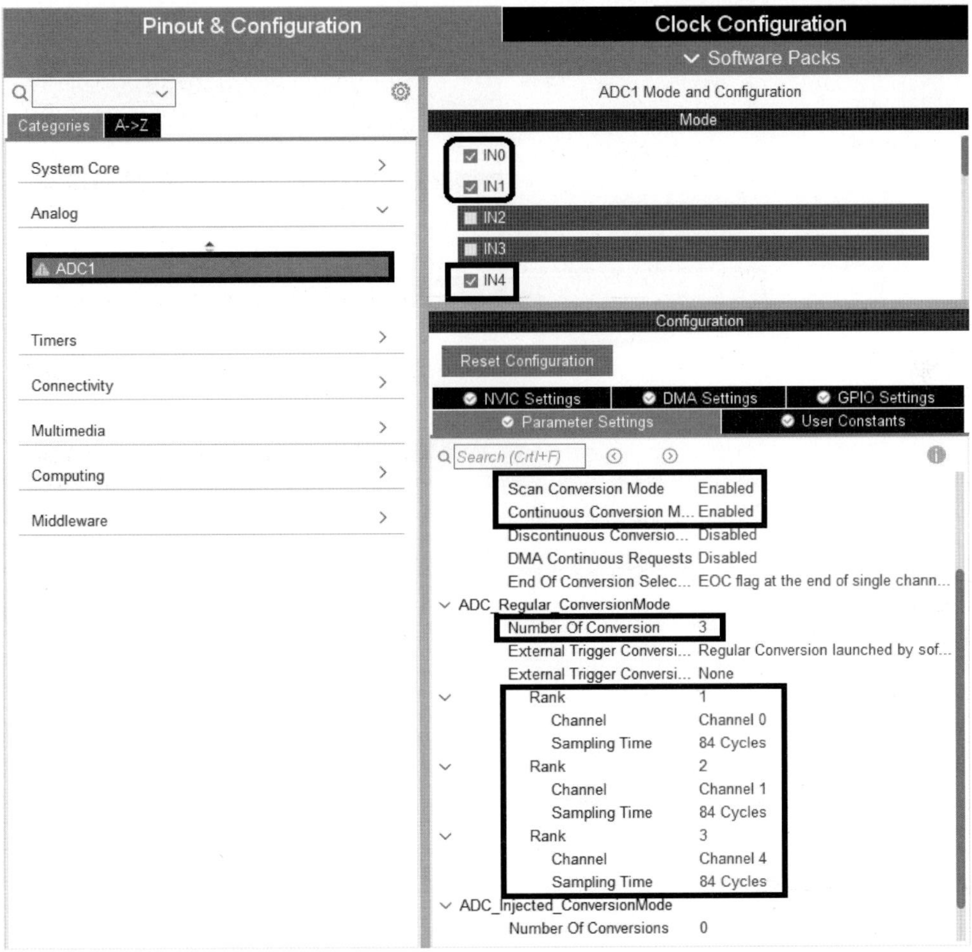

[그림 6.6] ADC1 채널 3개 설정

4.2. Generate Code 생성

[그림 6.7]은 코드를 생성하기 위한 화면이다. 그림과 같이 클릭해서 프로그램을 생성한다.

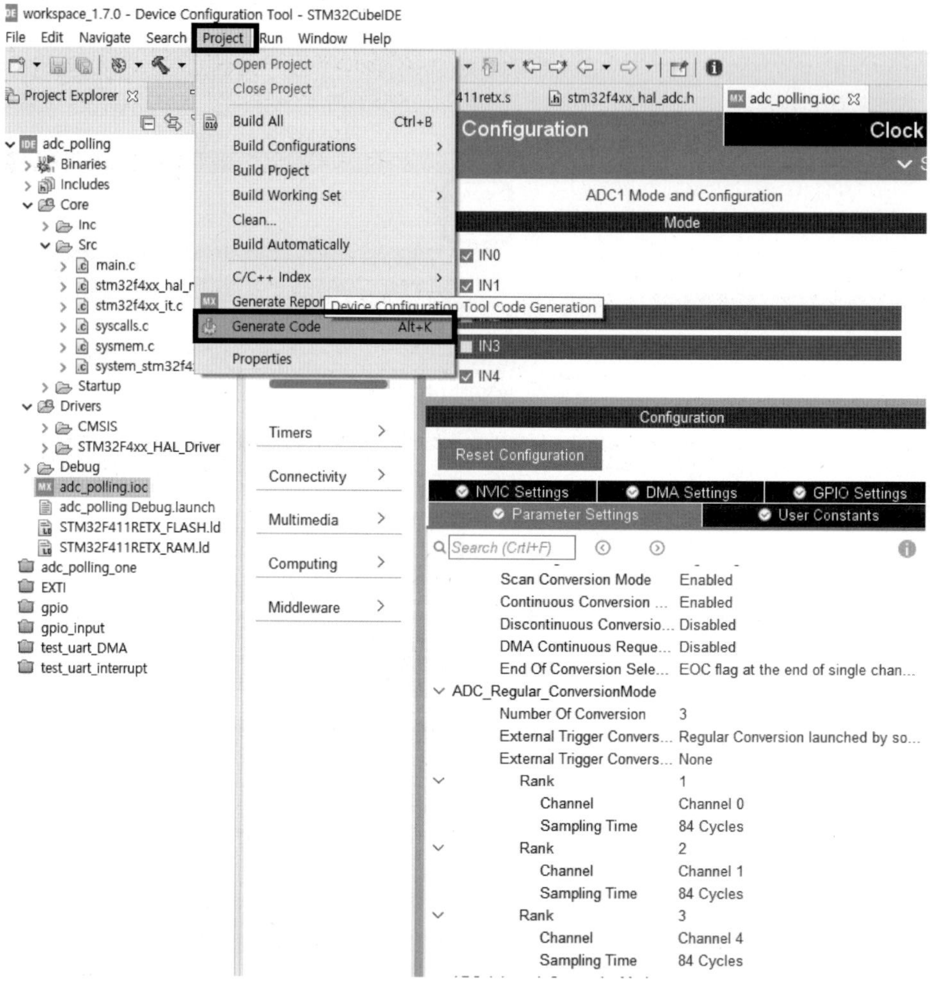

[그림 6.7] ADC1 채널 3개 설정

4.3. Source Code 작성하기

ADC를 Polling 방식으로 채널을 선택해서 사용하려면 MX_ADC1_Init(void) 함수의 ZZZ 수정이 필요하다.

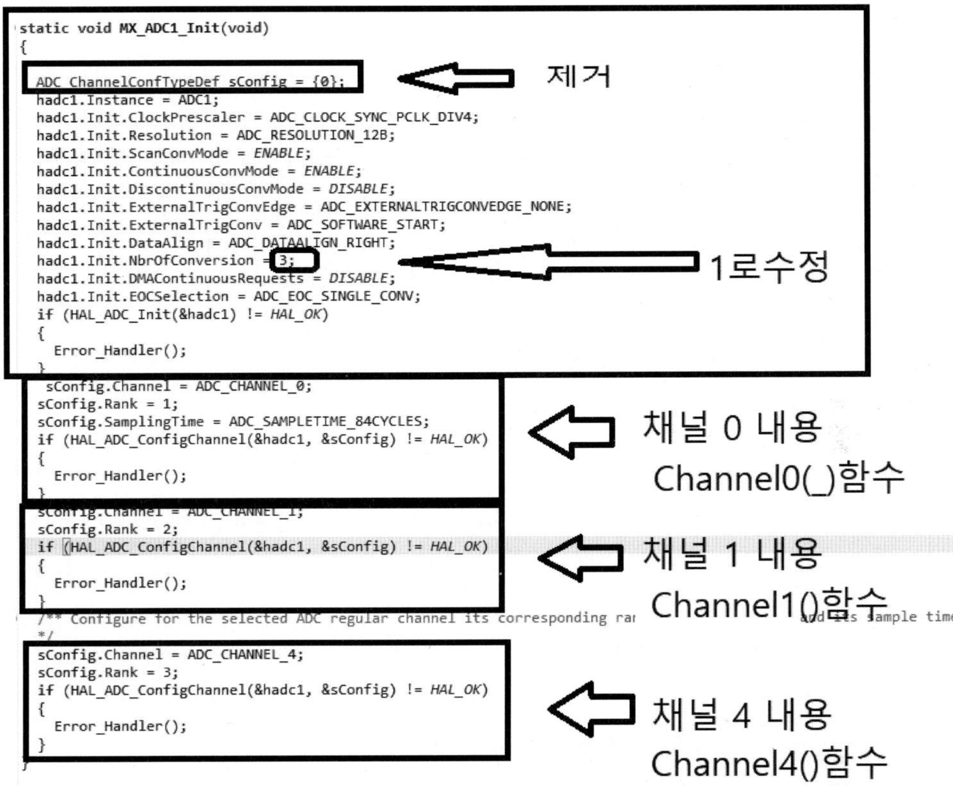

static void MX_ADC1_Init(void)
{
ADC_ChannelConfTypeDef sConfig = {0};
hadc1.Instance = ADC1;
hadc1.Init.ClockPrescaler = ADC_CLOCK_SYNC_PCLK_DIV4;
hadc1.Init.Resolution = ADC_RESOLUTION_12B;
hadc1.Init.ScanConvMode = ENABLE;
hadc1.Init.ContinuousConvMode = ENABLE;
hadc1.Init.DiscontinuousConvMode = DISABLE;
hadc1.Init.ExternalTrigConvEdge = ADC_EXTERNALTRIGCONVEDGE_NONE;
hadc1.Init.ExternalTrigConv = ADC_SOFTWARE_START;
hadc1.Init.DataAlign = ADC_DATAALIGN_RIGHT;
hadc1.Init.NbrOfConversion = 3;
hadc1.Init.DMAContinuousRequests = DISABLE;
hadc1.Init.EOCSelection = ADC_EOC_SINGLE_CONV;
if (HAL_ADC_Init(&hadc1) != HAL_OK)
{
Error_Handler();
}
sConfig.Channel = ADC_CHANNEL_0;
sConfig.Rank = 1;
sConfig.SamplingTime = ADC_SAMPLETIME_84CYCLES;
if (HAL_ADC_ConfigChannel(&hadc1, &sConfig) != HAL_OK)
{
Error_Handler();
}
sConfig.Rank = 2;
if (HAL_ADC_ConfigChannel(&hadc1, &sConfig) != HAL_OK)
{
Error_Handler();
}
sConfig.Channel = ADC_CHANNEL_4;
sConfig.Rank = 3;
if (HAL_ADC_ConfigChannel(&hadc1, &sConfig) != HAL_OK)
{
Error_Handler();
}
제거
1로수정
채널 0 내용
Channel0(_)함수
채널 1 내용
Channel1()함수
채널 4 내용
Channel4()함수

[그림 6.8] ADC1 채널 3개 설정 내용(수정 전)

[그림 6.8]을 다음과 같은 4개의 함수로 분리한다.

```
> static void MX_ADC1_Init(void)
> uint32_t Adc_Channel0(void)
> uint32_t Adc_Channel1(void)
> uint32_t Adc_Channel4(void)
```

수정된 MX_ADC1_Init(void) 함수는 아래 [그림 6.9]와 같다.

```
static void MX_ADC1_Init(void)
{

  hadc1.Instance = ADC1;
  hadc1.Init.ClockPrescaler = ADC_CLOCK_SYNC_PCLK_DIV4;
  hadc1.Init.Resolution = ADC_RESOLUTION_12B;
  hadc1.Init.ScanConvMode = ENABLE;
  hadc1.Init.ContinuousConvMode = ENABLE;
  hadc1.Init.DiscontinuousConvMode = DISABLE;
  hadc1.Init.ExternalTrigConvEdge = ADC_EXTERNALTRIGCONVEDGE_NONE;
  hadc1.Init.ExternalTrigConv = ADC_SOFTWARE_START;
  hadc1.Init.DataAlign = ADC_DATAALIGN_RIGHT;
  hadc1.Init.NbrOfConversion = 1;
  hadc1.Init.DMAContinuousRequests = DISABLE;
  hadc1.Init.EOCSelection = ADC_EOC_SINGLE_CONV;
  if (HAL_ADC_Init(&hadc1) != HAL_OK)
  {
    Error_Handler();
  }

}
```

[그림 6.9] MX_ADC1_Init(void) (수정 후)

[그림 6.10]은 변수 지정과 printf() 함수를 사용하기 위한 함수이다.

```
/* USER CODE BEGIN PFP */
ADC_ChannelConfTypeDef sConfig = {0};
/* USER CODE END PFP */

/* Private user code ---------------------------------
/* USER CODE BEGIN 0 */
uint32_t adc[3];
float volt[3];

int __io_putchar(int ch)
{
HAL_UART_Transmit(&huart2, (uint8_t *)&ch, 1, 0xFFFF);
return ch;
}
```

[그림 6.10] 변수 지정 및 printf() 함수 설정

[그림 6.11]은 MX_ADC1_Init(void) 내용을 채널에 맞도록 3개의 함수로 분리 수정한 내용이다.

```
uint32_t Adc_Channel0(void){
    uint32_t Temp_Adc;
      sConfig.Channel = ADC_CHANNEL_0;
      sConfig.Rank = 1;
      sConfig.SamplingTime = ADC_SAMPLETIME_84CYCLES;
      if (HAL_ADC_ConfigChannel(&hadc1, &sConfig) != HAL_OK)
      {
        Error_Handler();
      }
      HAL_ADC_Start(&hadc1);
      HAL_ADC_PollForConversion(&hadc1, 100);
      Temp_Adc=HAL_ADC_GetValue(&hadc1);
      HAL_ADC_Stop(&hadc1);
      return Temp_Adc;
}
uint32_t Adc_Channel1(void){
    uint32_t Temp_Adc;
      sConfig.Channel = ADC_CHANNEL_1;
      sConfig.Rank = 1;
      sConfig.SamplingTime = ADC_SAMPLETIME_84CYCLES;
      if (HAL_ADC_ConfigChannel(&hadc1, &sConfig) != HAL_OK)
      {
        Error_Handler();
      }
      HAL_ADC_Start(&hadc1);
      HAL_ADC_PollForConversion(&hadc1, 100);
      Temp_Adc=HAL_ADC_GetValue(&hadc1);
      HAL_ADC_Stop(&hadc1);
      return Temp_Adc;
}
uint32_t Adc_Channel4(void){
    uint32_t Temp_Adc;
      sConfig.Channel = ADC_CHANNEL_4;
      sConfig.Rank = 1;
      sConfig.SamplingTime = ADC_SAMPLETIME_84CYCLES;
      if (HAL_ADC_ConfigChannel(&hadc1, &sConfig) != HAL_OK)
      {
        Error_Handler();
      }
      HAL_ADC_Start(&hadc1);
      HAL_ADC_PollForConversion(&hadc1, 100);
      Temp_Adc=HAL_ADC_GetValue(&hadc1);
      HAL_ADC_Stop(&hadc1);
      return Temp_Adc;
}
```

[그림 6.11] 채널 3개 함수

[그림 6.12]는 채널별로 ADC 값을 읽어서 전압으로 환산한 내용이다.

```
  SystemClock_Config();

  /* USER CODE BEGIN SysInit */

  /* USER CODE END SysInit */

  /* Initialize all configured peripherals */
  MX_GPIO_Init();
  MX_USART2_UART_Init();
  MX_ADC1_Init();
  /* USER CODE BEGIN 2 */

  /* USER CODE END 2 */

  /* Infinite loop */
  /* USER CODE BEGIN WHILE */
  while (1)
  {
      adc[0]=Adc_Channel0();
      adc[1]=Adc_Channel1();
      adc[2]=Adc_Channel4();

      volt[0]=(float)adc[0]*3.3/4096.0;
      volt[1]=(float)adc[1]*3.3/4096.0;
      volt[2]=(float)adc[2]*3.3/4096.0;

      printf("Volt[0]=%f,Volt[1]=%f,Volt[2]=%f\r\n",volt[0],volt[1],volt[2]);
      HAL_Delay(100);
    /* USER CODE END WHILE */

    /* USER CODE BEGIN 3 */
  }
  /* USER CODE END 3 */
}

/**
  * @brief System Clock Configuration
  * @retval None
  */
```

[그림 6.12] 메인 코드 내용

4.4. 디버그 및 다운로드하기

메인 함수에 프로그램이 작성되었다면 다음으로 디버그는 [그림 6.13]의 를 클릭하거나 단축키 F11을 누르면 디버깅과 프로그램 다운로드가 완료된다.

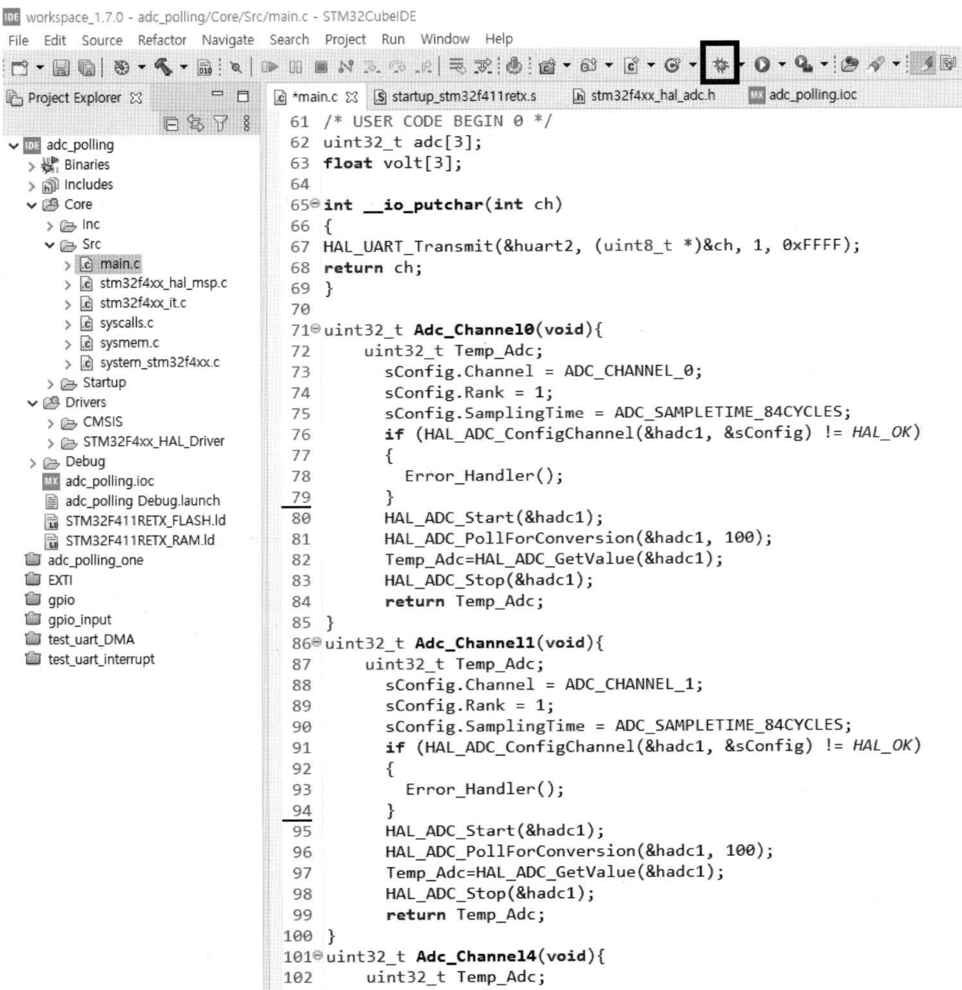
workspace_1.7.0 - adc_polling/Core/Src/main.c - STM32CubeIDE
File Edit Source Refactor Navigate Search Project Run Window Help
Project Explorer
*main.c
startup_stm32f411retx.s
stm32f4xx_hal_adc.h
adc_polling.ioc
adc_polling
Binaries
Includes
Core
Inc
Src
main.c
stm32f4xx_hal_msp.c
stm32f4xx_it.c
syscalls.c
sysmem.c
system_stm32f4xx.c
Startup
Drivers
CMSIS
STM32F4xx_HAL_Driver
Debug
adc_polling.ioc
adc_polling Debug.launch
STM32F411RETX_FLASH.ld
STM32F411RETX_RAM.ld
adc_polling_one
EXTI
gpio
gpio_input
test_uart_DMA
test_uart_interrupt
/* USER CODE BEGIN 0 */
uint32_t adc[3];
float volt[3];
int __io_putchar(int ch)
{
HAL_UART_Transmit(&huart2, (uint8_t *)&ch, 1, 0xFFFF);
return ch;
}
uint32_t Adc_Channel0(void){
uint32_t Temp_Adc;
sConfig.Channel = ADC_CHANNEL_0;
sConfig.Rank = 1;
sConfig.SamplingTime = ADC_SAMPLETIME_84CYCLES;
if (HAL_ADC_ConfigChannel(&hadc1, &sConfig) != HAL_OK)
{
Error_Handler();
}
HAL_ADC_Start(&hadc1);
HAL_ADC_PollForConversion(&hadc1, 100);
Temp_Adc=HAL_ADC_GetValue(&hadc1);
HAL_ADC_Stop(&hadc1);
return Temp_Adc;
}
uint32_t Adc_Channel1(void){
uint32_t Temp_Adc;
sConfig.Channel = ADC_CHANNEL_1;
sConfig.Rank = 1;
sConfig.SamplingTime = ADC_SAMPLETIME_84CYCLES;
if (HAL_ADC_ConfigChannel(&hadc1, &sConfig) != HAL_OK)
{
Error_Handler();
}
HAL_ADC_Start(&hadc1);
HAL_ADC_PollForConversion(&hadc1, 100);
Temp_Adc=HAL_ADC_GetValue(&hadc1);
HAL_ADC_Stop(&hadc1);
return Temp_Adc;
}
uint32_t Adc_Channel4(void){
uint32_t Temp_Adc;

[그림 6.13] 프로그램 디버그

4.5. 보드 프로그램 실행하기

[그림 6.14] 상단 중앙에 아이콘 ■ 클릭하면 디버그 모드가 종료되어 보드의 프로그램이 실행된다.

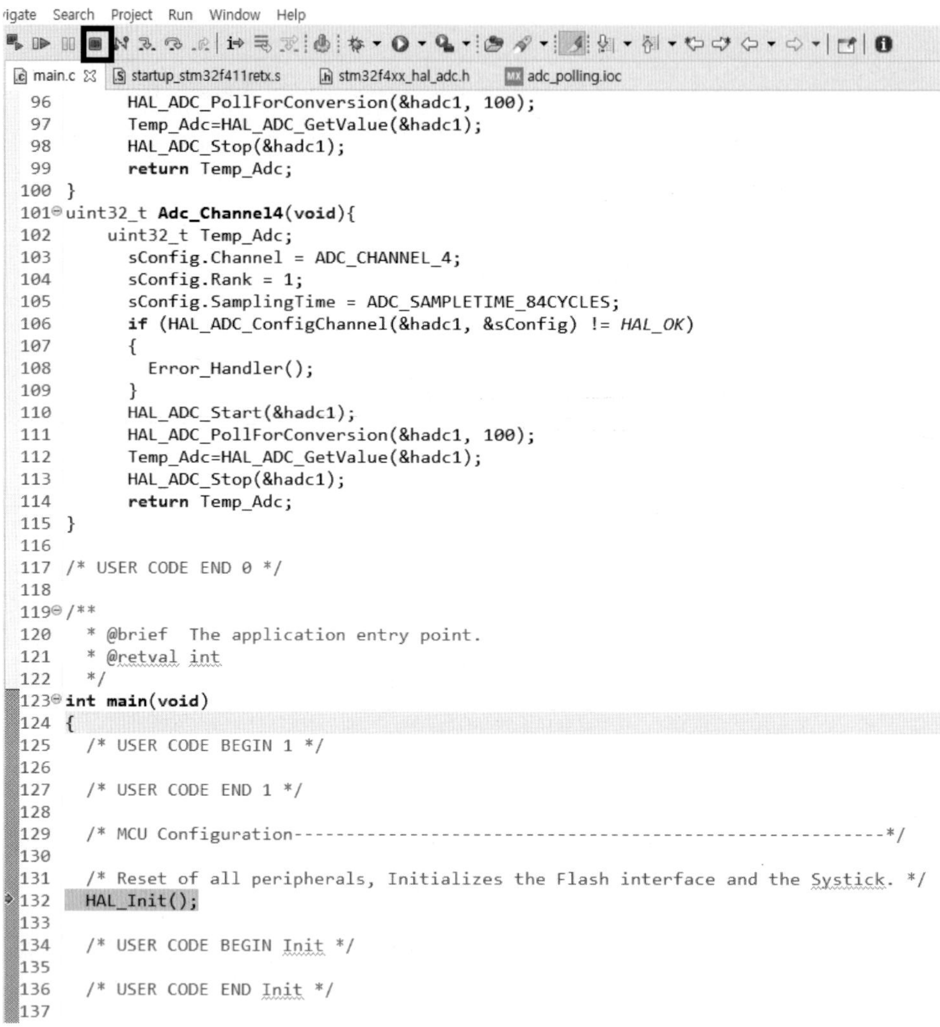

igate Search Project Run Window Help
main.c
startup_stm32f411retx.s
stm32f4xx_hal_adc.h
adc_polling.ioc
        HAL_ADC_PollForConversion(&hadc1, 100);
        Temp_Adc=HAL_ADC_GetValue(&hadc1);
        HAL_ADC_Stop(&hadc1);
        return Temp_Adc;
}
uint32_t Adc_Channel4(void){
    uint32_t Temp_Adc;
        sConfig.Channel = ADC_CHANNEL_4;
        sConfig.Rank = 1;
        sConfig.SamplingTime = ADC_SAMPLETIME_84CYCLES;
        if (HAL_ADC_ConfigChannel(&hadc1, &sConfig) != HAL_OK)
        {
          Error_Handler();
        }
        HAL_ADC_Start(&hadc1);
        HAL_ADC_PollForConversion(&hadc1, 100);
        Temp_Adc=HAL_ADC_GetValue(&hadc1);
        HAL_ADC_Stop(&hadc1);
        return Temp_Adc;
}

/* USER CODE END 0 */

/**
  * @brief The application entry point.
  * @retval int
  */
int main(void)
{
  /* USER CODE BEGIN 1 */

  /* USER CODE END 1 */

  /* MCU Configuration--*/

  /* Reset of all peripherals, Initializes the Flash interface and the Systick. */
  HAL_Init();

  /* USER CODE BEGIN Init */

  /* USER CODE END Init */

[그림 6.14] 디버그 모드 프로그램 실행

4.6. 프로그램 실행 확인

시리얼 통신 프로그램을 확인하기 쉬운 방법은 Tera term이나 Hercules, docklight 통신 터미널 [그림 6.15]을 이용하면 된다. 인터넷에서 hercules terminal 검색하면 다운로드 받을 수 있다.

관련 사이트: https://www.hw-group.com/software/hercules-setup-utility
https://tera-term.softonic.kr/
https://docklight.de/downloads/

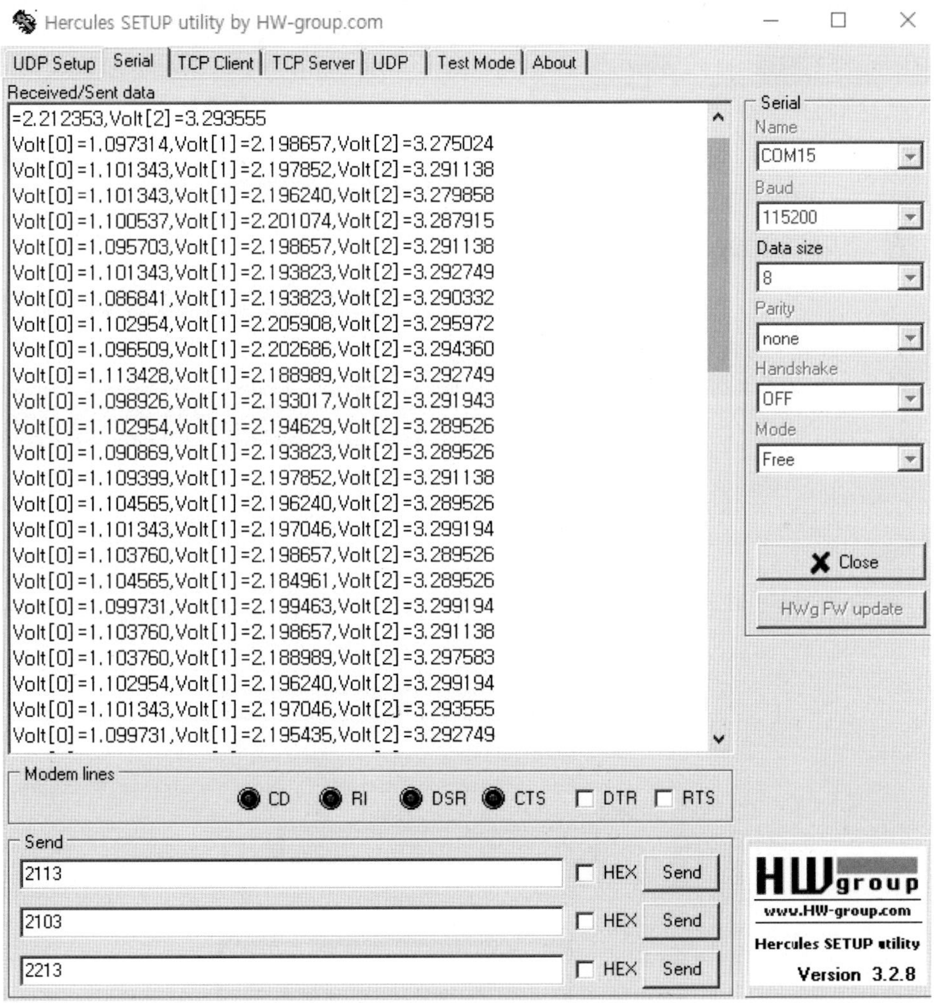

[그림 6.15] 시리얼 통신으로 결과 확인

5. Interrupt 방식

5.1. ADC 설정

System Core에서 RCC 및 SYS 및 TX RX 단자는 이전과 같이 설정한다. [그림 6.2]와 같이 구성한다. [그림 6.16], [그림 6.17]은 Channel0을 인터럽트 방식으로 선언을 나타내고 있다.

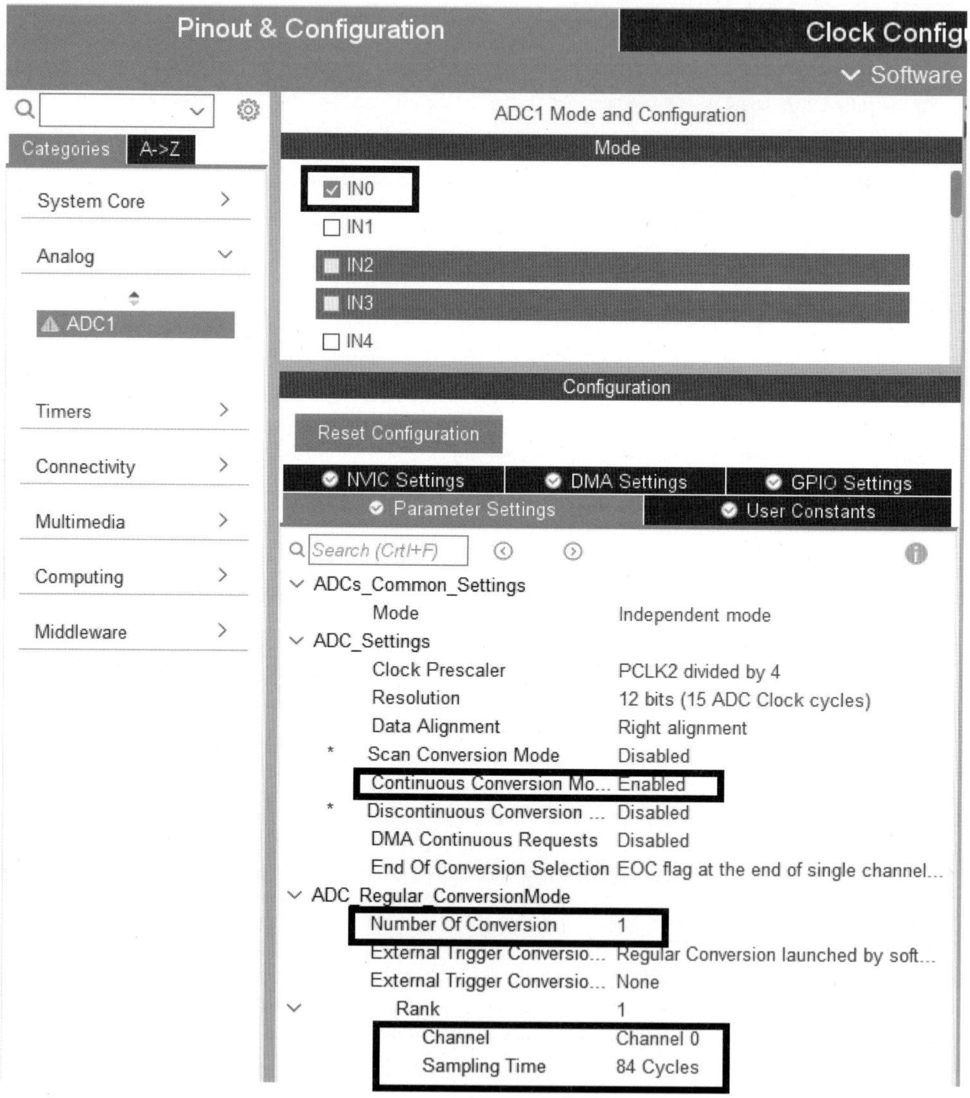

[그림 6.16] ADC1 채널 설정

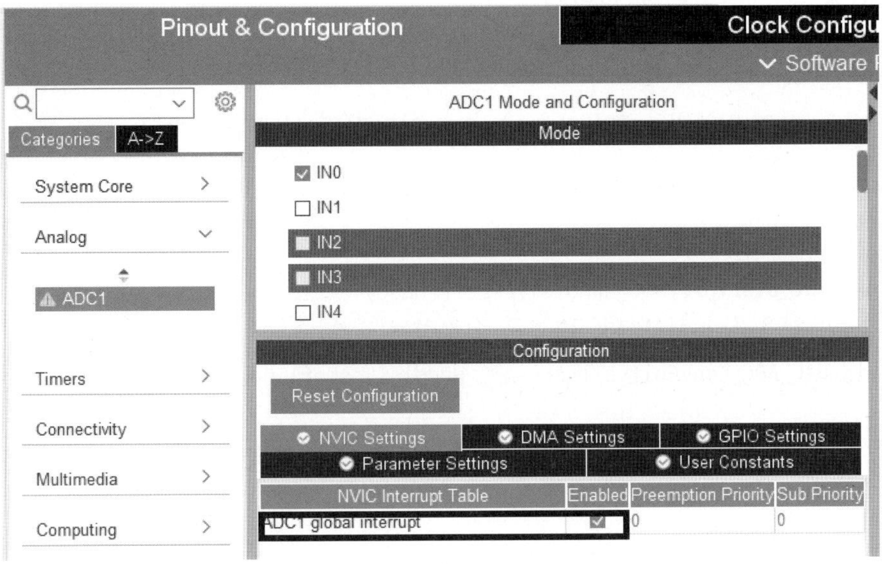

[그림 6.17] ADC1 인터럽트 설정

5.2. Generate Code 생성

[그림 6.18]은 코드를 생성하기 위한 화면이다. 그림과 같이  클릭해서 프로그램을 생성한다.

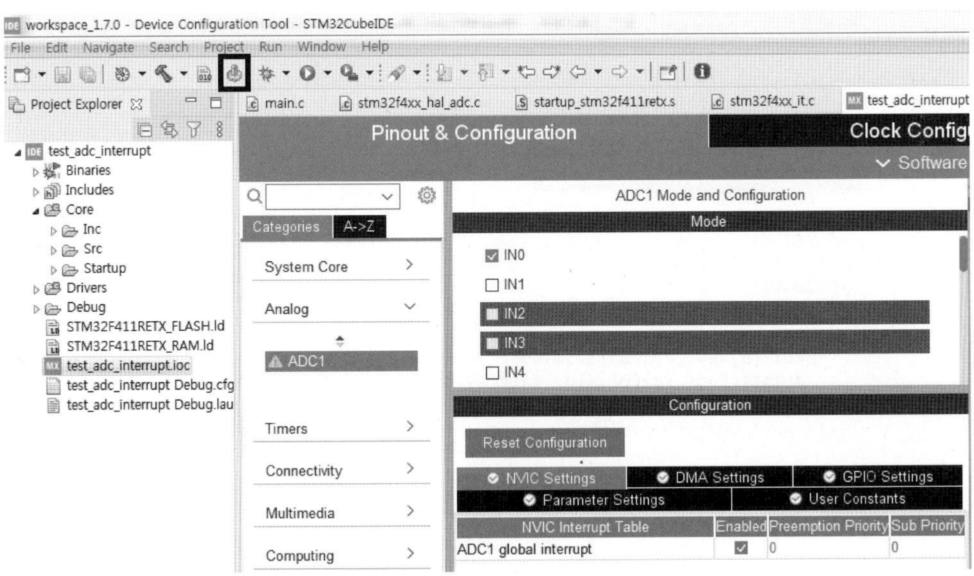

[그림 6.18] ADC1 코드 생성

5.3. Source Code 작성하기

ADC를 인터럽트 방식으로 사용하려면 관련 함수를 아는 것이 중요하다. 사용하는 방법은 어렵지 않다. 메인 함수에 ADC 인터럽트 시작을 해주고 관련 콜백 함수를 이용하면 된다. [그림 6.19]는 메인 프로그램이다.

> **HAL_ADC_Start_IT**(ADC_HandleTypeDef* hadc)
인터럽트를 Enable하고 ADC 변환을 시작하는 함수
>void **HAL_ADC_ConvCpltCallback**(ADC_HandleTypeDef* hadc)
{
    ADC 변환이 완료되면 인터럽트 콜뱍 함수가 실행됨
}

```
/* Private user code ---------------------------------------
/* USER CODE BEGIN 0 */
uint32_t adc_val[3];
float volt[3];
void HAL_ADC_ConvCpltCallback(ADC_HandleTypeDef* hadc)
{
    adc_val[0]=HAL_ADC_GetValue(&hadc1);
}

int __io_putchar(int ch)
{
HAL_UART_Transmit(&huart2, (uint8_t *)&ch, 1, 0xFFFF);
return ch;
}

/* USER CODE END 0 */

  /* USER CODE BEGIN 2 */
 HAL_ADC_Start_IT(&hadc1);
  /* USER CODE END 2 */

  /* Infinite loop */
  /* USER CODE BEGIN WHILE */
  while (1)
  {

      volt[0]=adc_val[0]*3.3/4096;
      printf("adc_val[0]=%ld, volt[0]=%f[V]\r\n",adc_val[0],volt[0]);
      HAL_Delay(100);
    /* USER CODE END WHILE */

    /* USER CODE BEGIN 3 */
  }
  /* USER CODE END 3 */
}
```

[그림 6.19] 메인 코드 내용

5.4. 디버그 및 다운로드하기

메인 함수에 프로그램이 작성되었다면 다음으로 디버그는 [그림 6.20]의 ▣를 클릭하거나 단축키 F11을 누르면 디버깅과 프로그램 다운로드가 완료된다.

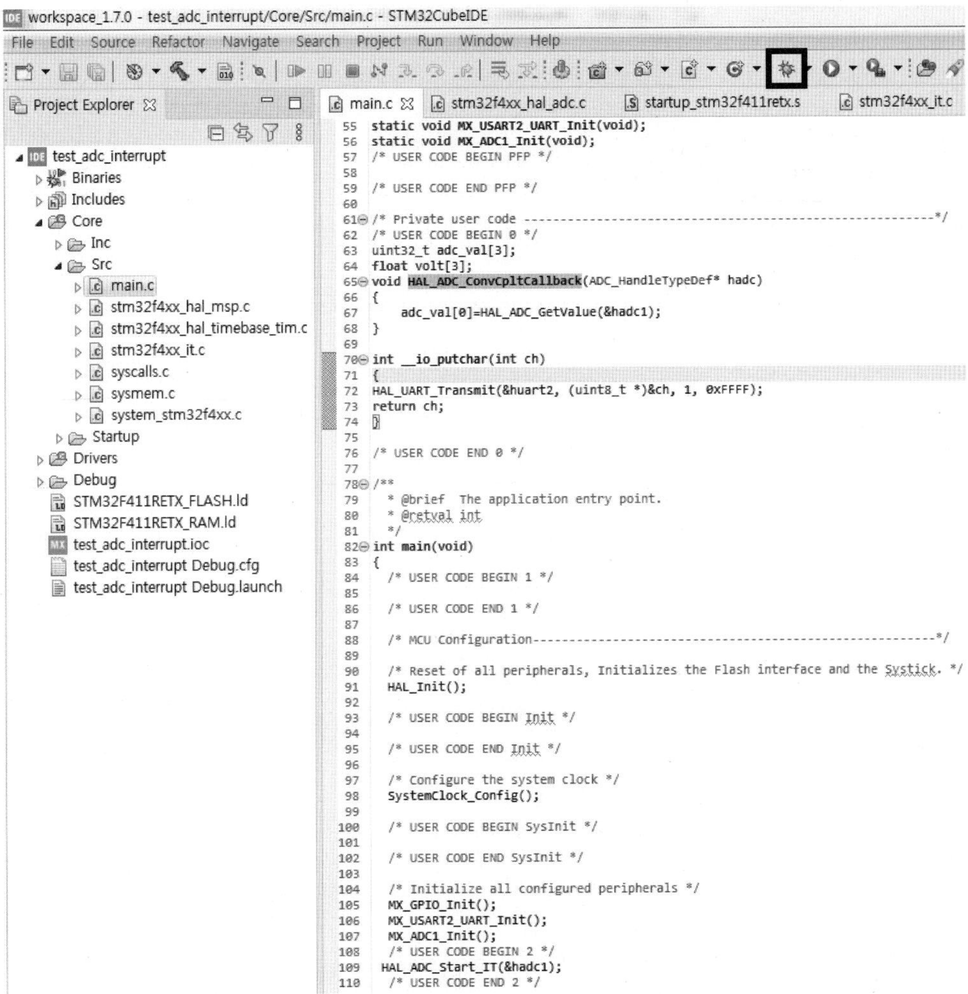

workspace_1.7.0 - test_adc_interrupt/Core/Src/main.c - STM32CubeIDE
File Edit Source Refactor Navigate Search Project Run Window Help
Project Explorer
test_adc_interrupt
Binaries
Includes
Core
Inc
Src
main.c
stm32f4xx_hal_msp.c
stm32f4xx_hal_timebase_tim.c
stm32f4xx_it.c
syscalls.c
sysmem.c
system_stm32f4xx.c
Startup
Drivers
Debug
STM32F411RETX_FLASH.ld
STM32F411RETX_RAM.ld
test_adc_interrupt.ioc
test_adc_interrupt Debug.cfg
test_adc_interrupt Debug.launch
main.c
stm32f4xx_hal_adc.c
startup_stm32f411retx.s
stm32f4xx_it.c
static void MX_USART2_UART_Init(void);
static void MX_ADC1_Init(void);
/* USER CODE BEGIN PFP */
/* USER CODE END PFP */
/* Private user code ---*/
/* USER CODE BEGIN 0 */
uint32_t adc_val[3];
float volt[3];
void HAL_ADC_ConvCpltCallback(ADC_HandleTypeDef* hadc)
{
adc_val[0]=HAL_ADC_GetValue(&hadc1);
}
int __io_putchar(int ch)
{
HAL_UART_Transmit(&huart2, (uint8_t *)&ch, 1, 0xFFFF);
return ch;
}
/* USER CODE END 0 */
/**
* @brief The application entry point.
* @retval int
*/
int main(void)
{
/* USER CODE BEGIN 1 */
/* USER CODE END 1 */
/* MCU Configuration--*/
/* Reset of all peripherals, Initializes the Flash interface and the Systick. */
HAL_Init();
/* USER CODE BEGIN Init */
/* USER CODE END Init */
/* Configure the system clock */
SystemClock_Config();
/* USER CODE BEGIN SysInit */
/* USER CODE END SysInit */
/* Initialize all configured peripherals */
MX_GPIO_Init();
MX_USART2_UART_Init();
MX_ADC1_Init();
/* USER CODE BEGIN 2 */
HAL_ADC_Start_IT(&hadc1);
/* USER CODE END 2 */

[그림 6.20] 프로그램 디버그

5.5. 보드 프로그램 실행하기

[그림 6.21] 상단 중앙에 아이콘 ■ 클릭하면 디버그 모드가 종료되어 보드의 프로그램이 실행된다.

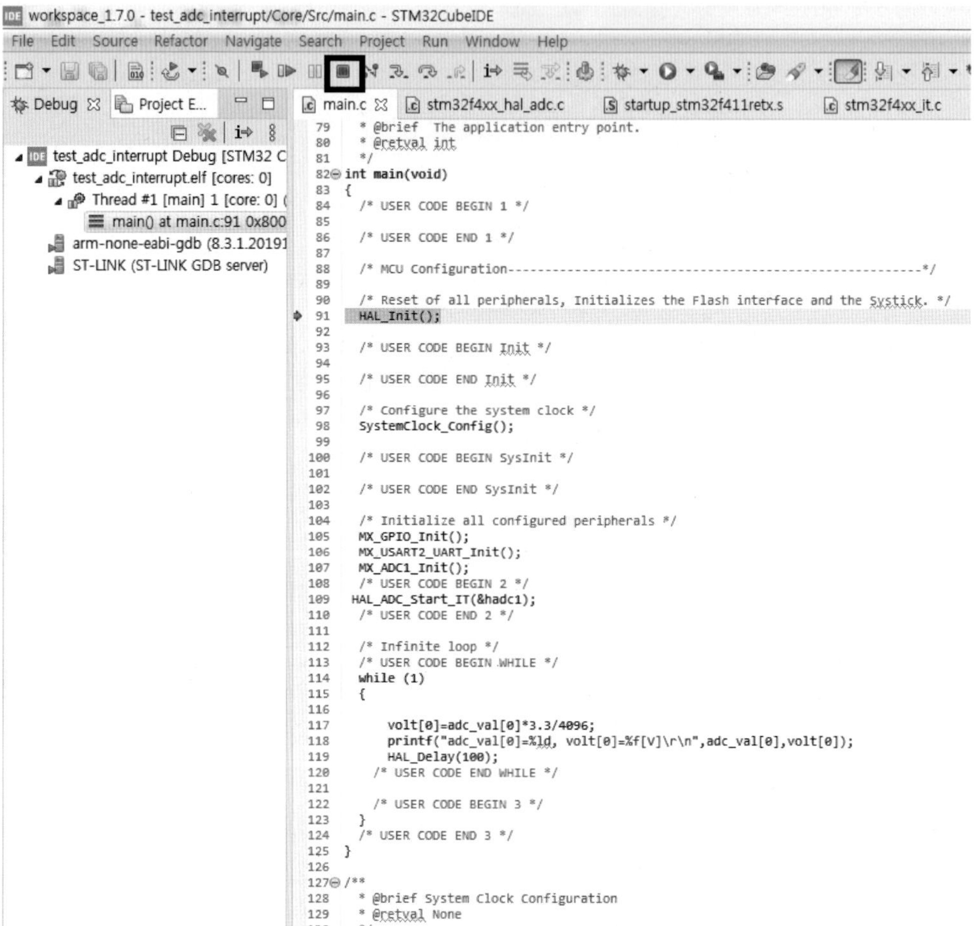
workspace_1.7.0 - test_adc_interrupt/Core/Src/main.c - STM32CubeIDE
File Edit Source Refactor Navigate Search Project Run Window Help
Debug
Project E...
main.c
stm32f4xx_hal_adc.c
startup_stm32f411retx.s
stm32f4xx_it.c
test_adc_interrupt Debug [STM32 C
test_adc_interrupt.elf [cores: 0]
Thread #1 [main] 1 [core: 0] (
main() at main.c:91 0x800
arm-none-eabi-gdb (8.3.1.20191
ST-LINK (ST-LINK GDB server)
* @brief The application entry point.
* @retval int
*/
int main(void)
{
/* USER CODE BEGIN 1 */
/* USER CODE END 1 */
/* MCU Configuration--*/
/* Reset of all peripherals, Initializes the Flash interface and the Systick. */
HAL_Init();
/* USER CODE BEGIN Init */
/* USER CODE END Init */
/* Configure the system clock */
SystemClock_Config();
/* USER CODE BEGIN SysInit */
/* USER CODE END SysInit */
/* Initialize all configured peripherals */
MX_GPIO_Init();
MX_USART2_UART_Init();
MX_ADC1_Init();
/* USER CODE BEGIN 2 */
HAL_ADC_Start_IT(&hadc1);
/* USER CODE END 2 */
/* Infinite loop */
/* USER CODE BEGIN WHILE */
while (1)
{
volt[0]=adc_val[0]*3.3/4096;
printf("adc_val[0]=%ld, volt[0]=%f[V]\r\n",adc_val[0],volt[0]);
HAL_Delay(100);
/* USER CODE END WHILE */
/* USER CODE BEGIN 3 */
}
/* USER CODE END 3 */
}
/**
* @brief System Clock Configuration
* @retval None
*/

[그림 6.21] 디버그 모드 프로그램 실행

5.6. 프로그램 실행 확인

시리얼 통신 프로그램을 확인하기 쉬운 방법은 Tera term이나 Hercules,docklight 통신 터미널 [그림 6.22]를 이용하면 된다. 인터넷에서 hercules terminal 검색하면 다운로드할 수 있다.

관련 사이트: https://www.hw-group.com/software/hercules-setup-utility

https://tera-term.softonic.kr/

https://docklight.de/downloads/

테스트 포인트를 옮겨가면서 측정해본다.

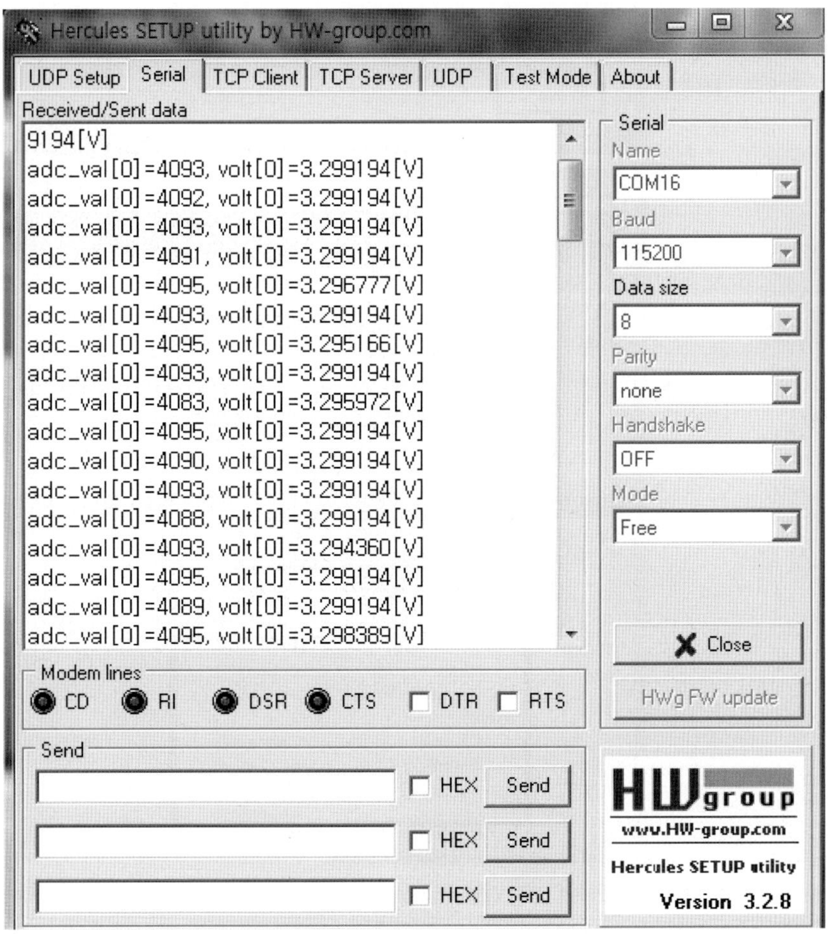

[그림 6.22] 시리얼 통신으로 결과 확인

6. DMA 방식

6.1. ADC 설정

System Core에서 RCC 및 SYS 및 TX RX 단자는 이전과 같이 설정한다. [그림 6.2]와 같이 구성한다. [그림 6.23], [그림 6.24]는 Channel0,1,4를 DMA 방식으로 선언하였다.

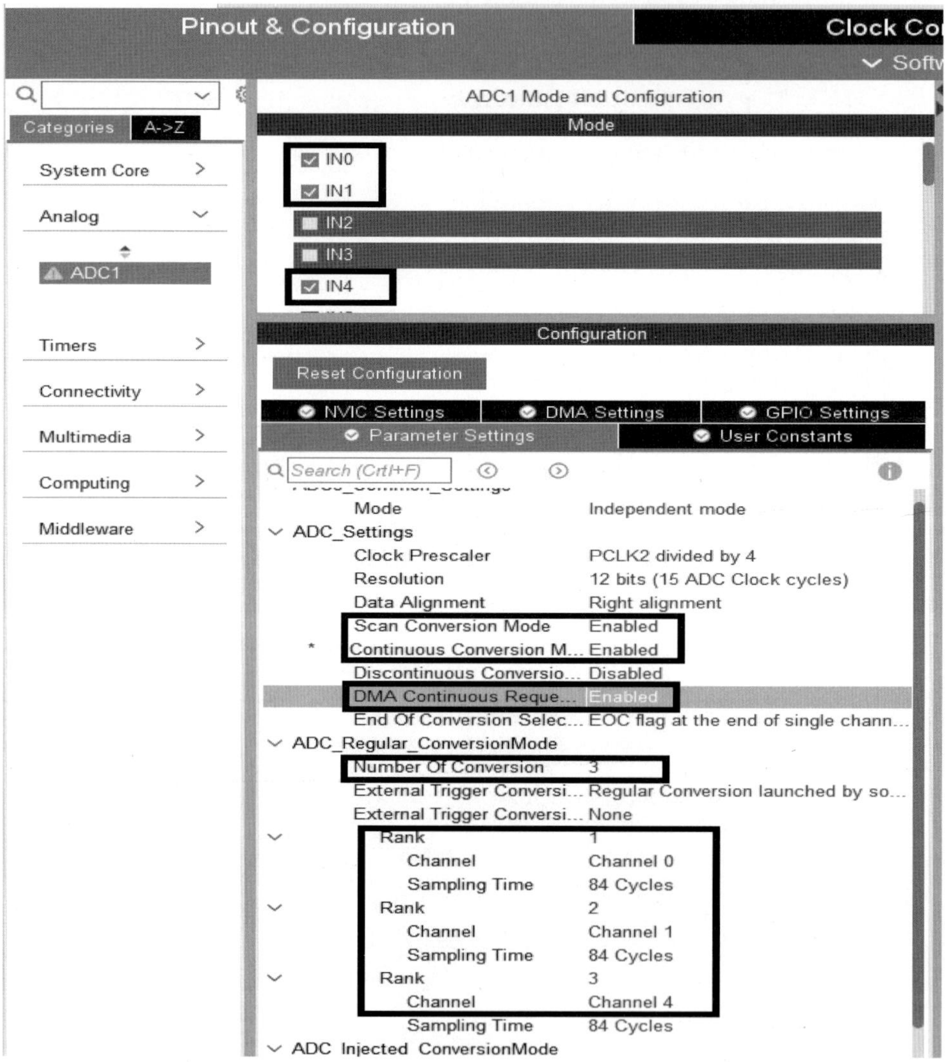

[그림 6.23] ADC1 채널 설정

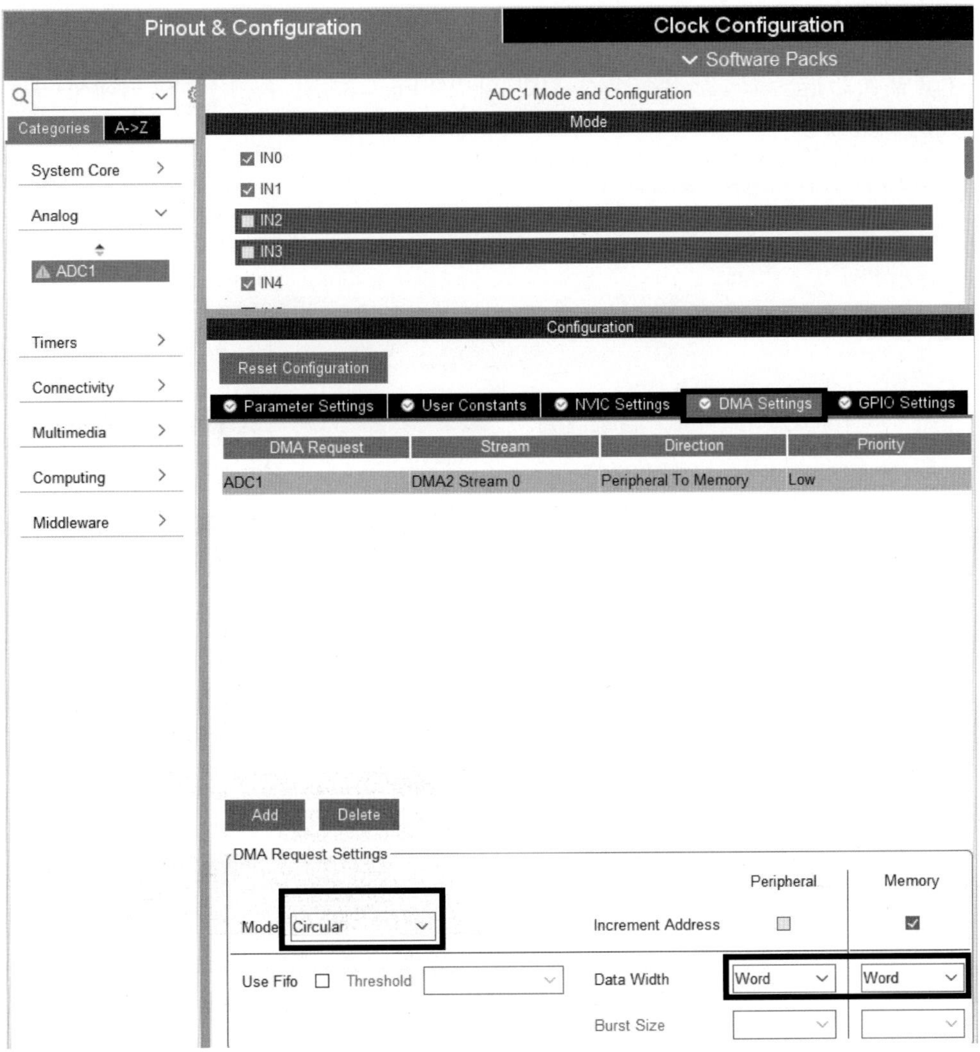

[그림 6.24] ADC1 DMA 설정

6.2. Generate Code 생성

[그림 6.25]는 코드를 생성하기 위한 화면이다. 그림과 같이 클릭해서 프로그램을 생성한다.

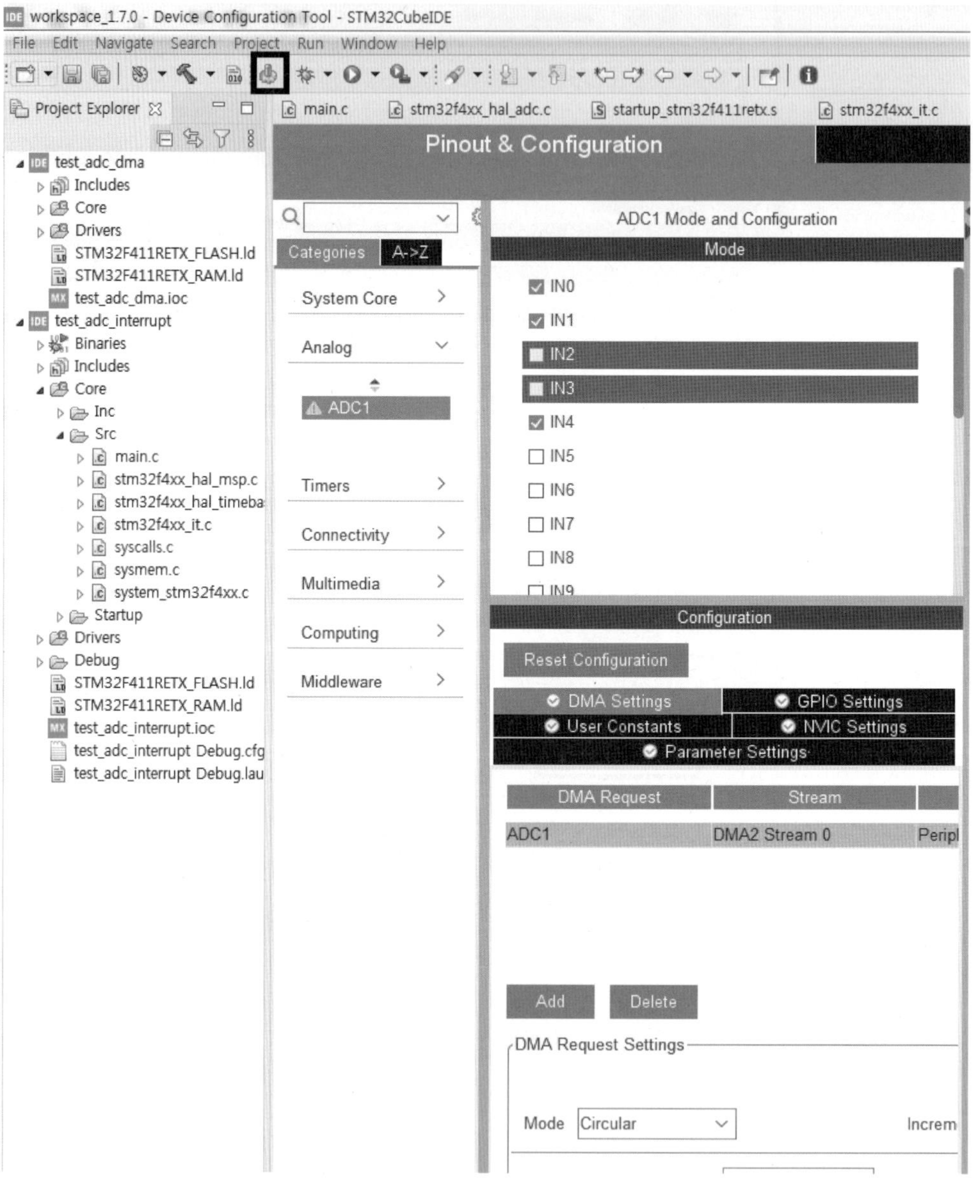

[그림 6.25] ADC1 코드 생성

6.3. Source Code 작성하기

ADC를 인터럽트 방식으로 사용하려면 관련 함수를 아는 것이 중요하다. 사용하는 방법은 어렵지 않다. 메인 함수에 ADC 인터럽트 시작을 해주고 관련 콜백 함수를 이용하면 된다. [그림 6.26]은 메인 프로그램이다.

>**HAL_ADC_Start_DMA**(ADC_HandleTypeDef* hadc)

인터럽트를 Enable하고 ADC 변환을 시작하는 함수

>**void HAL_ADC_ConvCpltCallback**(ADC_HandleTypeDef* hadc)

{

　　ADC 변환이 완료되면 인터럽트 콜백 함수가 실행됨

}

```
/* Private user code ---------------------------------------------------------*/
/* USER CODE BEGIN 0 */
uint32_t adc_val[3];
float volt[3];
int __io_putchar(int ch)
{
HAL_UART_Transmit(&huart2, (uint8_t *)&ch, 1, 0xFFFF);
return ch;
}
void HAL_ADC_ConvCpltCallback(ADC_HandleTypeDef* hadc)
{
        volt[0]=(float)adc_val[0]*3.3/4096.0;
        volt[1]=(float)adc_val[1]*3.3/4096.0;
        volt[2]=(float)adc_val[2]*3.3/4096.0;
}
/* USER CODE END 0 */


/* USER CODE BEGIN 2 */
HAL_ADC_Start_DMA(&hadc1, adc_val, 3);
/* USER CODE END 2 */

/* Infinite loop */
/* USER CODE BEGIN WHILE */
while (1)
{

    printf("volt[0]=%f[V],volt[1]=%f[V],volt[2]=%f[V]\r\n",volt[0],volt[1],volt[2]);
    HAL_Delay(300);

  /* USER CODE END WHILE */

  /* USER CODE BEGIN 3 */
}
/* USER CODE END 3 */
}
```

[그림 6.26] 메인 코드 내용

6.4. 디버그 및 다운로드하기

메인 함수에 프로그램이 작성되었다면 다음으로 디버그는 [그림 6.27]의 를 클릭하거나 단축키 F11을 누르면 디버깅과 프로그램 다운로드가 완료된다.

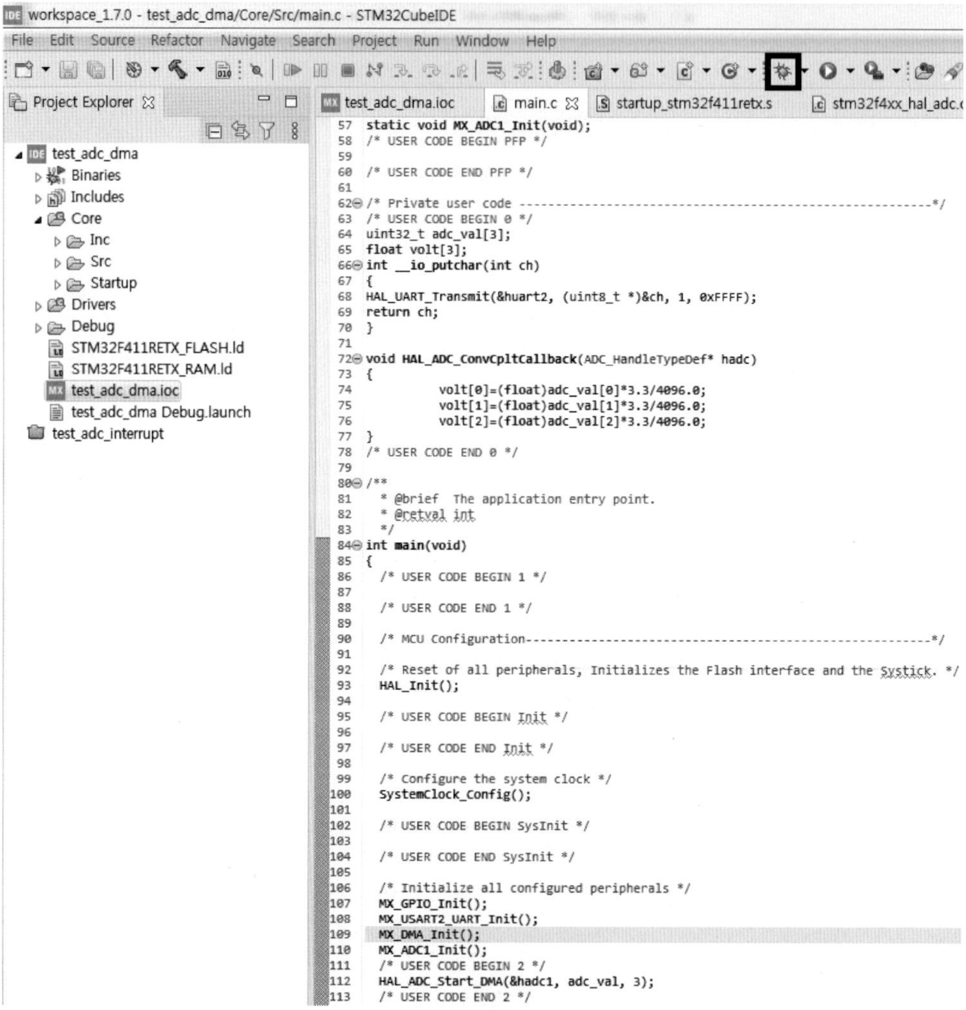

workspace_1.7.0 - test_adc_dma/Core/Src/main.c - STM32CubeIDE
File Edit Source Refactor Navigate Search Project Run Window Help
Project Explorer
test_adc_dma.ioc
main.c
startup_stm32f411retx.s
stm32f4xx_hal_adc.c
test_adc_dma
Binaries
Includes
Core
Inc
Src
Startup
Drivers
Debug
STM32F411RETX_FLASH.ld
STM32F411RETX_RAM.ld
test_adc_dma.ioc
test_adc_dma Debug.launch
test_adc_interrupt
static void MX_ADC1_Init(void);
/* USER CODE BEGIN PFP */
/* USER CODE END PFP */
/* Private user code ---*/
/* USER CODE BEGIN 0 */
uint32_t adc_val[3];
float volt[3];
int __io_putchar(int ch)
{
HAL_UART_Transmit(&huart2, (uint8_t *)&ch, 1, 0xFFFF);
return ch;
}
void HAL_ADC_ConvCpltCallback(ADC_HandleTypeDef* hadc)
{
volt[0]=(float)adc_val[0]*3.3/4096.0;
volt[1]=(float)adc_val[1]*3.3/4096.0;
volt[2]=(float)adc_val[2]*3.3/4096.0;
}
/* USER CODE END 0 */
/**
* @brief The application entry point.
* @retval int
*/
int main(void)
{
/* USER CODE BEGIN 1 */
/* USER CODE END 1 */
/* MCU Configuration--*/
/* Reset of all peripherals, Initializes the Flash interface and the Systick. */
HAL_Init();
/* USER CODE BEGIN Init */
/* USER CODE END Init */
/* Configure the system clock */
SystemClock_Config();
/* USER CODE BEGIN SysInit */
/* USER CODE END SysInit */
/* Initialize all configured peripherals */
MX_GPIO_Init();
MX_USART2_UART_Init();
MX_DMA_Init();
MX_ADC1_Init();
/* USER CODE BEGIN 2 */
HAL_ADC_Start_DMA(&hadc1, adc_val, 3);
/* USER CODE END 2 */

[그림 6.27] 프로그램 디버그

6.5. 보드 프로그램 실행하기

[그림 6.28] 상단 중앙에 아이콘 클릭하면 디버그 모드가 종료되어 보드의 프로그램이 실행된다.

workspace_1.7.0 - test_adc_dma/Core/Src/main.c - STM32CubeIDE
File Edit Source Refactor Navigate Search Project Run Window Help
Debug
Project E...
test_adc_dma.ioc
main.c
startup_stm32f411retx.s
stm32f4xx_hal_adc.c
test_adc_dma Debug [STM32 Corte
test_adc_dma.elf [cores: 0]
Thread #1 [main] 1 [core: 0] (
main() at main.c:93 0x800
arm-none-eabi-gdb (8.3.1.20191
ST-LINK (ST-LINK GDB server)
/* USER CODE END 0 */
/**
* @brief The application entry point.
* @retval int
*/
int main(void)
{
/* USER CODE BEGIN 1 */
/* USER CODE END 1 */
/* MCU Configuration--*/
/* Reset of all peripherals, Initializes the Flash interface and the Systick. */
HAL_Init();
/* USER CODE BEGIN Init */
/* USER CODE END Init */
/* Configure the system clock */
SystemClock_Config();
/* USER CODE BEGIN SysInit */
/* USER CODE END SysInit */
/* Initialize all configured peripherals */
MX_GPIO_Init();
MX_USART2_UART_Init();
MX_DMA_Init();
MX_ADC1_Init();
/* USER CODE BEGIN 2 */
HAL_ADC_Start_DMA(&hadc1, adc_val, 3);
/* USER CODE END 2 */
/* Infinite loop */
/* USER CODE BEGIN WHILE */
while (1)
{
printf("volt[0]=%f[V],volt[1]=%f[V],volt[2]=%f[V]\r\n",volt[0],volt[1],volt[2]);
HAL_Delay(300);
/* USER CODE END WHILE */
/* USER CODE BEGIN 3 */
}
/* USER CODE END 3 */
}
/**
* @brief System Clock Configuration

[그림 6.28] 디버그 모드 프로그램 실행

6.6. 프로그램 실행 확인

시리얼 통신 프로그램을 확인하기 쉬운 방법은 Tera term이나 Hercules, docklight 통신 터미널 [그림 6.22]를 이용하면 된다. 인터넷에서 hercules terminal 검색하면 다운로드할 수 있다.

관련 사이트: https://www.hw-group.com/software/hercules-setup-utility
https://tera-term.softonic.kr/
https://docklight.de/downloads/

테스트 포인트를 옮겨가면서 측정해본다.

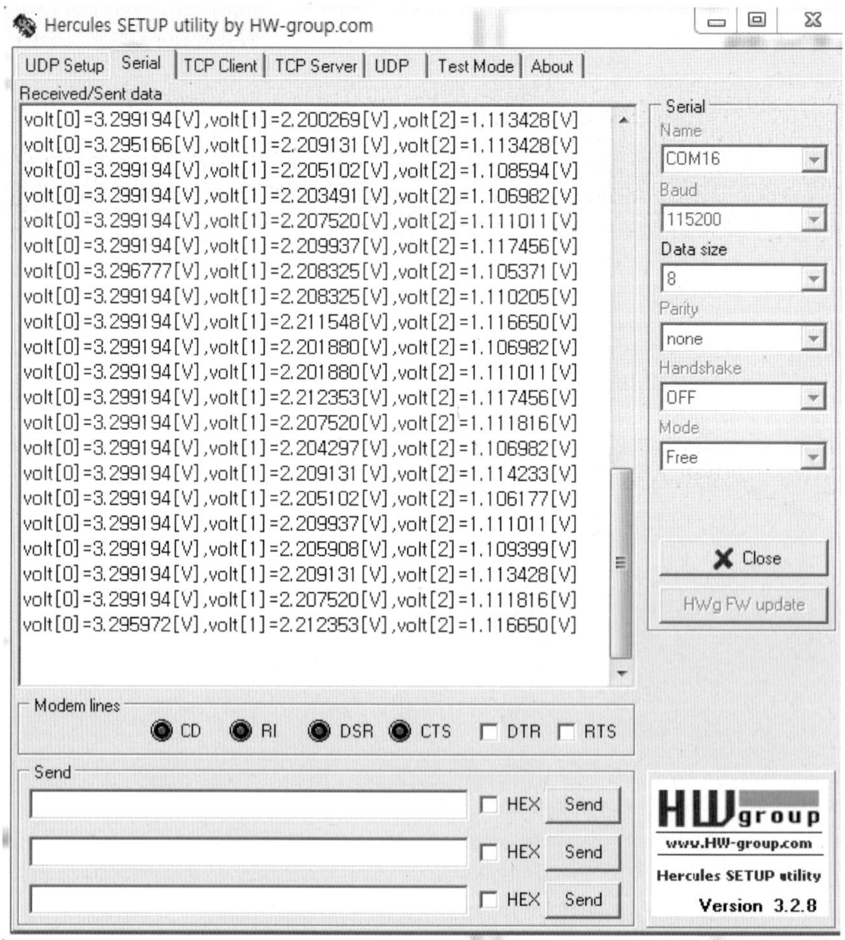

[그림 6.22] 시리얼 통신으로 결과 확인

CHAPTER 7

MCU Internal 온도 측정하기

1. 개요
2. Internal 온도 Sensor 계산
3. ADC 설정
4. Generate Code 생성
5. Source Code 작성하기
6. 디버그 및 다운로드하기
7. 보드 프로그램 실행하기
8. 프로그램 실행 확인

CHAPTER 07

Cortex-M4 기초 활용(ARM Core STM32)

MCU Internal 온도 측정하기

1. 개요

STM32는 CPU와 CPU 주변 온도를 측정할 수 있는 내부 온도 센서를 가지고 있다. 이는 ADC 채널 18에 연결되어 있으며 디지털 전압값으로 나타나고 범위는 –40도에서 125도, 정확도는 (+)1.5도이다.

2. Internal 온도 Sensor 계산

온도 센서는 ADC 채널 18에 고정되어 있으므로 그 채널 값을 읽어서 계산하면 된다. 관련된 식과 파라메터들은 [표 7.1]과 같다.

[표 7.1] Temperature sensor characteristics

Symbol	Parameter	Min	Typ	Max	Unit
$T_L^{(1)}$	V_{SENSE} linearity with temperature	-	±1	±2	°C
Avg_Slope(1)	Average slope	-	2.5	-	mV/°C
$V_{25}^{(1)}$	Voltage at 25 °C	-	0.76	-	V
$t_{START}^{(2)}$	Startup time	-	6	10	μs
$T_{S\_temp}^{(2)}$	ADC sampling time when reading the temperature (1 °C accuracy)	10	-	-	μs

온도 계산식은 다음과 같다.

Temperature (in ‘C) = {(V25 – VSENSE) / Avg_Slope} + 25.

VSENSE= (adc value of channel 18)*3.3/4096

Temperate=(0.76–VSENSE)/0.0025+25;

V25=Vsense at 25 degrees (typical value: 0.76).

Avg_Slope=the average slope of temperature versus Vsense curve (in mv/C or uv/C) (typical value is 2.5mV/C).

3. ADC 설정

System Core에서 RCC 및 SYS 및 TX RX 단자는 이전과 같이 설정한다. [그림 7.1]~[그림 7.2]는 채널 설정과 인터럽트 방식으로 선언하였다.

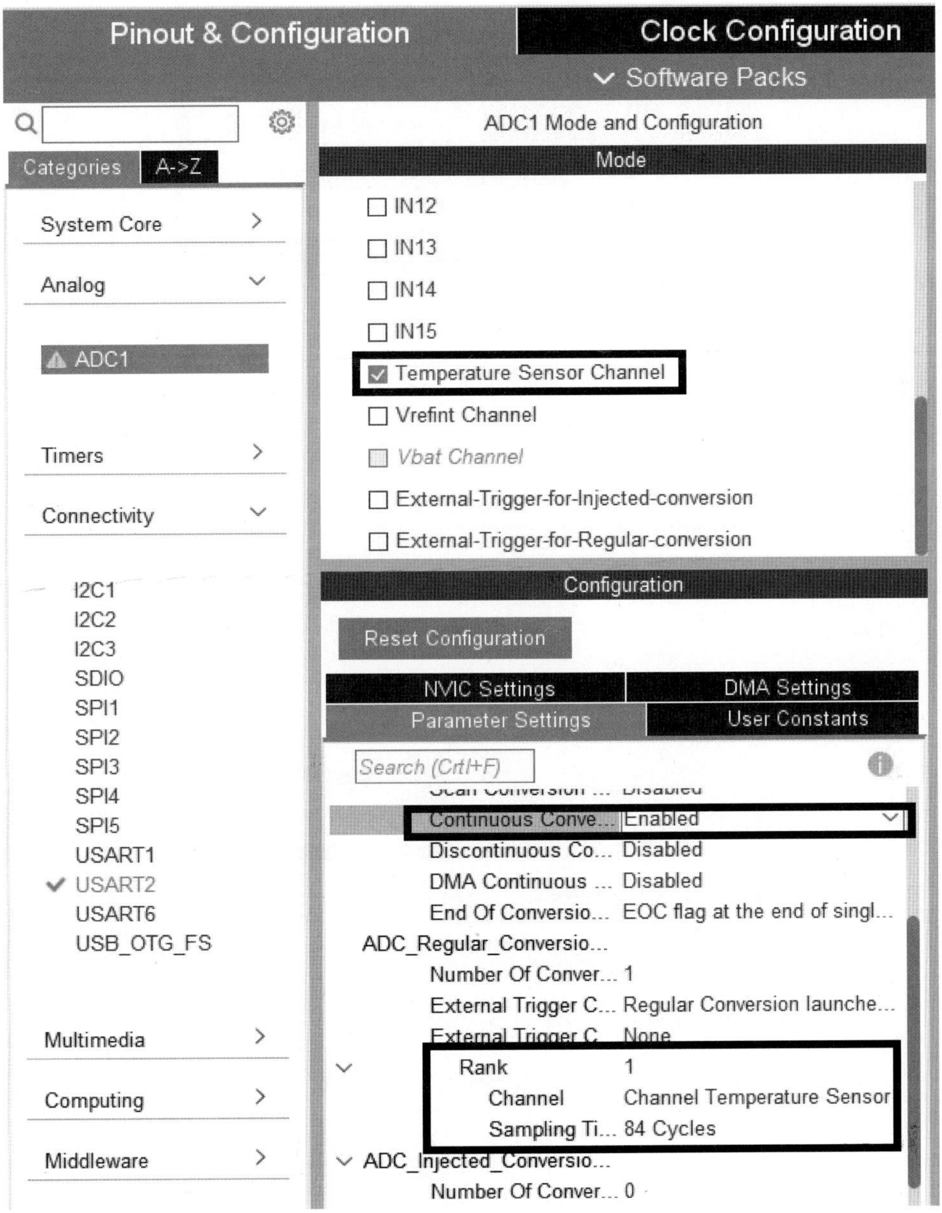

[그림 7.1] ADC 채널 설정

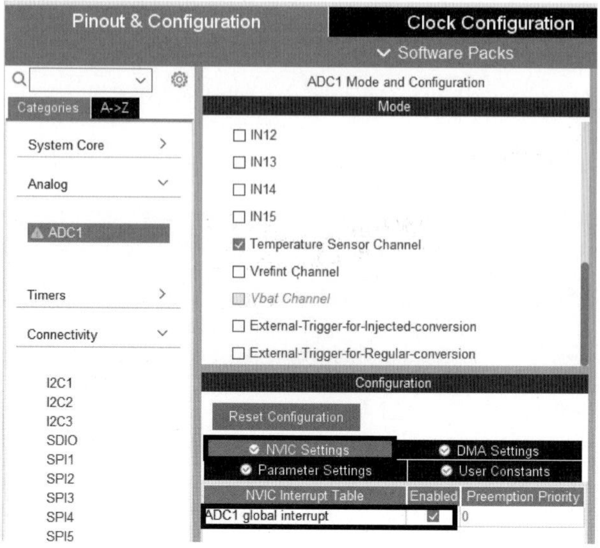

[그림 7.2] ADC1 인터럽트 설정

4. Generate Code 생성

[그림 7.3]은 코드를 생성하기 위한 화면이다. 그림과 같이 클릭해서 프로그램을 생성한다.

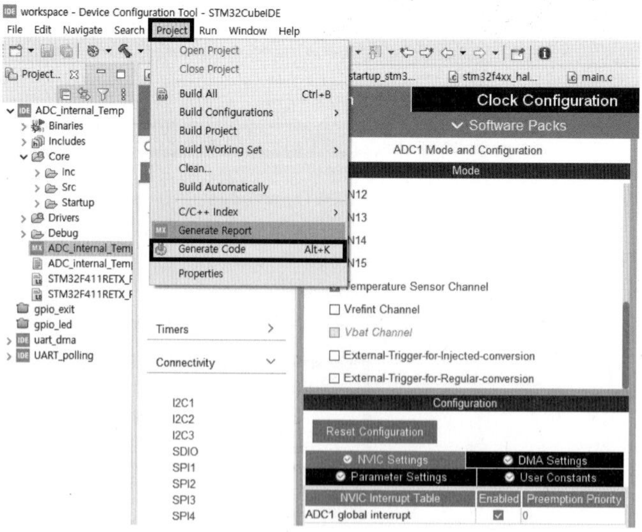

[그림 7.3] ADC1 코드 생성

5. Source Code 작성하기

ADC를 인터럽트 방식으로 사용하려면 관련 함수를 아는 것이 중요하다. 사용하는 방법은 어렵지 않다. 메인 함수에 ADC 인터럽트 시작을 해주고 관련 콜백 함수를 이용하면 된다. [그림 7.4]는 메인 프로그램이다.

>**HAL_ADC_Start_DMA**(ADC_HandleTypeDef* hadc)
인터럽트를 Enable하고 ADC 변환을 시작하는 함수
>**void HAL_ADC_ConvCpltCallback**(ADC_HandleTypeDef* hadc)
{
    ADC 변환이 완료되면 인터럽트 콜백 함수가 실행됨
}

```
/* USER CODE BEGIN 0 */
uint32_t adc_data;
float Temp;

void HAL_ADC_ConvCpltCallback(ADC_HandleTypeDef* hadc)
{
    adc_data= HAL_ADC_GetValue(&hadc1);
    Temp= (float)adc_data*3.3/4096.0;
    Temp=(0.76-Temp)/0.0025 +25;

    HAL_ADC_Start_IT(&hadc1);
}

int __io_putchar(int ch)
{
HAL_UART_Transmit(&huart2, (uint8_t *)&ch, 1, 0xFFFF);
 return ch;
}
/* USER CODE END 0 */
```

```
/* USER CODE BEGIN 2 */
HAL_ADC_Start_IT(&hadc1);
/* USER CODE END 2 */

/* Infinite loop */
/* USER CODE BEGIN WHILE */
while (1)
{
    printf("Temp=%f\r\n",Temp);
    HAL_Delay(200);
  /* USER CODE END WHILE */

  /* USER CODE BEGIN 3 */
}
/* USER CODE END 3 */
```

[그림 7.4] 메인 코드 내용

6. 디버그 및 다운로드하기

메인 함수에 프로그램이 작성되었다면 다음으로 디버그는 [그림 7.5]의 를 클릭하거나 단축키 F11을 누르면 디버깅과 프로그램 다운로드가 완료된다.

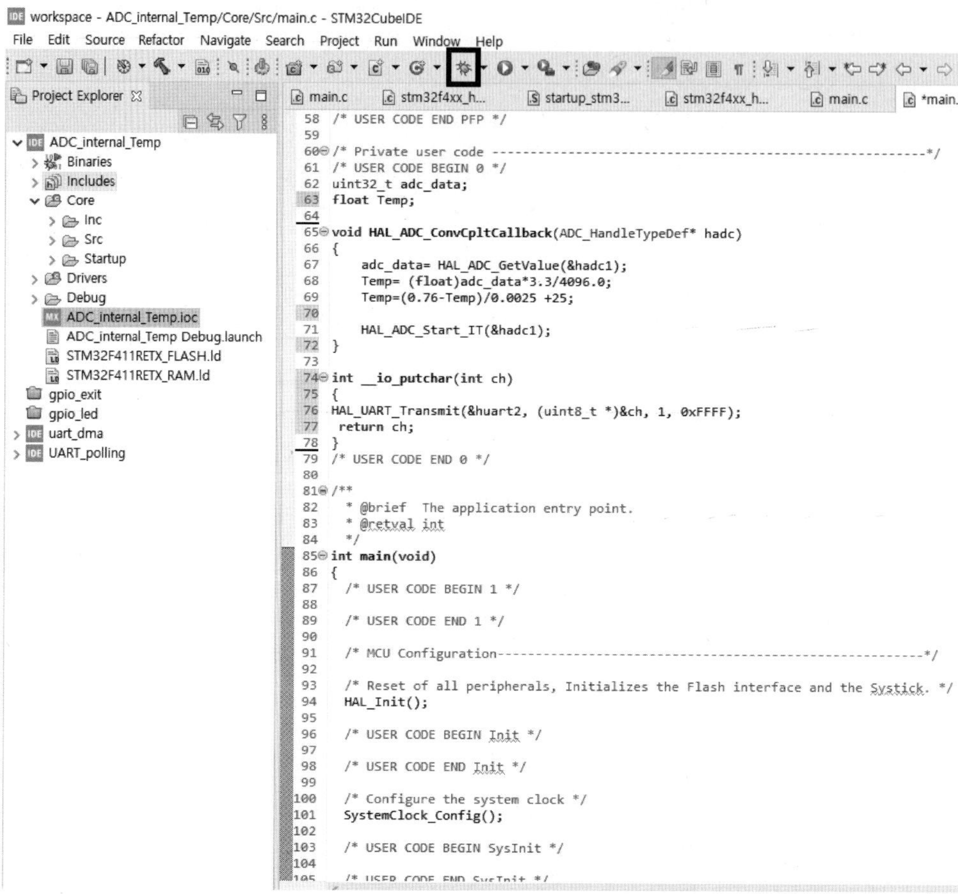
workspace - ADC_internal_Temp/Core/Src/main.c - STM32CubeIDE
File Edit Source Refactor Navigate Search Project Run Window Help
Project Explorer
ADC_internal_Temp
Binaries
Includes
Core
Inc
Src
Startup
Drivers
Debug
ADC_internal_Temp.ioc
ADC_internal_Temp Debug.launch
STM32F411RETX_FLASH.ld
STM32F411RETX_RAM.ld
gpio_exit
gpio_led
uart_dma
UART_polling
main.c
stm32f4xx_h...
startup_stm3...
stm32f4xx_h...
main.c
*main.c
/* USER CODE END PFP */
/* Private user code ---*/
/* USER CODE BEGIN 0 */
uint32_t adc_data;
float Temp;
void HAL_ADC_ConvCpltCallback(ADC_HandleTypeDef* hadc)
{
adc_data= HAL_ADC_GetValue(&hadc1);
Temp= (float)adc_data*3.3/4096.0;
Temp=(0.76-Temp)/0.0025 +25;
HAL_ADC_Start_IT(&hadc1);
}
int __io_putchar(int ch)
{
HAL_UART_Transmit(&huart2, (uint8_t *)&ch, 1, 0xFFFF);
return ch;
}
/* USER CODE END 0 */
/**
* @brief The application entry point.
* @retval int
*/
int main(void)
{
/* USER CODE BEGIN 1 */
/* USER CODE END 1 */
/* MCU Configuration--*/
/* Reset of all peripherals, Initializes the Flash interface and the Systick. */
HAL_Init();
/* USER CODE BEGIN Init */
/* USER CODE END Init */
/* Configure the system clock */
SystemClock_Config();
/* USER CODE BEGIN SysInit */

[그림 7.5] 프로그램 디버그

7. 보드 프로그램 실행하기

[그림 7.6] 상단 중앙에 아이콘 클릭하면 디버그 모드가 종료되어 보드의 프로그램이 실행된다.

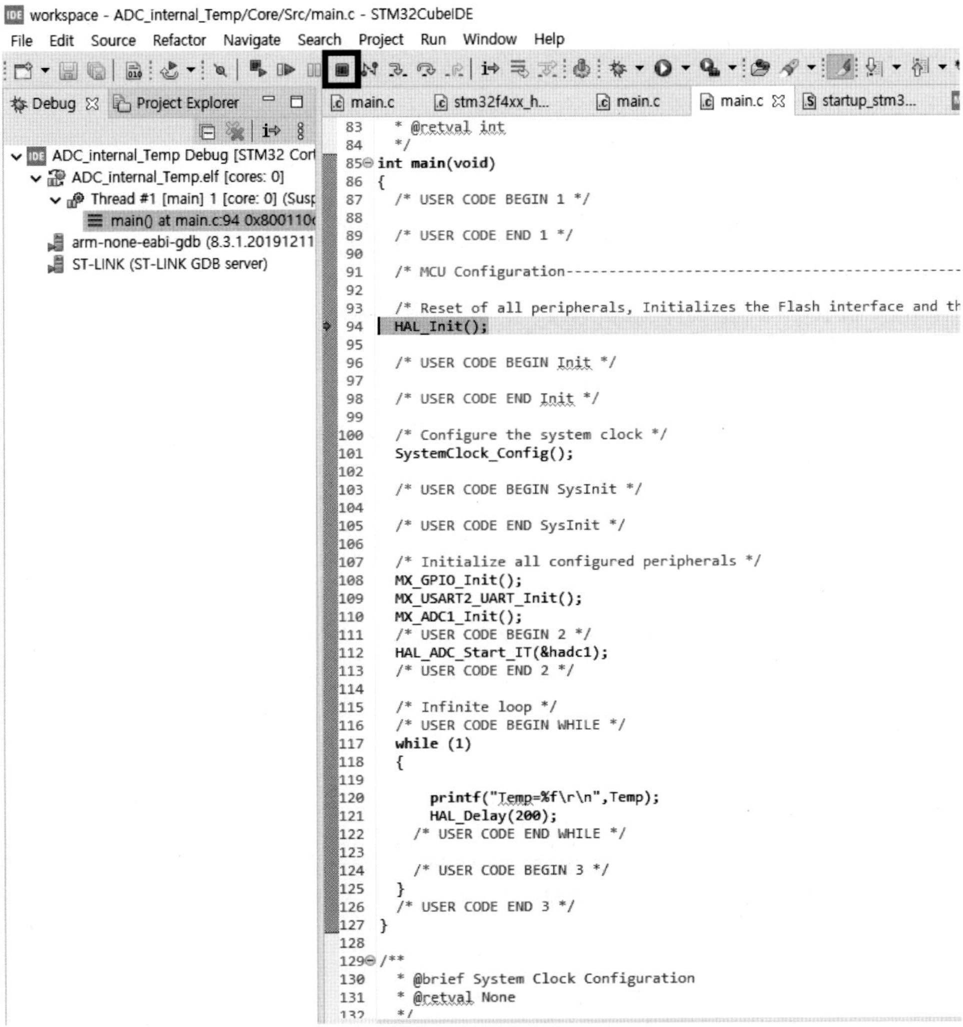

workspace - ADC_internal_Temp/Core/Src/main.c - STM32CubeIDE
File Edit Source Refactor Navigate Search Project Run Window Help
Debug
Project Explorer
main.c
stm32f4xx_h...
main.c
main.c
startup_stm3...
ADC_internal_Temp Debug [STM32 Cor
ADC_internal_Temp.elf [cores: 0]
Thread #1 [main] 1 [core: 0] (Susp
main() at main.c:94 0x800110
arm-none-eabi-gdb (8.3.1.20191211
ST-LINK (ST-LINK GDB server)
83 * @retval int
84 */
85 int main(void)
86 {
87 /* USER CODE BEGIN 1 */
88
89 /* USER CODE END 1 */
90
91 /* MCU Configuration--
92
93 /* Reset of all peripherals, Initializes the Flash interface and th
94 HAL_Init();
95
96 /* USER CODE BEGIN Init */
97
98 /* USER CODE END Init */
99
100 /* Configure the system clock */
101 SystemClock_Config();
102
103 /* USER CODE BEGIN SysInit */
104
105 /* USER CODE END SysInit */
106
107 /* Initialize all configured peripherals */
108 MX_GPIO_Init();
109 MX_USART2_UART_Init();
110 MX_ADC1_Init();
111 /* USER CODE BEGIN 2 */
112 HAL_ADC_Start_IT(&hadc1);
113 /* USER CODE END 2 */
114
115 /* Infinite loop */
116 /* USER CODE BEGIN WHILE */
117 while (1)
118 {
119
120 printf("Temp=%f\r\n",Temp);
121 HAL_Delay(200);
122 /* USER CODE END WHILE */
123
124 /* USER CODE BEGIN 3 */
125 }
126 /* USER CODE END 3 */
127 }
128
129 /**
130 * @brief System Clock Configuration
131 * @retval None

[그림 7.6] 디버그 모드 프로그램 실행

8. 프로그램 실행 확인

시리얼 통신 프로그램을 확인하기 쉬운 방법은 Tera term이나 Hercules, docklight 통신 터미널 [그림 7.7]을 이용하면 된다. 인터넷에서 hercules terminal 검색하면 다운로드할 수 있다.

관련 사이트: https://www.hw-group.com/software/hercules-setup-utility

https://tera-term.softonic.kr/

https://docklight.de/downloads/

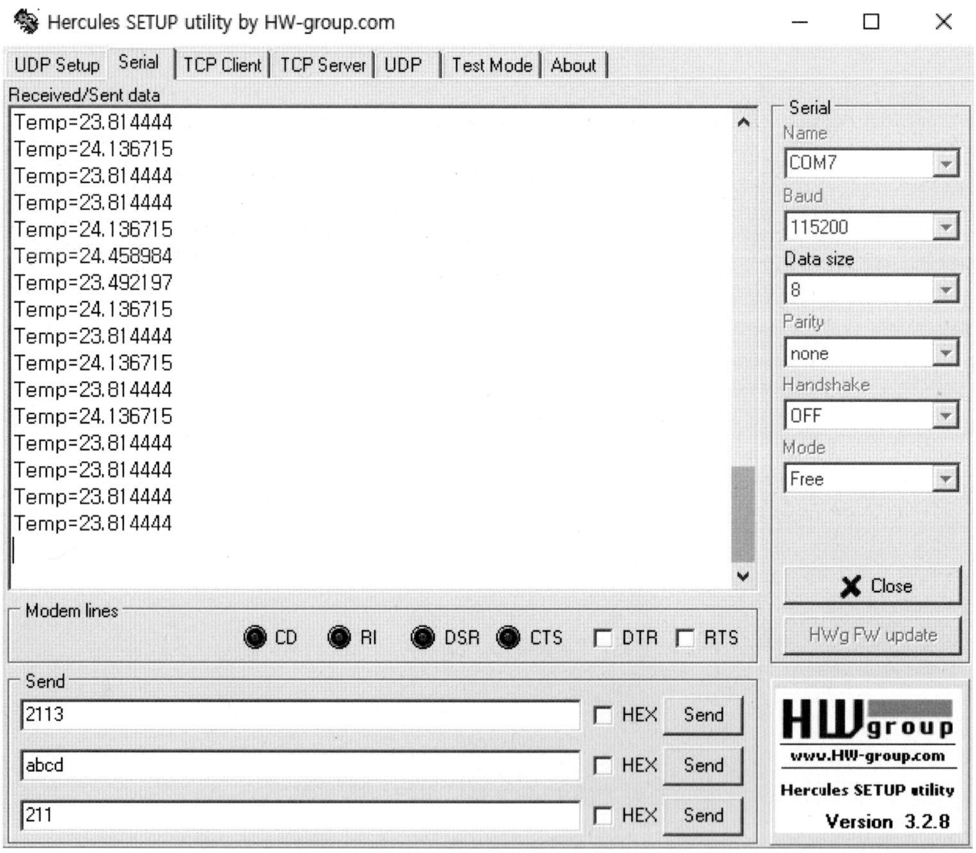

[그림 7.7] 시리얼 통신으로 결과 확인

Cortex-M_4 기초 활용(ARM Core STM32)

CHAPTER 8

Timer 활용하기 1 (타이머 기초, RTC)

1. 개요
2. 타이머 실습 위한 하드웨어 설정
3. 클럭 설정
4. 타이머 1개 사용하기
5. 타이머 2개 사용하기
6. RTC(Real-Time Clock) 사용하기(Date,Time)
7. RTC(Real-Time Clock) 사용하기(Alarm)

CHAPTER 08

Cortex-M4 기초 활용(ARM Core STM32)

Timer 활용하기 1 (타이머 기초, RTC)

1. 개요

STM32의 타이머는 [그림 8.1]~[그림 8.2]와 같이 여러 기능과 함께 다양한 유형의 타이머가 있다. 이를 참조하여 사용하면 된다. ▷범용 타이머(General-purpose timer): 출력 비교, 원 펄스 모드, 입력 캡처, 센서 인터페이스 등을 위한 응용 분야에 사용됨. ▷고급 타이머(Advanced timer): 많은 기능이 있으며, 일반 용도 외에 여기에는 모터 제어 및 디지털 전원과 관련된 여러 기능이 포함되어 있음. ▷기본 타이머(base timer): 기본 타이머에는 입력 출력이 없으며 타임베이스 타이머로 사용되거나 DAC 주변 장치를 트리거함. ▷1 또는 2 타이머(One or Two timer): 제한된 수의 범용 타이머로 사용 채널. ▷상보 출력이 있는 1 또는 2 타이머(One or Two timer): 이전 유형과 동일하지만 한 채널에 데드 타임 생성기가 있음.

Timer type		STM32F0 series	STM32F101 /102/103/ 105/107 lines	STM32F100 value line	STM32L1 series	STM32F2 and STM32F4 series	STM32F30x and STM32F3x8	STM32F37x lines
Advanced		TIM1	TIM1	TIM1	-	TIM1	TIM1	-
		-	TIM8	-	-	TIM8	TIM8	-
							TIM20(1)	
General purpose	**16-bit**	-	TIM2	TIM2	TIM2	-	TIM2	TIM2
		TIM3	TIM3	TIM3	TIM3	TIM3	TIM3	TIM3
		-	TIM4	TIM4	TIM4	TIM4	TIM4	TIM4
		-	TIM5	TIM5	-	-	-	TIM5
		-	-	-	-	-	-	TIM19
	32-bit	TIM2	-	-	-	TIM2	TIM2	TIM2
		-	-	-	-	TIM5		TIM5
Basic		TIM6	TIM6	TIM6	TIM6	TIM6	TIM6	TIM6
		-	TIM7	TIM7	TIM7	TIM7	TIM7	TIM7
		-	-	-	-	-	-	TIM18
1-channel		-	TIM10	-	TIM10	TIM10	-	-
		-	TIM11	-	TIM11	TIM11	-	-
		-	TIM13	TIM13		TIM13	-	TIM13
		TIM14	TIM14	TIM14		TIM14	-	TIM14
2-channel		-	TIM9		TIM9	TIM9	-	-
		-	TIM12	TIM12	-	TIM12	-	TIM12
1-channel with one complementary output		TIM15	-	TIM15	-	-	TIM15	TIM15
		-	-	-	-	-	-	-
2-channel with one complementary output		TIM16	-	TIM16	-	-	TIM16	TIM16
		TIM17	-	TIM17	-	-	TIM17	TIM17

[그림 8.1] STM32 timer

Timer type	Counter resolution	Counter type	DMA	Channels	Comp. channels	Synchronization	
						Master config.	Slave config.
Advanced	16 bit	up, down and center aligned	Yes	4	3	Yes	Yes
General-purpose	16 bit 32 bit(1)	up, down and center aligned	Yes	4	0	Yes	Yes
Basic	16 bit	up	Yes	0	0	Yes	No
1-channel	16 bit	up	No	1	0	Yes (OC signal)	No
2-channel	16 bit	up	No	2	0	Yes	Yes
1-channel with one complementary output	16 bit	up	Yes	1	1	Yes (OC signal)	No
2-channel with one complementary output	16 bit	up	Yes	2	1	No	Yes

[그림 8.2] STM32 타이머 기능

2. 타이머 실습 위한 하드웨어 설정

내부 클럭을 사용하고 별도의 외부 트리거 신호를 사용하지 않으면 하드웨어 없이 동작시킬 수 있다. 따라서 [그림 8.3]과 같이 LED 제어 용도 타이머를 사용한다.

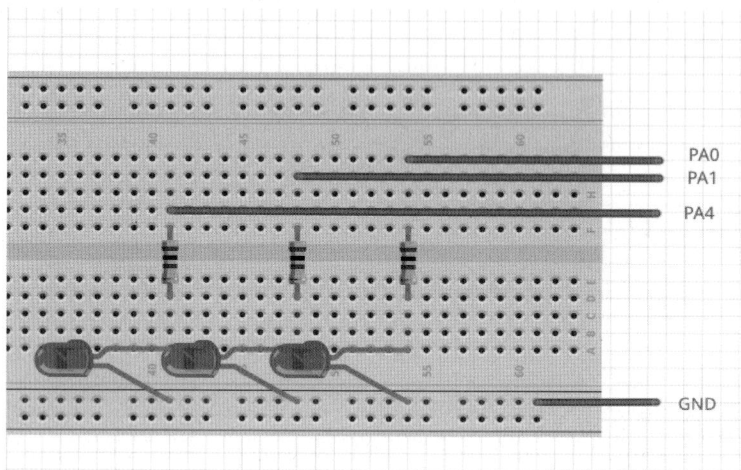

[그림 8.3] LED 제어

3. 클럭 설정

[그림 8.4]~[그림 8.6]은 클럭 설정 과정을 나타낸다. 메인 클럭은 외부 클럭(Crystal/Ceramic Resonator)과 내부 클럭[HSI RC(BYPASS Clock Source)] 둘 중에 하나를 이용하여야 한다. 내부 클럭보다는 외부 클럭이 정확하며, 본 보드에서는 Crystal/Ceramic Resonator는 연결되어 있지 않아 사용하지 않으며 ST-LINK MCO 출력 8MHz를 사용한다.

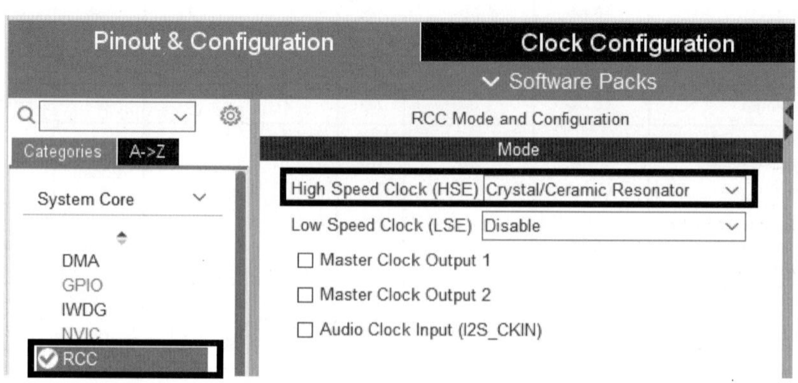

[그림 8.4] 클럭 설정

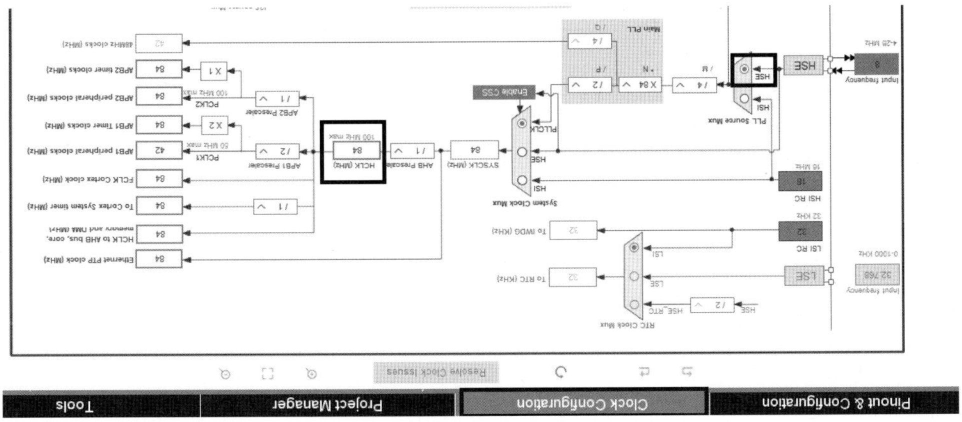

[그림 8.5] 클럭 설정

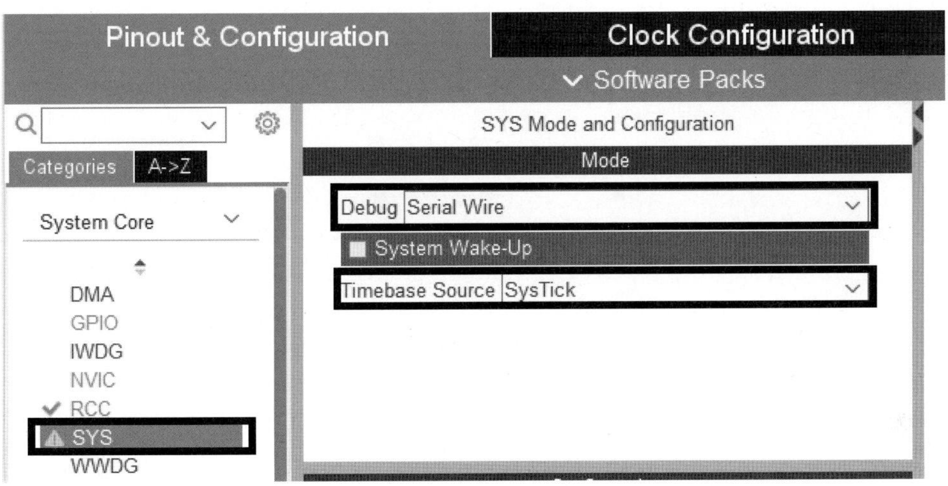

[그림 8.6] SYS 설정

4. 타이머 1개 사용하기

4.1. 타이머 설정

Advanced Timer1은 16비트 4채널 타이머로 APB2에 속하며 84Mhz로 동작한다. 기본으로 LED 제어를 위한 GPIO 설정 및 시리얼 통신을 설정하고 확인을 한 후에 타이머를 설정한다. [그림 8.7]을 참조해서 [그림 8.8]~[그림 8.10]번을 순차적으로 설정한다.

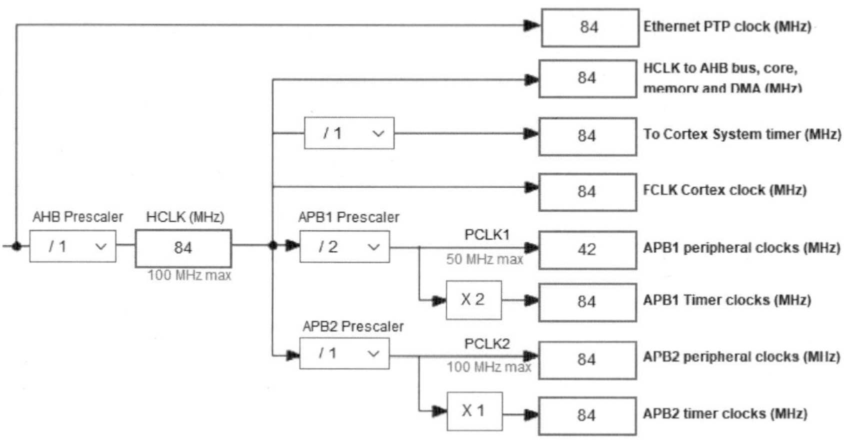

[그림 8.7] 클럭 설정(APB1 =84Mhz)

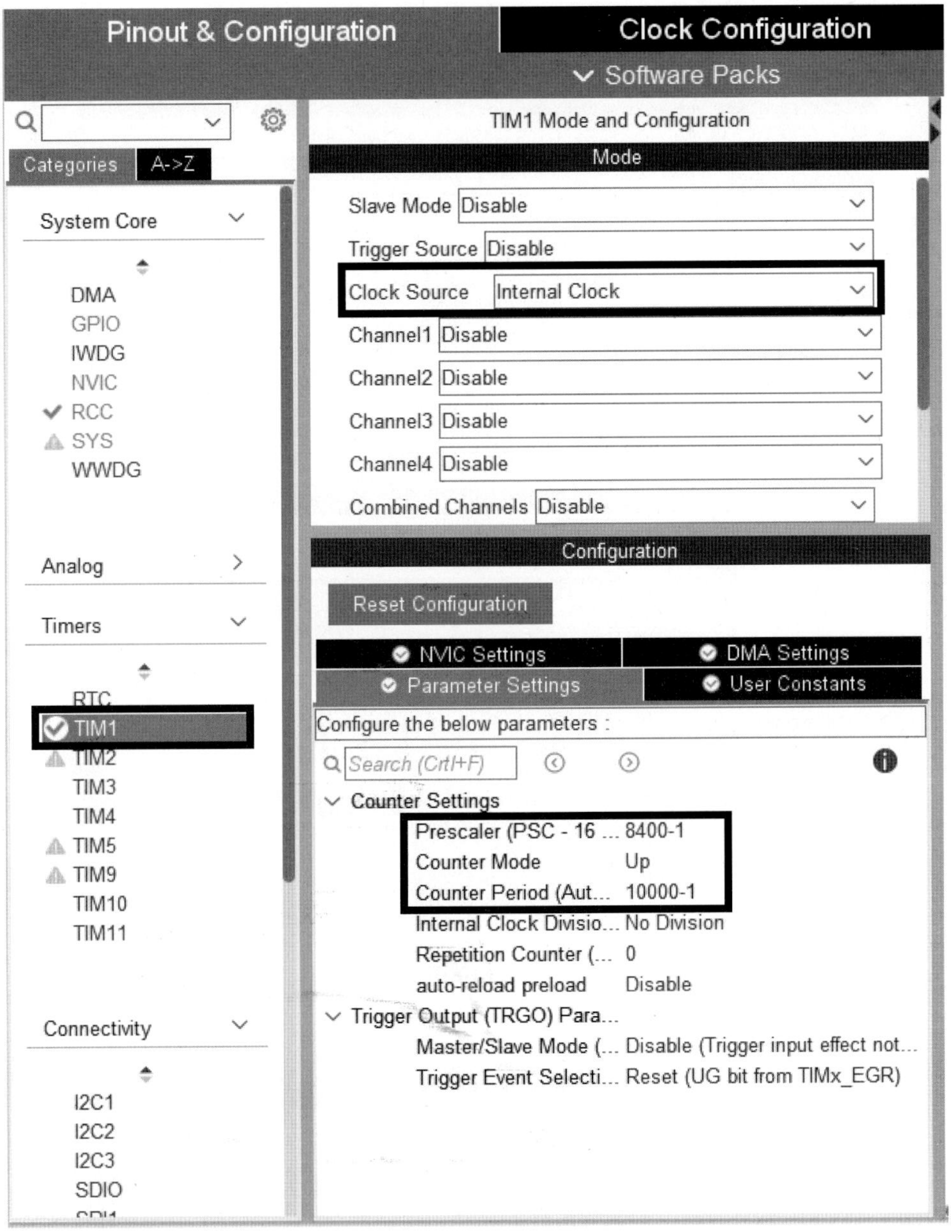
Pinout & Configuration
Clock Configuration
Software Packs
Categories
A->Z
System Core
DMA
GPIO
IWDG
NVIC
RCC
SYS
WWDG
Analog
Timers
RTC
TIM1
TIM2
TIM3
TIM4
TIM5
TIM9
TIM10
TIM11
Connectivity
I2C1
I2C2
I2C3
SDIO
TIM1 Mode and Configuration
Mode
Slave Mode Disable
Trigger Source Disable
Clock Source Internal Clock
Channel1 Disable
Channel2 Disable
Channel3 Disable
Channel4 Disable
Combined Channels Disable
Configuration
Reset Configuration
NVIC Settings
DMA Settings
Parameter Settings
User Constants
Configure the below parameters :
Search (Crtl+F)
Counter Settings
Prescaler (PSC - 16 ... 8400-1
Counter Mode Up
Counter Period (Aut... 10000-1
Internal Clock Divisio... No Division
Repetition Counter (... 0
auto-reload preload Disable
Trigger Output (TRGO) Para...
Master/Slave Mode (... Disable (Trigger input effect not...
Trigger Event Selecti... Reset (UG bit from TIMx_EGR)

[그림 8.8] 타이머 1 설정

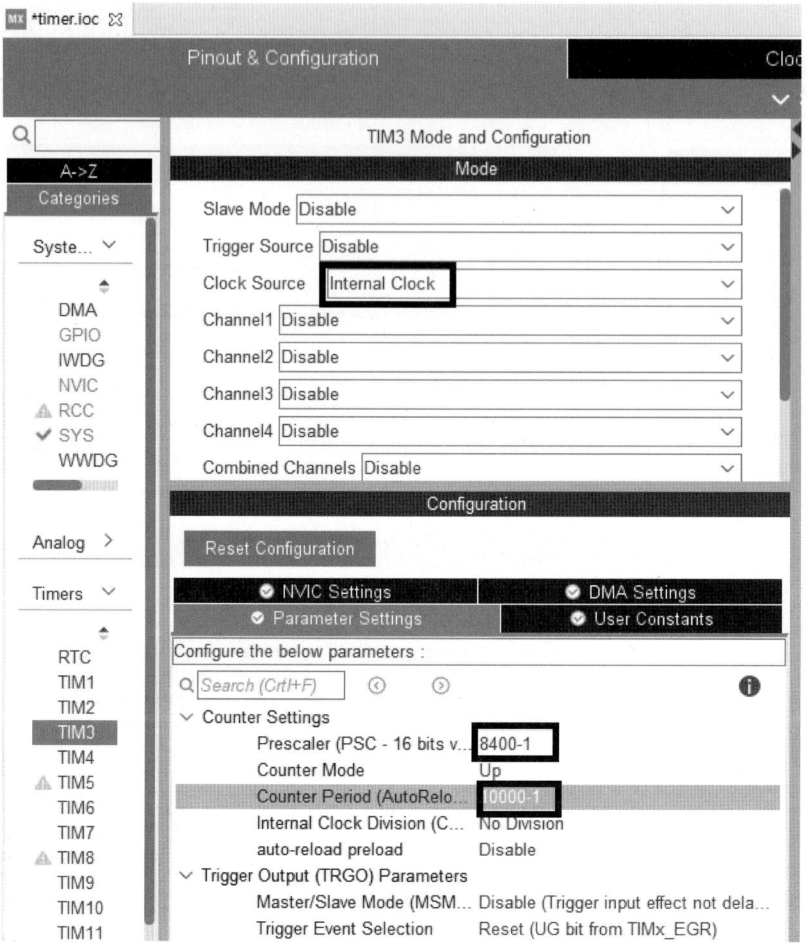

[그림 8.9] 타이머 3 설정

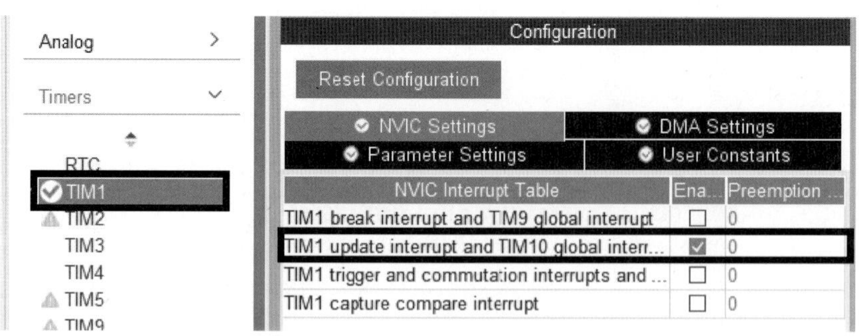

[그림 8.10] 타이머 1 설정(인터럽트)

4.2. Generate Code 생성

[그림 8.11]은 코드를 생성하기 위한 화면이다. 그림과 같이 클릭해서 프로그램을 생성한다.

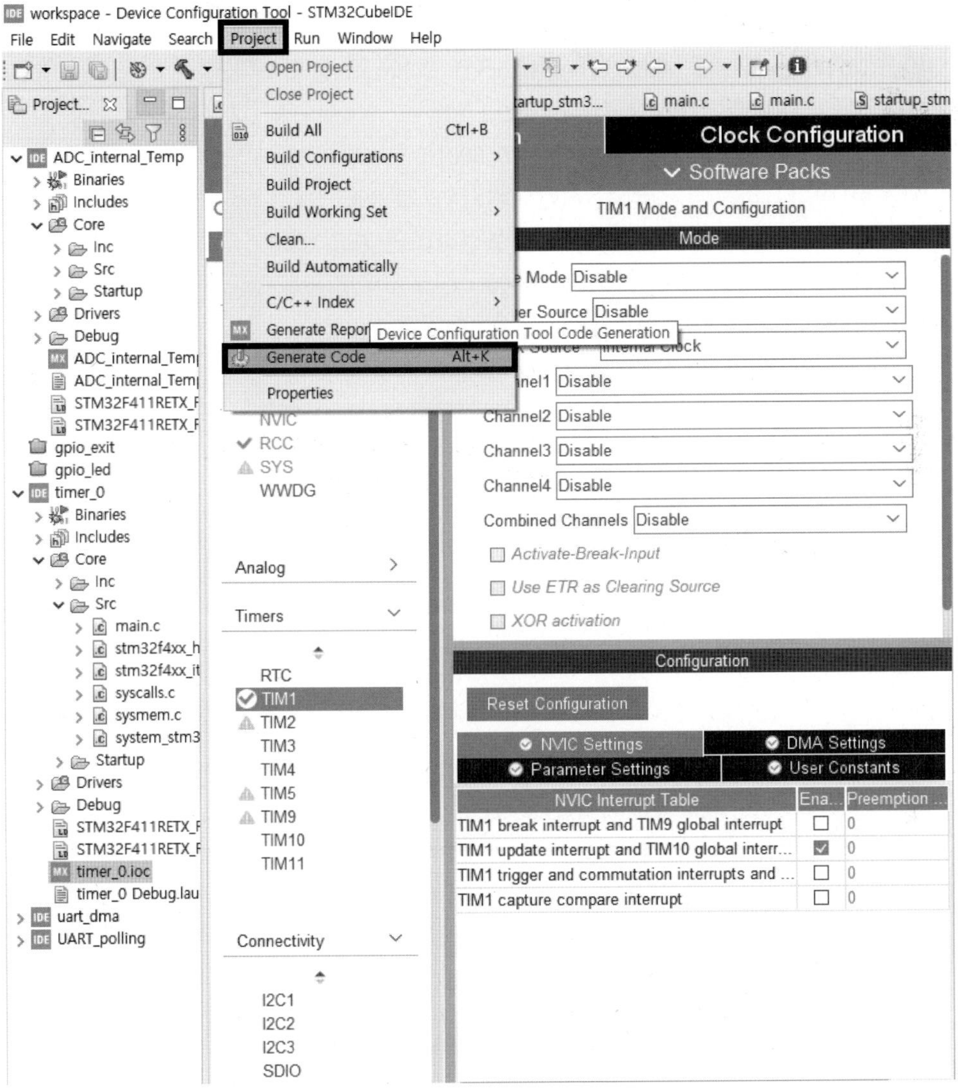

[그림 8.11] 프로그램 코드를 생성

4.3. Source Code 작성하기

LED를 1초 간격으로 on/off 하는 프로그램을 작성해 보자. 이를 위해서 타이머 설정은 다음과 같다. 프로그램 과정은 [그림 8.12]와 같이 프로그램하면 된다.

STM32F 411RE는 APB1과 APB2 timer clockrk 84Mhz로 동일하다.
>APB2 clock: 84Mhz , Prescale= 8,400,
주파수 계산: f= 84,000,000/8,400= 10,000[Hz],
1 카운터(증가) 하는데 걸리는 시간 T=1/f=1/10,000,
1초 설정 위한 ARR=10,000, 0.5초로 설정하려면 ARR =5000
Auto reload Register(ARR)설정 : 10000
ARR 값(16bit) 설정범위: 0~65535
>시작함수: HAL_TIM_Base_Start_IT(&htim3);
>Callback 함수 프로그램:

```
void HAL_TIM_PeriodElapsedCallback(TIM_HandleTypeDef *htim)

{
    타이머 인터럽트 콜백 함수
}
```

```
/* Private user code ---------------------------------------------------------*/
/* USER CODE BEGIN 0 */
void HAL_TIM_PeriodElapsedCallback(TIM_HandleTypeDef *htim){

    HAL_GPIO_TogglePin(GPIOA, GPIO_PIN_0);
}
/* USER CODE END 0 */


/* Initialize all configured peripherals */
MX_GPIO_Init();
MX_USART2_UART_Init();
MX_TIM1_Init();
/* USER CODE BEGIN 2 */
HAL_TIM_Base_Start_IT(&htim1);
/* USER CODE END 2 */

/* Infinite loop */
/* USER CODE BEGIN WHILE */
while (1)
{
  /* USER CODE END WHILE */

  /* USER CODE BEGIN 3 */
}
/* USER CODE END 3 */
}
```

[그림 8.12] 메인 코드 작성

4.4. 디버그 및 다운로드하기

메인 함수에 프로그램이 작성되었다면 다음으로 디버그는 [그림 8.13]의 를 클릭하거나 단축키 F11을 누르면 디버깅과 프로그램 다운로드가 완료된다. 그림을 보면 타이머의 스케일러 값과 주기(ARR) 값이 잘 설정되어 있는 것을 알 수 있다.

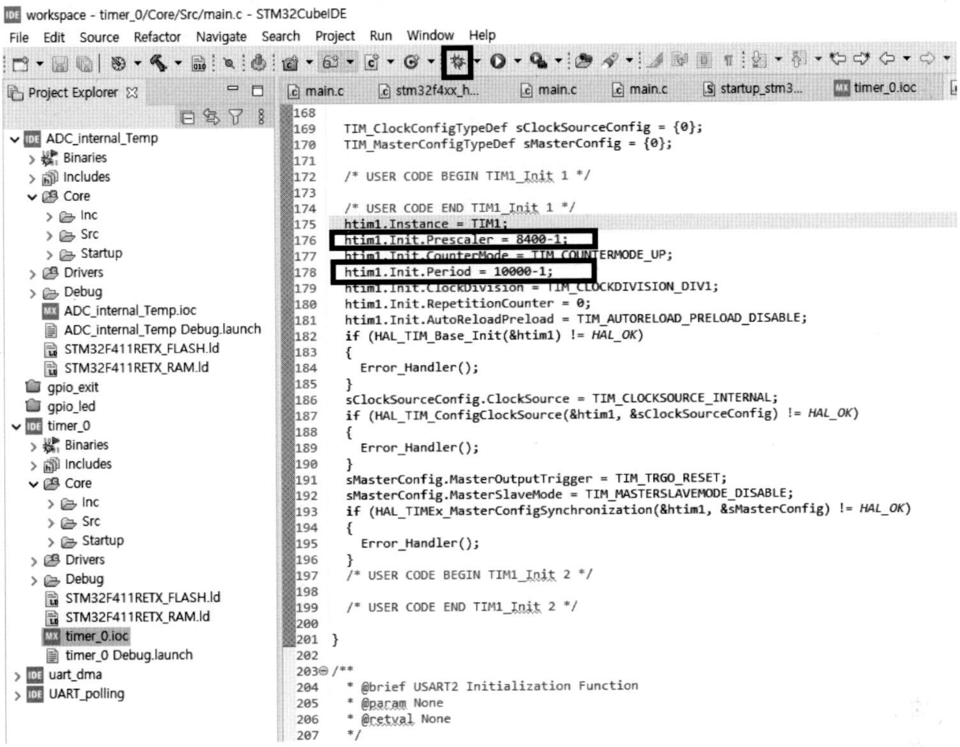
workspace - timer_0/Core/Src/main.c - STM32CubeIDE
File Edit Source Refactor Navigate Search Project Run Window Help
Project Explorer
ADC_internal_Temp
Binaries
Includes
Core
Inc
Src
Startup
Drivers
Debug
ADC_internal_Temp.ioc
ADC_internal_Temp Debug.launch
STM32F411RETX_FLASH.ld
STM32F411RETX_RAM.ld
gpio_exit
gpio_led
timer_0
Binaries
Includes
Core
Inc
Src
Startup
Drivers
Debug
STM32F411RETX_FLASH.ld
STM32F411RETX_RAM.ld
timer_0.ioc
timer_0 Debug.launch
uart_dma
UART_polling
main.c
stm32f4xx_h...
main.c
main.c
startup_stm3...
timer_0.ioc
TIM_ClockConfigTypeDef sClockSourceConfig = {0};
TIM_MasterConfigTypeDef sMasterConfig = {0};
/* USER CODE BEGIN TIM1_Init 1 */
/* USER CODE END TIM1_Init 1 */
htim1.Instance = TIM1;
htim1.Init.Prescaler = 8400-1;
htim1.Init.CounterMode = TIM_COUNTERMODE_UP;
htim1.Init.Period = 10000-1;
htim1.Init.ClockDivision = TIM_CLOCKDIVISION_DIV1;
htim1.Init.RepetitionCounter = 0;
htim1.Init.AutoReloadPreload = TIM_AUTORELOAD_PRELOAD_DISABLE;
if (HAL_TIM_Base_Init(&htim1) != HAL_OK)
{
Error_Handler();
}
sClockSourceConfig.ClockSource = TIM_CLOCKSOURCE_INTERNAL;
if (HAL_TIM_ConfigClockSource(&htim1, &sClockSourceConfig) != HAL_OK)
{
Error_Handler();
}
sMasterConfig.MasterOutputTrigger = TIM_TRGO_RESET;
sMasterConfig.MasterSlaveMode = TIM_MASTERSLAVEMODE_DISABLE;
if (HAL_TIMEx_MasterConfigSynchronization(&htim1, &sMasterConfig) != HAL_OK)
{
Error_Handler();
}
/* USER CODE BEGIN TIM1_Init 2 */
/* USER CODE END TIM1_Init 2 */
}
/**
* @brief USART2 Initialization Function
* @param None
* @retval None
*/

[그림 8.13] 프로그램 디버그

4.5. 보드 프로그램 실행하기

[그림 8.14] 상단 중앙에 아이콘 ■ 클릭하면 디버그 모드가 종료되어 보드의 프로그램이 실행된다.

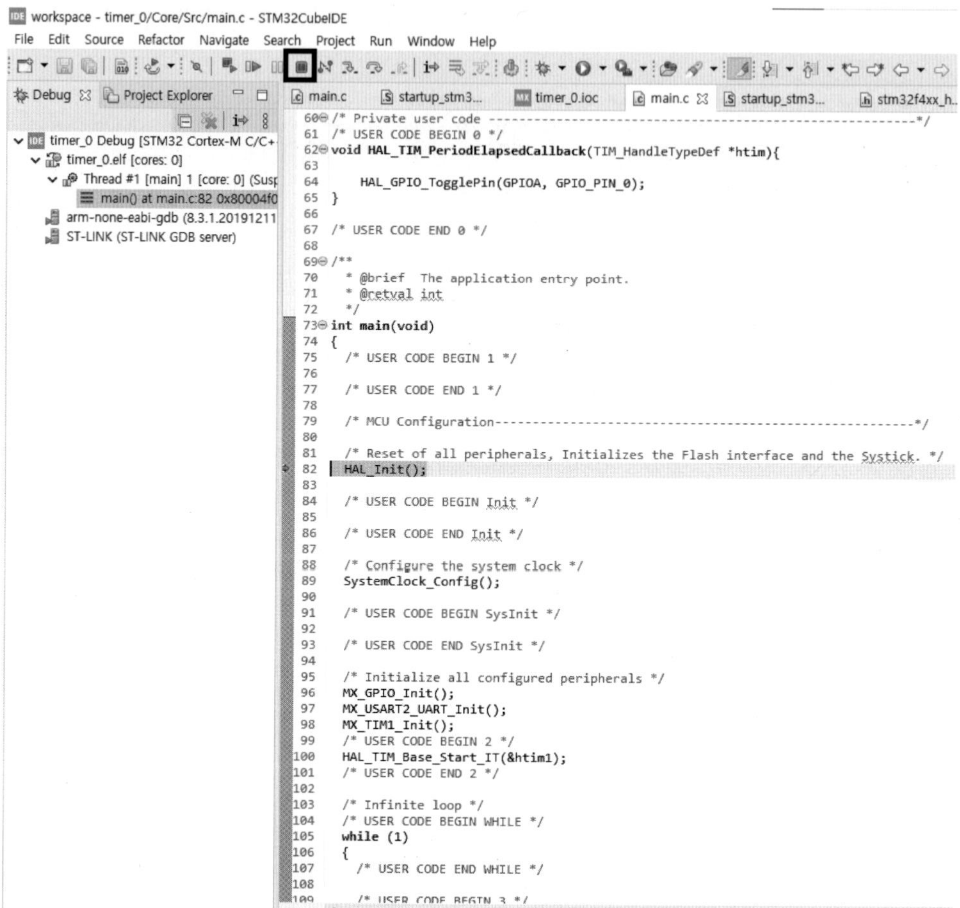
workspace - timer_0/Core/Src/main.c - STM32CubeIDE
File Edit Source Refactor Navigate Search Project Run Window Help
Debug
Project Explorer
main.c
startup_stm3...
timer_0.ioc
main.c
startup_stm3...
stm32f4xx_h...
timer_0 Debug [STM32 Cortex-M C/C+
timer_0.elf [cores: 0]
Thread #1 [main] 1 [core: 0] (Susp
main() at main.c:82 0x80004f0
arm-none-eabi-gdb (8.3.1.20191211
ST-LINK (ST-LINK GDB server)
/* Private user code ---*/
/* USER CODE BEGIN 0 */
void HAL_TIM_PeriodElapsedCallback(TIM_HandleTypeDef *htim){
HAL_GPIO_TogglePin(GPIOA, GPIO_PIN_0);
}
/* USER CODE END 0 */
/**
* @brief The application entry point.
* @retval int
*/
int main(void)
{
/* USER CODE BEGIN 1 */
/* USER CODE END 1 */
/* MCU Configuration--*/
/* Reset of all peripherals, Initializes the Flash interface and the Systick. */
HAL_Init();
/* USER CODE BEGIN Init */
/* USER CODE END Init */
/* Configure the system clock */
SystemClock_Config();
/* USER CODE BEGIN SysInit */
/* USER CODE END SysInit */
/* Initialize all configured peripherals */
MX_GPIO_Init();
MX_USART2_UART_Init();
MX_TIM1_Init();
/* USER CODE BEGIN 2 */
HAL_TIM_Base_Start_IT(&htim1);
/* USER CODE END 2 */
/* Infinite loop */
/* USER CODE BEGIN WHILE */
while (1)
{
/* USER CODE END WHILE */

[그림 8.14] 디버그 모드 프로그램 실행

4.6. 프로그램 수정하기

LED가 1초 간격으로 토글 됨을 확인할 수 있다. 만약 0.5초 간격으로 토글을 하려면 다음 [그림 8.15]와 같이 조정하고 디거깅을 하면 된다.

```
  /* Initialize all configured peripherals */
  MX_GPIO_Init();
  MX_USART2_UART_Init();
  MX_TIM1_Init();
  /* USER CODE BEGIN 2 */
  HAL_TIM_Base_Start_IT(&htim1);
  /* USER CODE END 2 */

  /* Infinite loop */
  /* USER CODE BEGIN WHILE */
  while (1)
  {
    /* USER CODE END WHILE */
  TIM1->ARR=5000;

    /* USER CODE BEGIN 3 */
  }
  /* USER CODE END 3 */
}
```

[그림 8.15] 프로그램 수정

5. 타이머 2개 사용하기

5.1. 타이머 설정

범용 타이머 3,4는 16bit 4채널 타이머로 APB1에 속하며 84Mhz로 동작한다. 기본으로 LED 제어를 위한 GPIO 설정 및 시리얼 통신을 설정하고 확인을 한 후에 타이머를 설정한다. 타이머 3에는 1초, 타이머 4에는 0.5초를 설정해서 LED를 토글하는 프로그램을 작성하기 위한 설정 과정을 [그림 8.16]~[그림 8.17]에 순차적으로 설정한다.

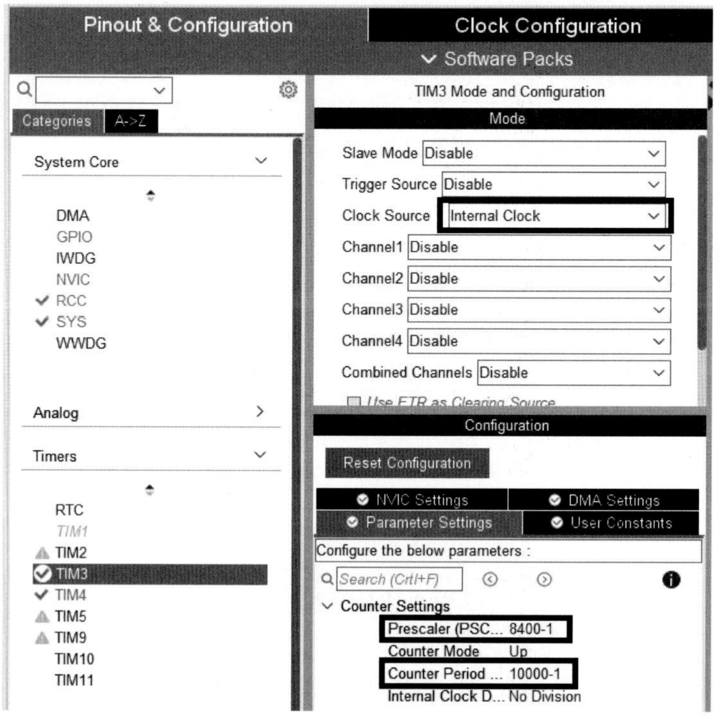

[그림 8.16] 타이머 3 설정

[그림 8.17] 타이머 3 인터럽트설정

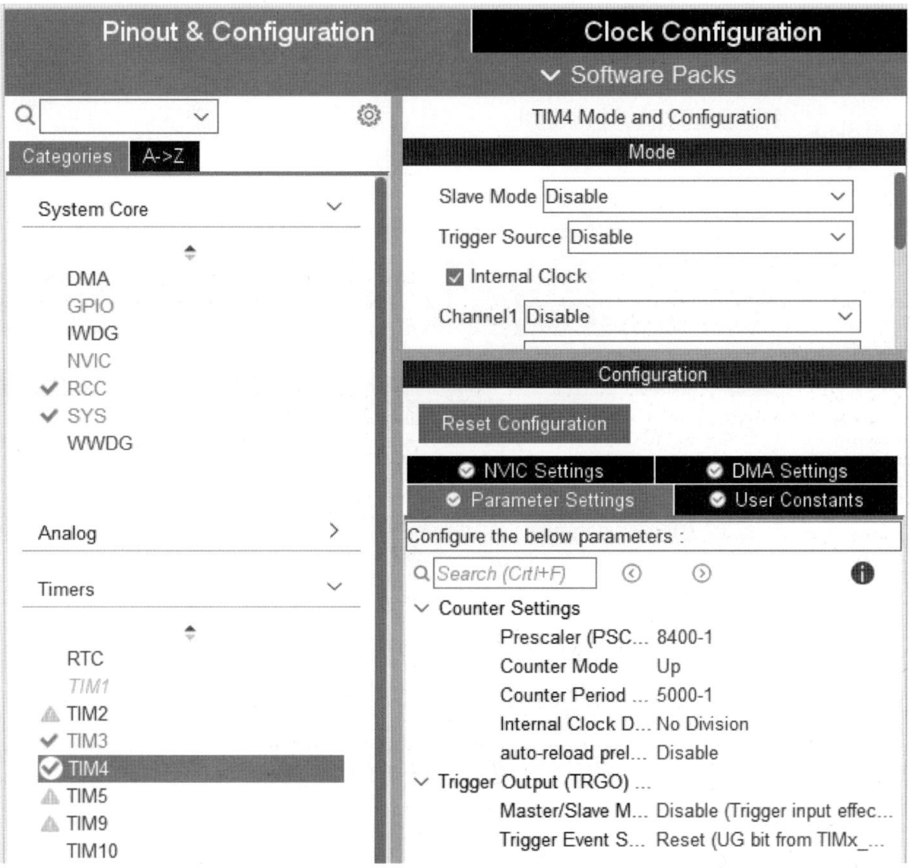

[그림 8.18] 타이머 4 설정

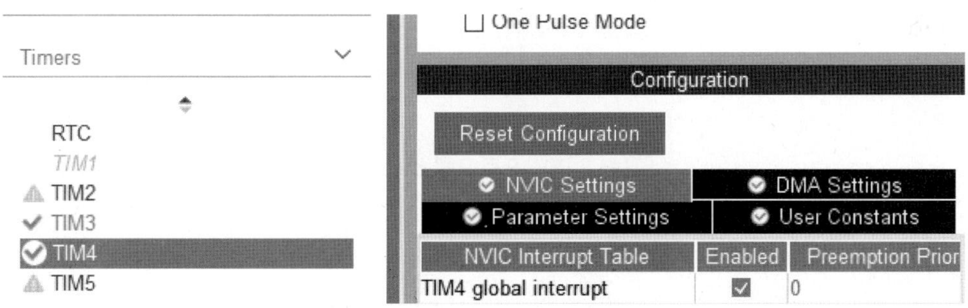

[그림 8.19] 타이머 4 설정(인터럽트)

5.2. Generate Code 생성

[그림 8.20]은 코드를 생성하기 위한 화면이다. 그림과 같이 클릭해서 프로그램을 생성한다.

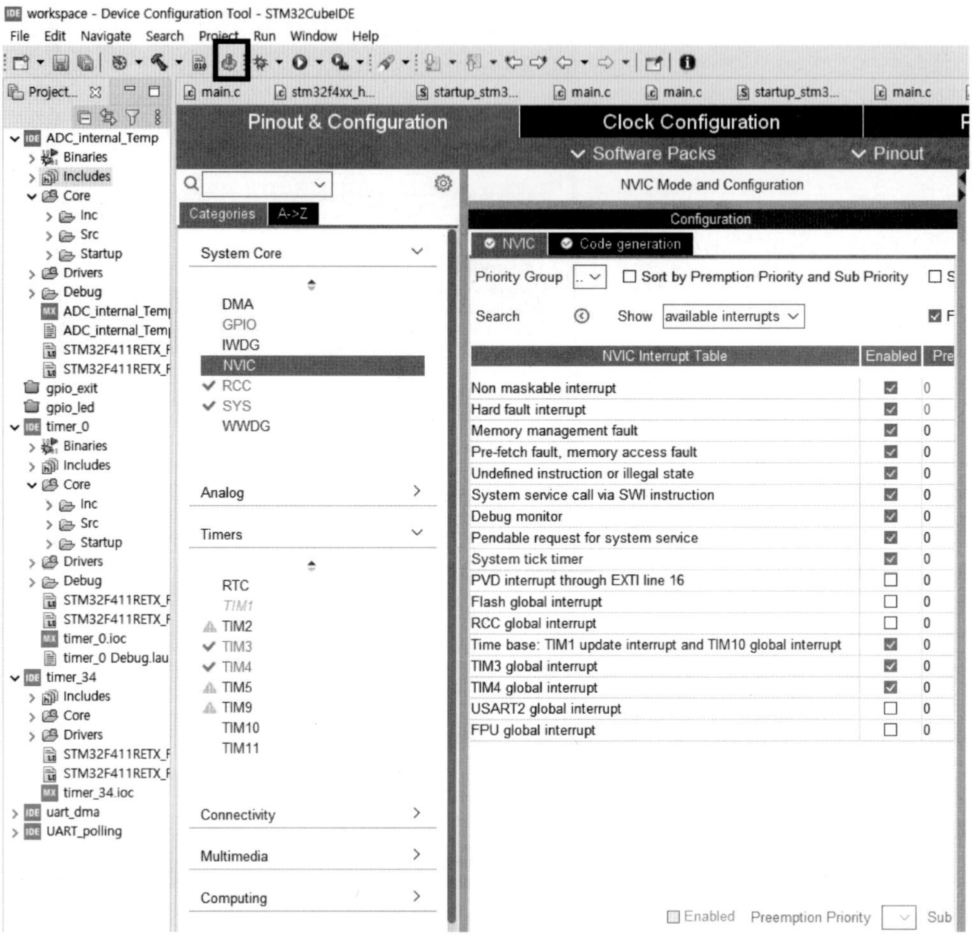

[그림 8.20] 프로그램 코드를 생성

5.3. Source Code 작성하기

SYS 설정에서 SysStick 이외의 타이머를 선택할 경우 여기서는 (TIM1) 타이머 콜백 함수가 main.c 파일에 자동 생성되므로 별도로 만들 필요가 없으며, 해당 함수로 이동해서 프로그램을 하면 된다. 타이머 3, 4를 이용 PA0 LED를 1초 간격, PA1 LED를 0.5초 간격으로 on/off 하는 프로그램을 작성 해보자. 이를 위해서 타이머 설정은 다음과 같다.

```
>Prescale=8,400,
주파수 계산: f= 84,000,000/8,400= 10,000[Hz],
1 카운터(증가) 하는데 걸리는 시간 T=1/f=1/10,000,
타이머 3: ARR=10,000(1초), 타이머 4: ARR =5000(0.5초)
>시작함수:
 HAL_TIM_Base_Start_IT(&htim3);
 HAL_TIM_Base_Start_IT(&htim4);
>Callback 함수 프로그램:
void HAL_TIM_PeriodElapsedCallback(TIM_HandleTypeDef *htim)
{
     타이머 인터럽트 콜백 함수
}
```

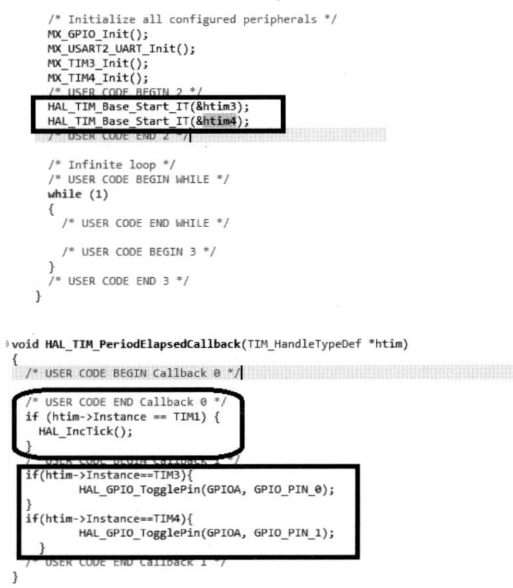

```
/* Initialize all configured peripherals */
MX_GPIO_Init();
MX_USART2_UART_Init();
MX_TIM3_Init();
MX_TIM4_Init();
HAL_TIM_Base_Start_IT(&htim3);
HAL_TIM_Base_Start_IT(&htim4);

/* Infinite loop */
/* USER CODE BEGIN WHILE */
while (1)
{
  /* USER CODE END WHILE */

  /* USER CODE BEGIN 3 */
}
/* USER CODE END 3 */
}

void HAL_TIM_PeriodElapsedCallback(TIM_HandleTypeDef *htim)
{
  /* USER CODE BEGIN Callback 0 */

  /* USER CODE END Callback 0 */
  if (htim->Instance == TIM1) {
    HAL_IncTick();
  }
if(htim->Instance==TIM3){
        HAL_GPIO_TogglePin(GPIOA, GPIO_PIN_0);
}
if(htim->Instance==TIM4){
        HAL_GPIO_TogglePin(GPIOA, GPIO_PIN_1);
  }
}
```

[그림 8.21] 메인 코드 작성

5.4. 디버그 및 다운로드하기

메인 함수에 프로그램이 작성되었다면 다음으로 디버그는 [그림 8.22]의 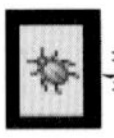를 클릭하거나 단축키 F11을 누르면 디버깅과 프로그램 다운로드가 완료된다.

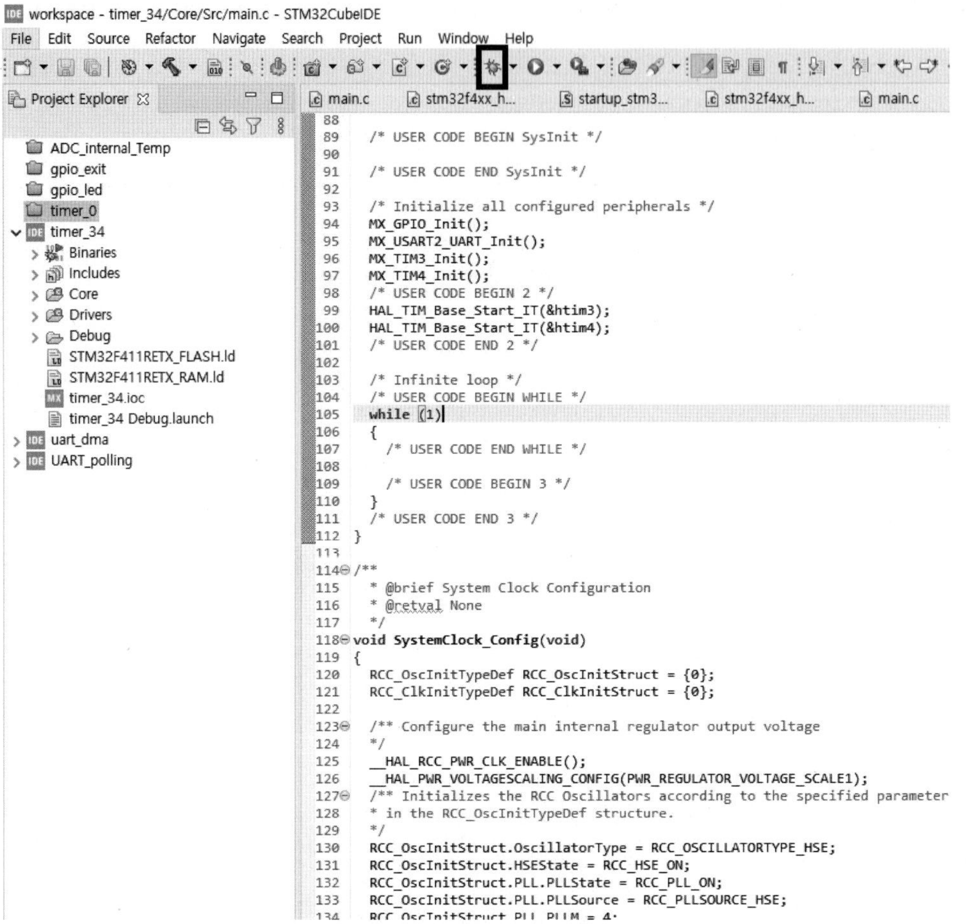

[그림 8.22] 프로그램 디버그

5.5. 보드 프로그램 실행하기

[그림 8.23] 상단 중앙에 아이콘 클릭하면 디버그 모드가 종료되어 보드의 프로그램이 실행된다.

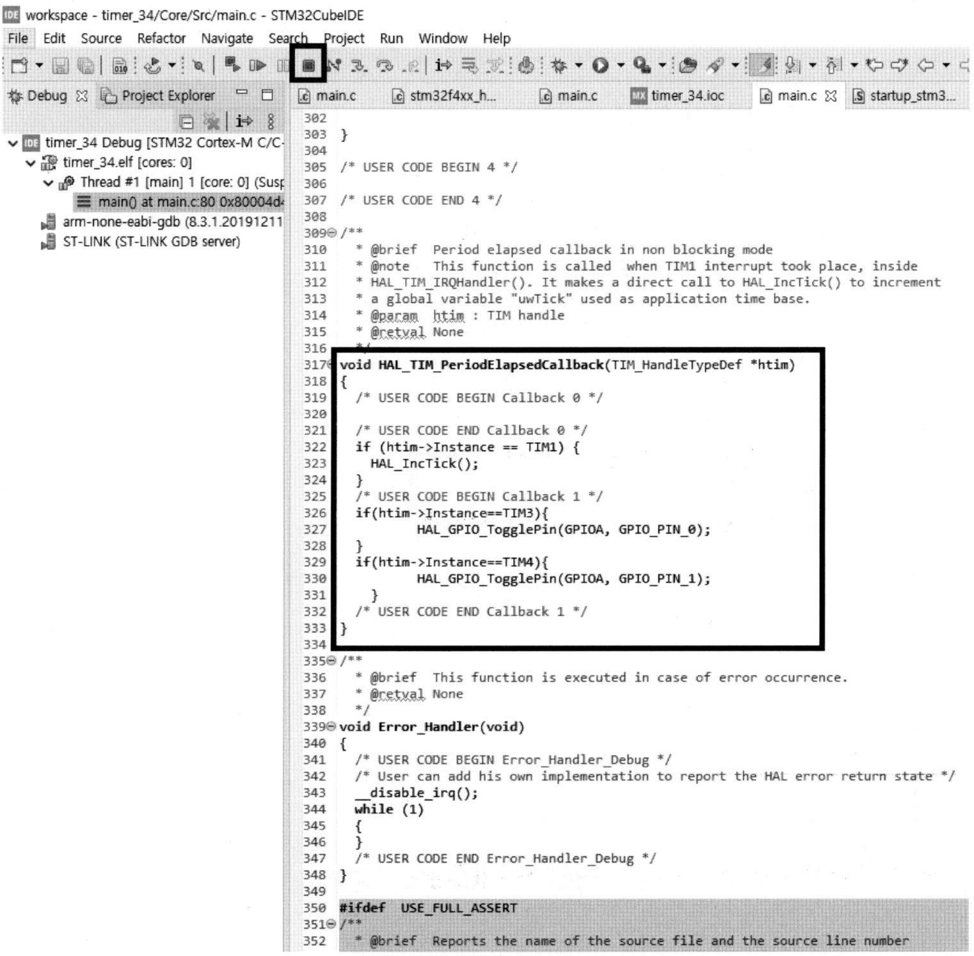
workspace - timer_34/Core/Src/main.c - STM32CubeIDE
File Edit Source Refactor Navigate Search Project Run Window Help
Debug
Project Explorer
main.c
stm32f4xx_h...
main.c
timer_34.ioc
main.c
startup_stm3...
timer_34 Debug [STM32 Cortex-M C/C
timer_34.elf [cores: 0]
Thread #1 [main] 1 [core: 0] (Susp
main() at main.c:80 0x80004d
arm-none-eabi-gdb (8.3.1.20191211
ST-LINK (ST-LINK GDB server)
}
/* USER CODE BEGIN 4 */
/* USER CODE END 4 */
/**
* @brief Period elapsed callback in non blocking mode
* @note This function is called when TIM1 interrupt took place, inside
* HAL_TIM_IRQHandler(). It makes a direct call to HAL_IncTick() to increment
* a global variable "uwTick" used as application time base.
* @param htim : TIM handle
* @retval None
*/
void HAL_TIM_PeriodElapsedCallback(TIM_HandleTypeDef *htim)
{
/* USER CODE BEGIN Callback 0 */
/* USER CODE END Callback 0 */
if (htim->Instance == TIM1) {
HAL_IncTick();
}
/* USER CODE BEGIN Callback 1 */
if(htim->Instance==TIM3){
HAL_GPIO_TogglePin(GPIOA, GPIO_PIN_0);
}
if(htim->Instance==TIM4){
HAL_GPIO_TogglePin(GPIOA, GPIO_PIN_1);
}
/* USER CODE END Callback 1 */
}
/**
* @brief This function is executed in case of error occurrence.
* @retval None
*/
void Error_Handler(void)
{
/* USER CODE BEGIN Error_Handler_Debug */
/* User can add his own implementation to report the HAL error return state */
__disable_irq();
while (1)
{
}
/* USER CODE END Error_Handler_Debug */
}
#ifdef USE_FULL_ASSERT
/**
* @brief Reports the name of the source file and the source line number

[그림 8.23] 디버그 모드 프로그램 실행

5.6. 프로그램 수정하기

만약 0.1, 0.2초 간격으로 토글을 하려면 다음 [그림 8.24]와 같이 조정하고 디거깅을 하면 된다.

```
  /* Initialize all configured peripherals */
  MX_GPIO_Init();
  MX_USART2_UART_Init();
  MX_TIM3_Init();
  MX_TIM4_Init();
  /* USER CODE BEGIN 2 */
  HAL_TIM_Base_Start_IT(&htim3);
  HAL_TIM_Base_Start_IT(&htim4);
  /* USER CODE END 2 */

  /* Infinite loop */
  /* USER CODE BEGIN WHILE */
  while (1)
  {
      TIM3->ARR=1000;
      TIM4->ARR=2000;
    /* USER CODE END WHILE */

    /* USER CODE BEGIN 3 */
  }
  /* USER CODE END 3 */
}
```

[그림 8.24] 프로그램 수정

6. RTC(Real-time clock) 사용하기(Date,Time)

RTC는 독립된 BCD 카운터 이며 전용 레지터들은 연, 월, 일, 요일, 시, 분, 초를 포함하고 있다. 또한, 프로그램 가능한 알람 기능과 정지 및 대기 보드로부터 Wakeup 기능을 갖는 프로그램 가능 주기적인 인터럽트 기능을 제공한다. [그림 8.25]는 연월을 이용하기 위한 기본 설정이다.

6.1. 클럭 설정

RTC를 사용하기 위한 클럭 설정은 [그림 8.25], [그림 8.26]과 같고 외부에 의해서 32.768 주파수를 설정할 수 있다.

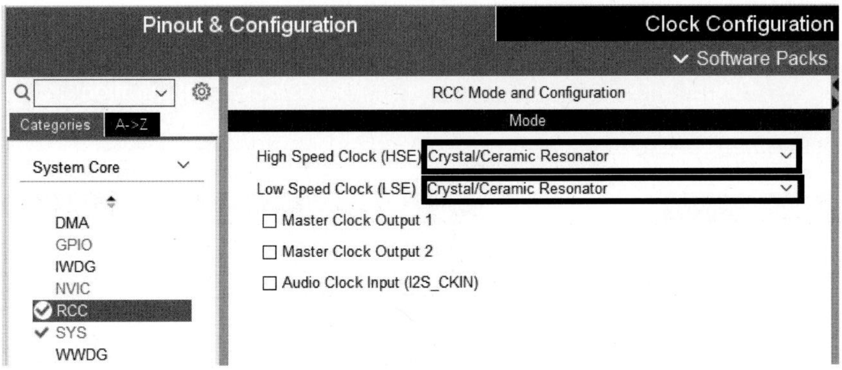

[그림 8.25] RTC 클럭 설정

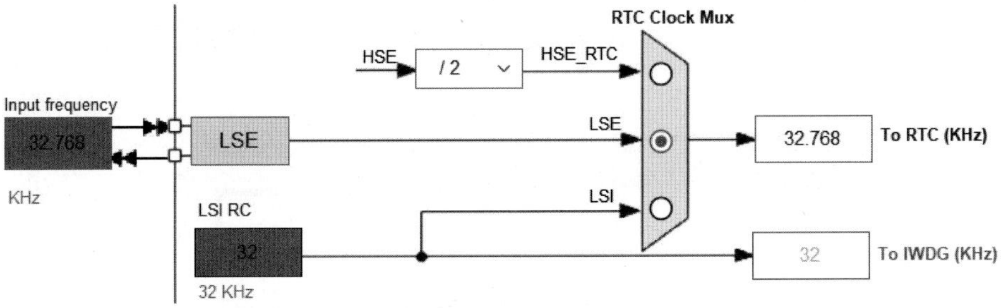

[그림 8.26] RTC 클럭 설정

6.2. RTC 설정하기

[그림 8.27]은 1초(32768/((127+1)*(255+1)) =1Hz)마다 MCU가 레지스터값을 갱신하기 위한 설정이다.

설정 시간: 21년 8월 31일 월요일 11시 59분 30초

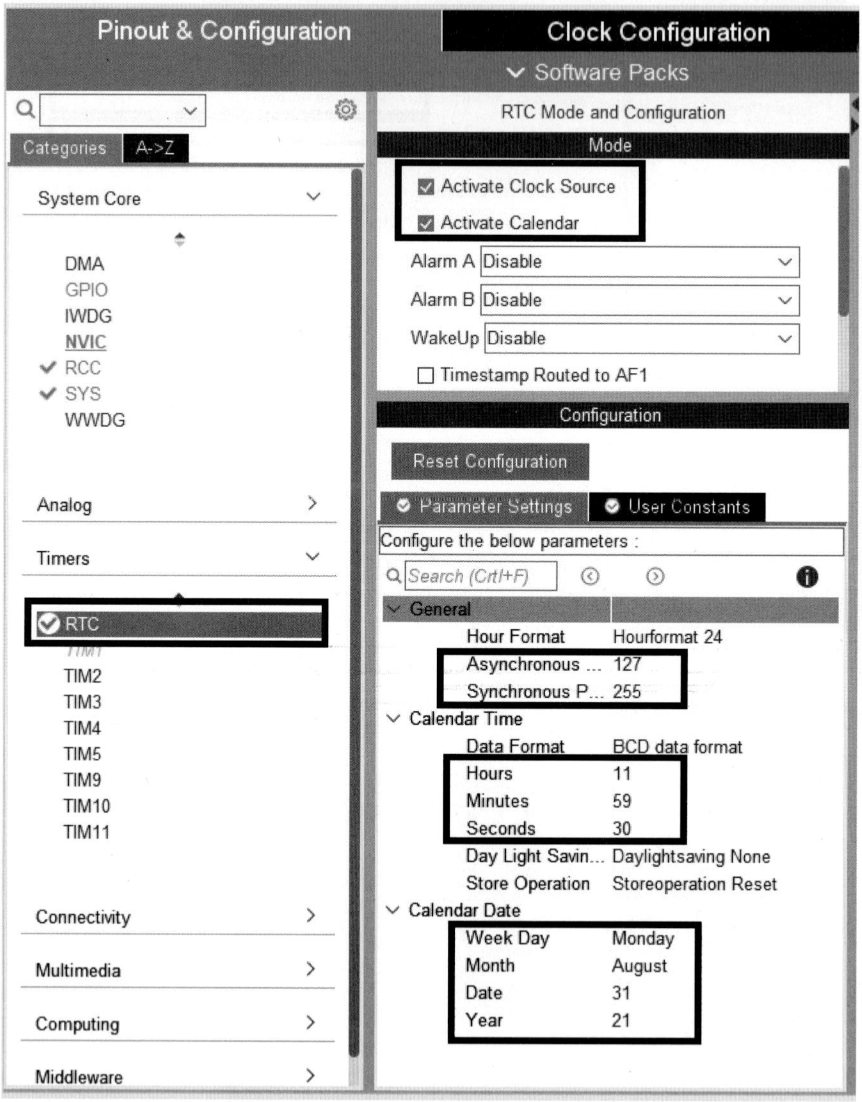

[그림 8.27] 시간 설정

6.2. Generate Code 생성

[그림 8.28]은 코드를 생성하기 위한 화면이다. 그림과 같이 클릭해서 프로그램을 생성한다.

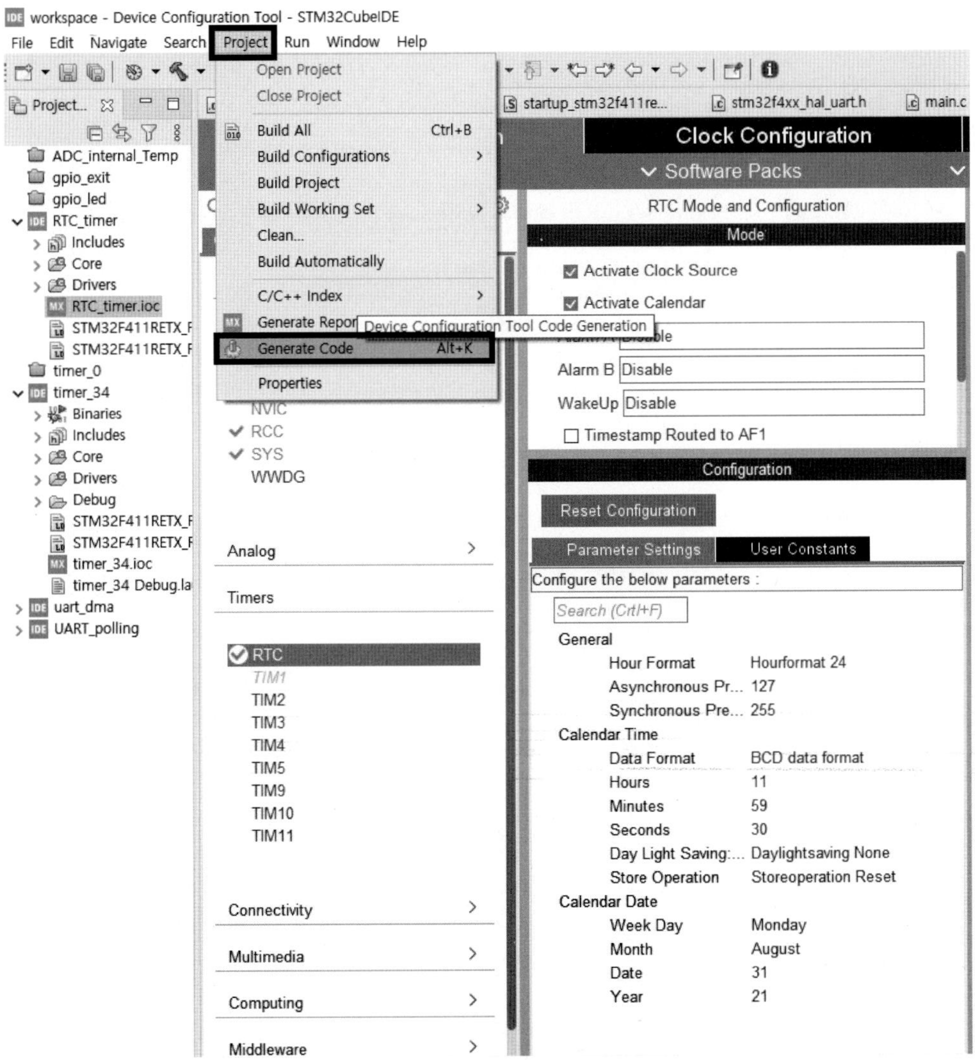

[그림 8.28] 프로그램 코드를 생성

6.3. Source Code 작성하기

RTC의 Time과 Date의 구조체 형식만 알면 쉽게 이용할 수 가 있다. [그림 8.29]~[그림 8.30]은 RTC 구조체 형식이며 [그림 8.31]은 메인 프로그램이다.

```c
typedef struct
{
  uint8_t Hours;            /*!< Specifies the RTC Time Hour.
                                 This parameter must be a number between Min_Data = 0 and Max_Data = 12 if the RTC_HourFormat_12 is selected
                                 This parameter must be a number between Min_Data = 0 and Max_Data = 23 if the RTC_HourFormat_24 is selected

  uint8_t Minutes;          /*!< Specifies the RTC Time Minutes.
                                 This parameter must be a number between Min_Data = 0 and Max_Data = 59 */

  uint8_t Seconds;          /*!< Specifies the RTC Time Seconds.
                                 This parameter must be a number between Min_Data = 0 and Max_Data = 59 */

  uint8_t TimeFormat;       /*!< Specifies the RTC AM/PM Time.
                                 This parameter can be a value of @ref RTC_AM_PM_Definitions */

  uint32_t SubSeconds;     /*!< Specifies the RTC_SSR RTC Sub Second register content.
                                 This parameter corresponds to a time unit range between [0-1] Second
                                 with [1 Sec / SecondFraction +1] granularity */

  uint32_t SecondFraction;  /*!< Specifies the range or granularity of Sub Second register content
                                 corresponding to Synchronous pre-scaler factor value (PREDIV_S)
                                 This parameter corresponds to a time unit range between [0-1] Second
                                 with [1 Sec / SecondFraction +1] granularity.
                                 This field will be used only by HAL_RTC_GetTime function */

  uint32_t DayLightSaving;  /*!< This interface is deprecated. To manage Daylight Saving Time,
                                 please use HAL_RTC_DST_xxx functions */

  uint32_t StoreOperation;  /*!< This interface is deprecated. To manage Daylight Saving Time,
                                 please use HAL_RTC_DST_xxx functions */
}RTC_TimeTypeDef;
```

[그림 8.29] RTC_Time의 구조체

```c
typedef struct
{
  uint8_t WeekDay;  /*!< Specifies the RTC Date WeekDay.
                         This parameter can be a value of @ref RTC_WeekDay_Definitions */

  uint8_t Month;    /*!< Specifies the RTC Date Month (in BCD format).
                         This parameter can be a value of @ref RTC_Month_Date_Definitions */

  uint8_t Date;     /*!< Specifies the RTC Date.
                         This parameter must be a number between Min_Data = 1 and Max_Data = 31 */

  uint8_t Year;     /*!< Specifies the RTC Date Year.
                         This parameter must be a number between Min_Data = 0 and Max_Data = 99 */

}RTC_DateTypeDef;
```

[그림 8.30] RTC_Date 구조체

```c
RTC_TimeTypeDef sTime;
RTC_DateTypeDef sDate;
/* USER CODE END 2 */

/* Infinite loop */
/* USER CODE BEGIN WHILE */
while (1)
{
    HAL_RTC_GetTime(&hrtc, &sTime, RTC_FORMAT_BIN);
    HAL_RTC_GetDate(&hrtc, &sDate, RTC_FORMAT_BIN);

printf("date=%d %d %d %d, time=%d %d %d \r\n",sDate.Year,sDate.Month,sDate.WeekDay,sDate.Date,sTime.Hours,sTime.Minutes,sTime.Seconds);
  /* USER CODE END WHILE */

  /* USER CODE BEGIN 3 */
}
/* USER CODE END 3 */
```

[그림 8.31] 메인 코드 작성

6.4. 디버그 및 다운로드하기

메인 함수에 프로그램이 작성되었다면 다음으로 디버그는 [그림 8.32]의 를 클릭하거나 단축키 F11을 누르면 디버깅과 프로그램 다운로드가 완료된다.

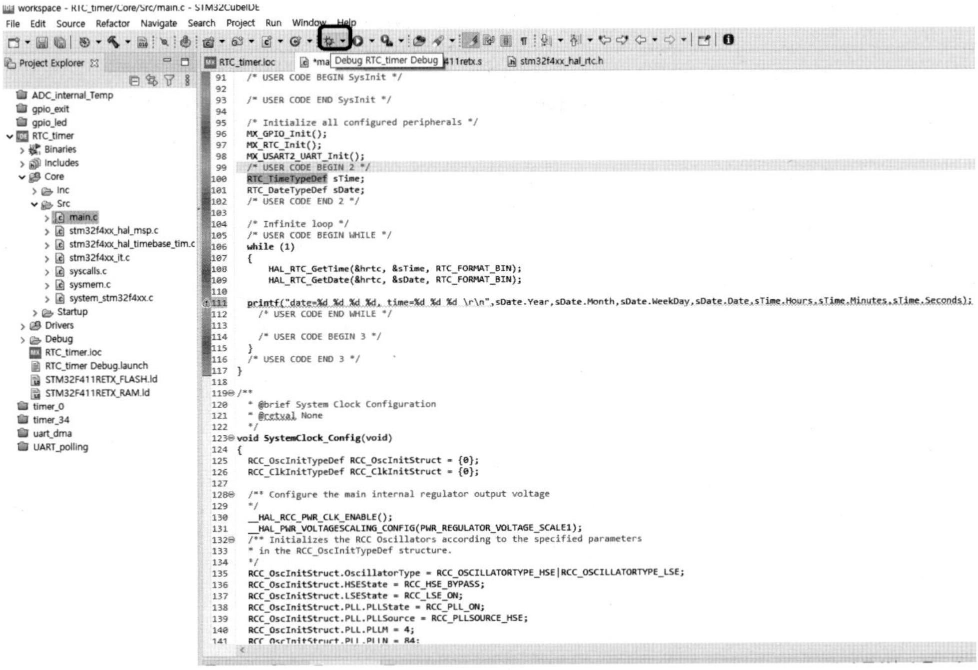

[그림 8.32] 프로그램 디버그

6.5. 보드 프로그램 실행하기

[그림 8.33] 상단 중앙에 아이콘 ■ 클릭하면 디버그 모드가 종료되어 보드의 프로그램이 실행된다.

workspace - RTC_timer/Core/Src/main.c - STM32CubeIDE
File Edit Source Refactor Navigate Search Project Run Window Help
Debug
Project Explorer
RTC_timer.ioc
main.c
startup_stm32f411retx.s
stm32f4xx_hal_rtc.h
RTC_timer Debug [STM32 Cortex-M C/C
RTC_timer.elf [cores: 0]
Thread #1 [main] 1 [core: 0] (Susp
main() at main.c:82 0x80005a
arm-none-eabi-gdb (8.3.1.20191211
ST-LINK (ST-LINK GDB server)
/* USER CODE END PV */
/* Private function prototypes ---
void SystemClock_Config(void);
static void MX_GPIO_Init(void);
static void MX_RTC_Init(void);
static void MX_USART2_UART_Init(void);
/* USER CODE BEGIN PFP */
/* USER CODE END PFP */
/* Private user code ---
/* USER CODE BEGIN 0 */
int __io_putchar(int ch)
{
HAL_UART_Transmit(&huart2, (uint8_t *)&ch, 1, 0xFFFF);
return ch;
}
/* USER CODE END 0 */
/**
* @brief The application entry point.
* @retval int
*/
int main(void)
{
/* USER CODE BEGIN 1 */
/* USER CODE END 1 */
/* MCU Configuration--
/* Reset of all peripherals, Initializes the Flash interface and the Syst
HAL_Init();
/* USER CODE BEGIN Init */
/* USER CODE END Init */
/* Configure the system clock */
SystemClock_Config();
Console
Problems
Executables
Debugger Console
Memory
RTC_timer Debug [STM32 Cortex-M C/C++ Application]
Verifying ...
Download verified successfully

[그림 8.33] 디버그 모드 프로그램 실행

6.6. 시리얼 통신 포트 결과확인

[그림 8.34]와 프로그램 결과를 시리얼로 모니터링한 결과이다.

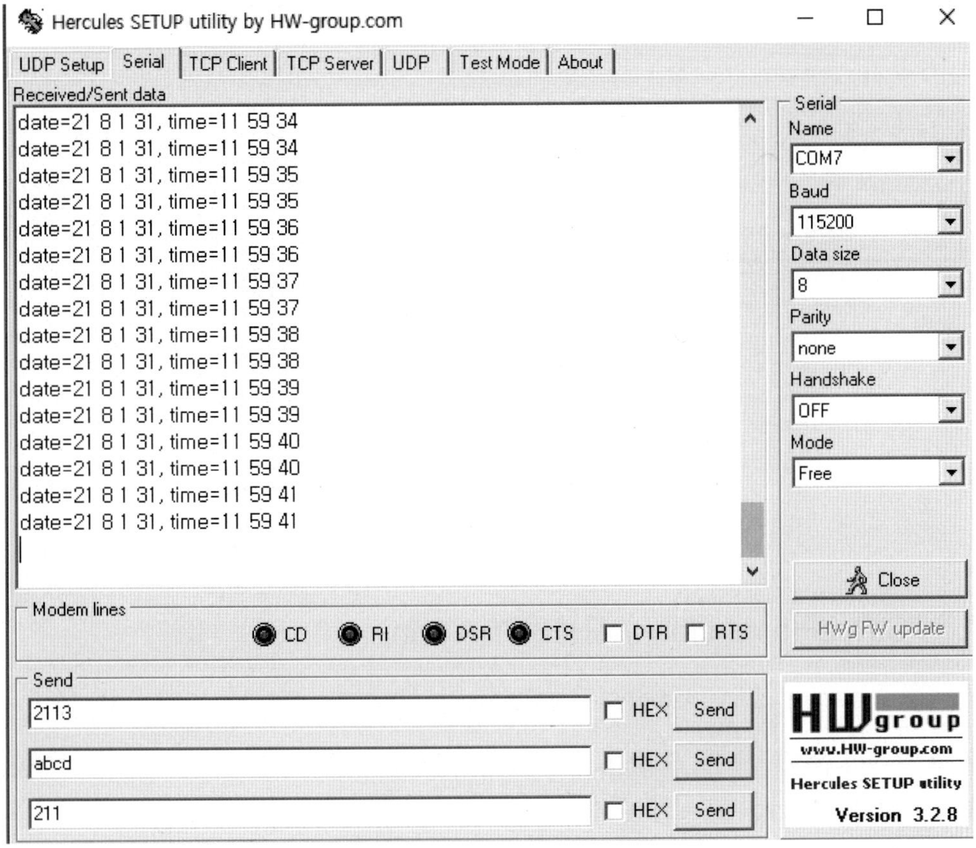

[그림 8.34] 프로그램 결과 확인

6.7. 프로그램 수정하기

[그림 8.35]와 같이 특정 날자 및 시간을 설정해서 출력해 본다.

```
RTC_TimeTypeDef sTime;
RTC_DateTypeDef sDate;

sTime.Hours=1;
sTime.Minutes=0;
sTime.Seconds=0;

sDate.Year=21;
sDate.Month=9;
sDate.WeekDay=2;
sDate.Date=1;
HAL_RTC_SetDate(&hrtc, &sDate, RTC_FORMAT_BIN);
HAL_RTC_SetTime(&hrtc, &sTime, RTC_FORMAT_BIN);
/* USER CODE END 2 */

/* Infinite loop */
/* USER CODE BEGIN WHILE */
while (1)
{
    HAL_RTC_GetTime(&hrtc, &sTime, RTC_FORMAT_BIN);
    HAL_RTC_GetDate(&hrtc, &sDate, RTC_FORMAT_BIN);

printf("date=%d %d %d %d, time=%d %d %d \r\n",sDate.Year,sDate.Month,sDate.WeekDay,sDate.Date,sTime.Hours,sTime.Minutes,sTime.Seconds);
HAL_Delay(500);
  /* USER CODE END WHILE */
```

[그림 8.35] 디버그 모드 프로그램 실행

7. RTC(Real–time clock) 사용하기(Alarm)

RTC는 독립된 BCD 카운터이며 전용 레지스터들은 연, 월, 일, 요일, 시, 분, 초를 포함하고 있다. 또한, 프로그램 가능한 알람 기능과 정지 및 대기 보드로부터 Wakeup 기능을 갖는 프로그램 가능 주기적인 인터럽트 기능을 제공한다.

7.1. 클럭 설정

RTC를 사용하기 위한 클럭 설정은 [그림 8.36], [그림8.37]과 같고 외부에 의해서 32.768 주파수를 설정할 수 있다.

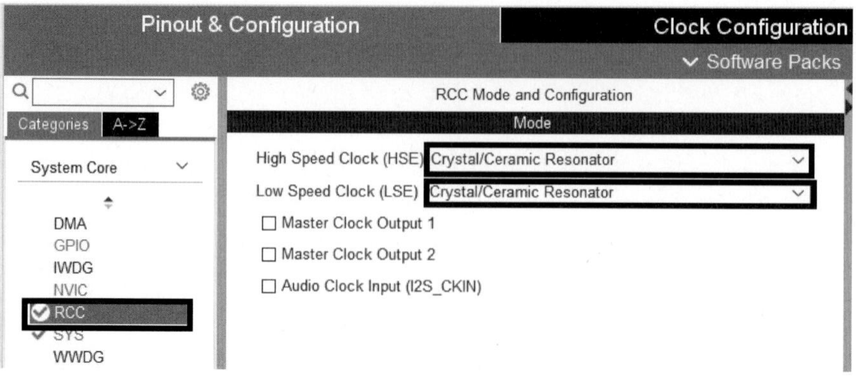

[그림 8.36] RTC 클럭 설정

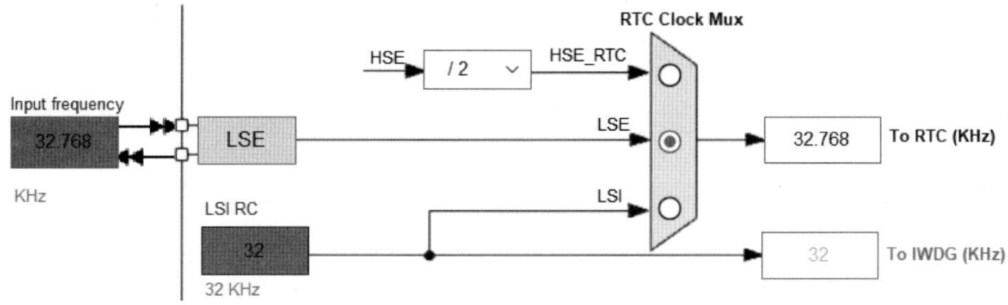

[그림 8.37 RTC] 클럭 설정

7.2. RTC 설정하기(Alarm)

[그림 8.38]은 1초(32768/((127+1)*(255+1)) =1Hz)마다 MCU가 레지스터값을 갱신하기 위한 설정이다. Alarm을 12:00:00초에 일어나게 하고 시간 설정을 21년 9월 1일 수요일 11시 59분 00초로 한다.

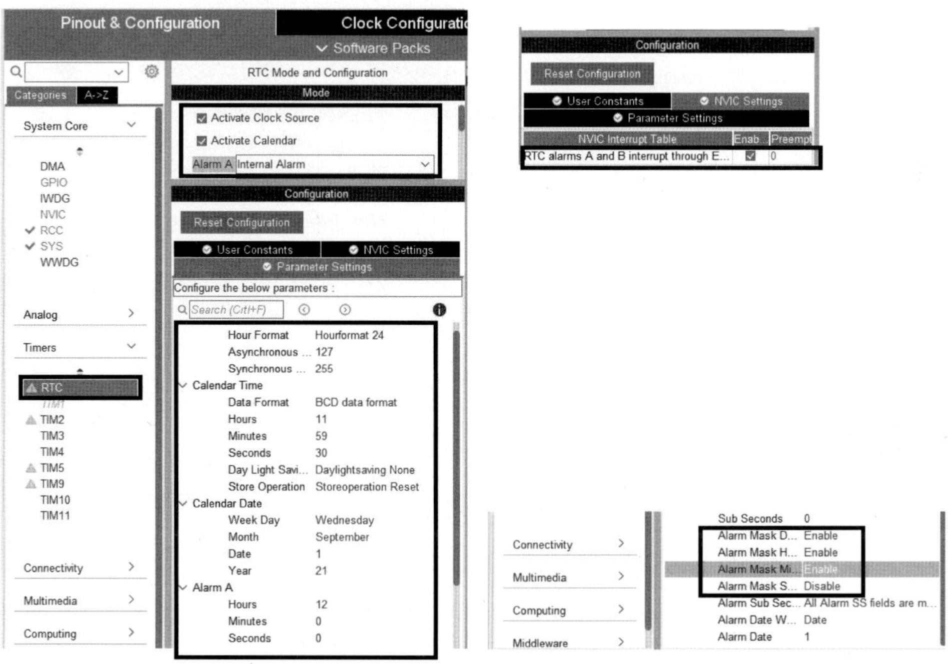

[그림 8.38] RTC 클럭 설정

7.3. Generate Code 생성

[그림 8.39]는 코드를 생성하기 위한 화면이다. 그림과 같이 클릭해서 프로그램을 생성한다.

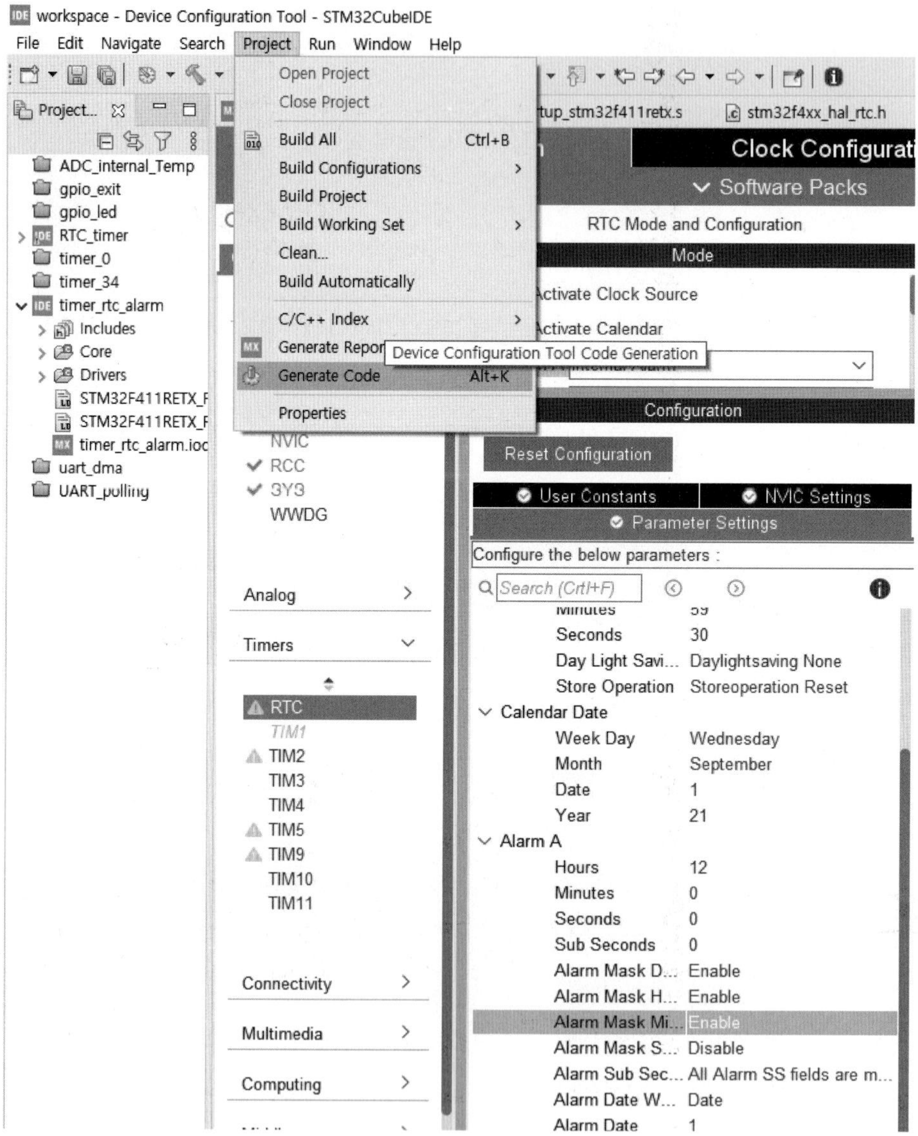

[그림 8.39] 프로그램 코드를 생성

7.4. Source Code 작성하기

RTC의 Alarm 구조체 형식은 [그림 8.40]에 나타냈다. 설정된 시간이 되면 알람 콜백 함수가 수행된다. [그림 8.39]를 보면 Seconds만 Mask Disable 되어 있어 1분마다 콜백 함수가 실행된다. [그림 8.41]은 메인 프로그램이다.

```
typedef struct
{
  RTC_TimeTypeDef AlarmTime;     /*!< Specifies the RTC Alarm Time members */

  uint32_t AlarmMask;            /*!< Specifies the RTC Alarm Masks.
                                      This parameter can be a value of @ref RTC_AlarmMask_Definitions */

  uint32_t AlarmSubSecondMask;   /*!< Specifies the RTC Alarm SubSeconds Masks.
                                      This parameter can be a value of @ref RTC_Alarm_Sub_Seconds_Masks_Definitions */

  uint32_t AlarmDateWeekDaySel;  /*!< Specifies the RTC Alarm is on Date or WeekDay.
                                     This parameter can be a value of @ref RTC_AlarmDateWeekDay_Definitions */

  uint8_t AlarmDateWeekDay;      /*!< Specifies the RTC Alarm Date/WeekDay.
                                      If the Alarm Date is selected, this parameter must be set to a value in the 1-31 range.
                                      If the Alarm WeekDay is selected, this parameter can be a value of @ref RTC_WeekDay_Definitions */

  uint32_t Alarm;                /*!< Specifies the alarm .
                                      This parameter can be a value of @ref RTC_Alarms_Definitions */
}RTC_AlarmTypeDef;
```

[그림 8.40] RTC_Alarm의 구조체

```
void HAL_RTC_AlarmAEventCallback(RTC_HandleTypeDef *hrtc)
{


}
```

```
uint8_t Alarm_count=0;
void HAL_RTC_AlarmAEventCallback(RTC_HandleTypeDef *hrtc)
{
    Alarm_count++;
   printf("Alarm count=%d \r\n",Alarm_count);

}

RTC_TimeTypeDef sTime;
RTC_DateTypeDef sDate;
RTC_AlarmTypeDef sAlarm;

/* USER CODE END 2 */

/* Infinite loop */
/* USER CODE BEGIN WHILE */
while (1)
{
  HAL_RTC_GetTime(&hrtc, &sTime, RTC_FORMAT_BIN);
  HAL_RTC_GetDate(&hrtc, &sDate, RTC_FORMAT_BIN);
  HAL_RTC_GetAlarm(&hrtc, &sAlarm, RTC_ALARM_A,RTC_FORMAT_BIN);

printf("date=%d %d %d %d, time=%d %d %d \r\n",sDate.Year,sDate.Month,sDate.WeekDay,sDate.Date,sTime.Hours,sTime.Minutes,sTime.Seconds);
HAL_Delay(5000);
/* USER CODE END WHILE */
/* USER CODE BEGIN 3 */
}
/* USER CODE END 3 */
```

[그림 8.41] 메인 코드 작성

7.5. 디버그 및 다운로드하기

메인 함수에 프로그램이 작성되었다면 다음으로 디버그는 [그림 8.42]의 를 클릭하거나 단축키 F11을 누르면 디버깅과 프로그램 다운로드가 완료된다.

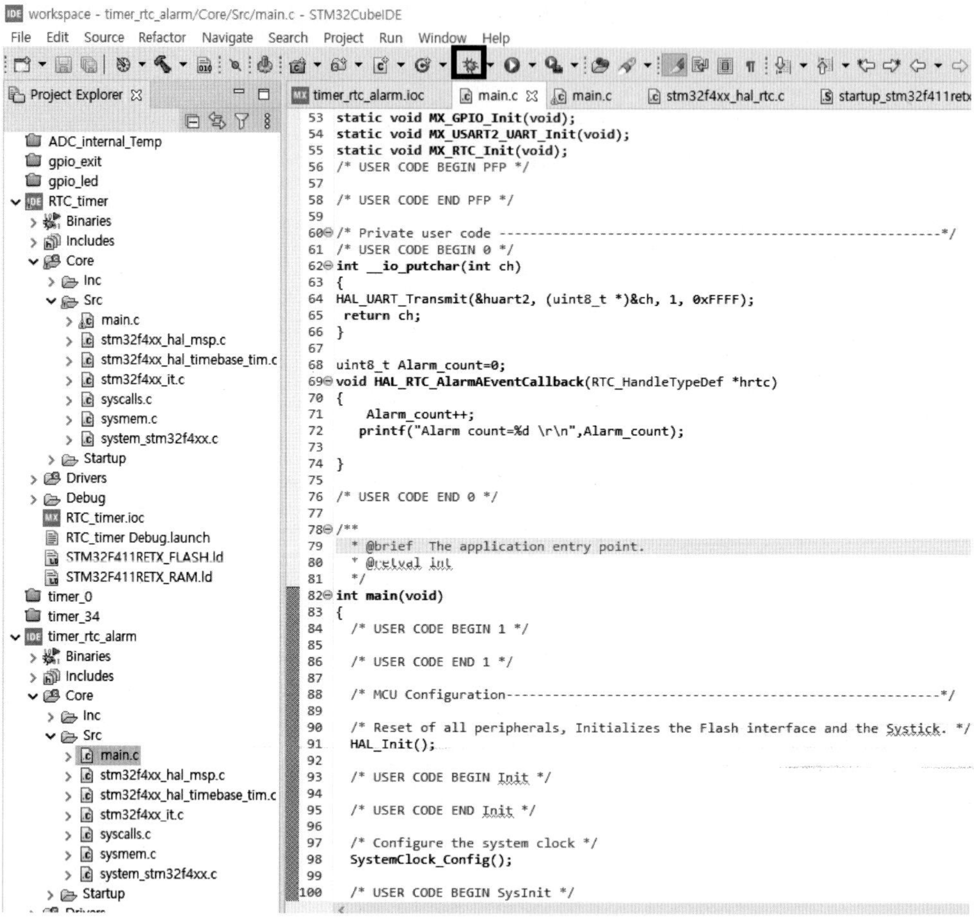

workspace - timer_rtc_alarm/Core/Src/main.c - STM32CubeIDE
File Edit Source Refactor Navigate Search Project Run Window Help
Project Explorer
timer_rtc_alarm.ioc
main.c
main.c
stm32f4xx_hal_rtc.c
startup_stm32f411retx
ADC_internal_Temp
gpio_exit
gpio_led
RTC_timer
Binaries
Includes
Core
Inc
Src
main.c
stm32f4xx_hal_msp.c
stm32f4xx_hal_timebase_tim.c
stm32f4xx_it.c
syscalls.c
sysmem.c
system_stm32f4xx.c
Startup
Drivers
Debug
RTC_timer.ioc
RTC_timer Debug.launch
STM32F411RETX_FLASH.ld
STM32F411RETX_RAM.ld
timer_0
timer_34
timer_rtc_alarm
Binaries
Includes
Core
Inc
Src
main.c
stm32f4xx_hal_msp.c
stm32f4xx_hal_timebase_tim.c
stm32f4xx_it.c
syscalls.c
sysmem.c
system_stm32f4xx.c
Startup
53 static void MX_GPIO_Init(void);
54 static void MX_USART2_UART_Init(void);
55 static void MX_RTC_Init(void);
56 /* USER CODE BEGIN PFP */
57
58 /* USER CODE END PFP */
59
60 /* Private user code ---*/
61 /* USER CODE BEGIN 0 */
62 int __io_putchar(int ch)
63 {
64 HAL_UART_Transmit(&huart2, (uint8_t *)&ch, 1, 0xFFFF);
65 return ch;
66 }
67
68 uint8_t Alarm_count=0;
69 void HAL_RTC_AlarmAEventCallback(RTC_HandleTypeDef *hrtc)
70 {
71 Alarm_count++;
72 printf("Alarm count=%d \r\n",Alarm_count);
73
74 }
75
76 /* USER CODE END 0 */
77
78 /**
79 * @brief The application entry point.
80 * @retval int
81 */
82 int main(void)
83 {
84 /* USER CODE BEGIN 1 */
85
86 /* USER CODE END 1 */
87
88 /* MCU Configuration--*/
89
90 /* Reset of all peripherals, Initializes the Flash interface and the Systick. */
91 HAL_Init();
92
93 /* USER CODE BEGIN Init */
94
95 /* USER CODE END Init */
96
97 /* Configure the system clock */
98 SystemClock_Config();
99
100 /* USER CODE BEGIN SysInit */

[그림 8.42] 프로그램 디버그

7.6. 보드 프로그램 실행하기

[그림 8.43] 상단 중앙에 아이콘 ■ 클릭하면 디버그 모드가 종료되어 보드의 프로그램이 실행된다.

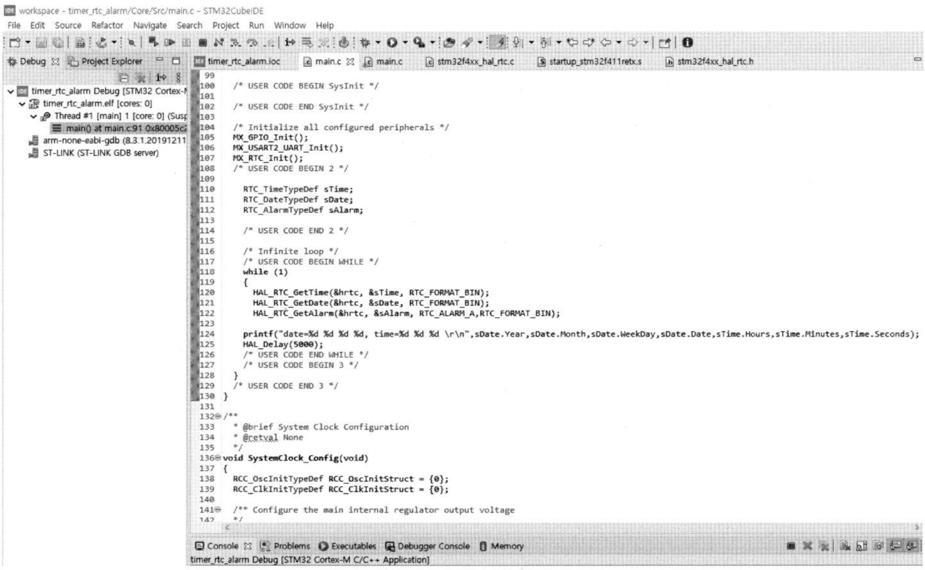
workspace - timer_rtc_alarm/Core/Src/main.c - STM32CubeIDE
File Edit Source Refactor Navigate Search Project Run Window Help
Debug
Project Explorer
timer_rtc_alarm.ioc
main.c
stm32f4xx_hal_rtc.c
startup_stm32f411retx.s
stm32f4xx_hal_rtc.h
timer_rtc_alarm Debug [STM32 Cortex-M C/C++ Application]
timer_rtc_alarm.elf [cores: 0]
Thread #1 [main] 1 [core: 0] (Suspended)
main() at main.c:91 0x80005c2
arm-none-eabi-gdb (8.3.1.20191211)
ST-LINK (ST-LINK GDB server)
/* USER CODE BEGIN SysInit */
/* USER CODE END SysInit */
/* Initialize all configured peripherals */
MX_GPIO_Init();
MX_USART2_UART_Init();
MX_RTC_Init();
/* USER CODE BEGIN 2 */
RTC_TimeTypeDef sTime;
RTC_DateTypeDef sDate;
RTC_AlarmTypeDef sAlarm;
/* USER CODE END 2 */
/* Infinite loop */
/* USER CODE BEGIN WHILE */
while (1)
{
HAL_RTC_GetTime(&hrtc, &sTime, RTC_FORMAT_BIN);
HAL_RTC_GetDate(&hrtc, &sDate, RTC_FORMAT_BIN);
HAL_RTC_GetAlarm(&hrtc, &sAlarm, RTC_ALARM_A,RTC_FORMAT_BIN);
printf("date=%d %d %d %d, time=%d %d %d \r\n",sDate.Year,sDate.Month,sDate.WeekDay,sDate.Date,sTime.Hours,sTime.Minutes,sTime.Seconds);
HAL_Delay(5000);
/* USER CODE END WHILE */
/* USER CODE BEGIN 3 */
}
/* USER CODE END 3 */
}
/**
* @brief System Clock Configuration
* @retval None
*/
void SystemClock_Config(void)
{
RCC_OscInitTypeDef RCC_OscInitStruct = {0};
RCC_ClkInitTypeDef RCC_ClkInitStruct = {0};
/** Configure the main internal regulator output voltage
Console
Problems
Executables
Debugger Console
Memory
timer_rtc_alarm Debug [STM32 Cortex-M C/C++ Application]

[그림 8.43] 디버그 모드 프로그램 실행

7.7. 시리얼 통신 포트 결과 확인

[그림 8.44]와 프로그램 결과를 시리얼로 모니터링한 결과이다.

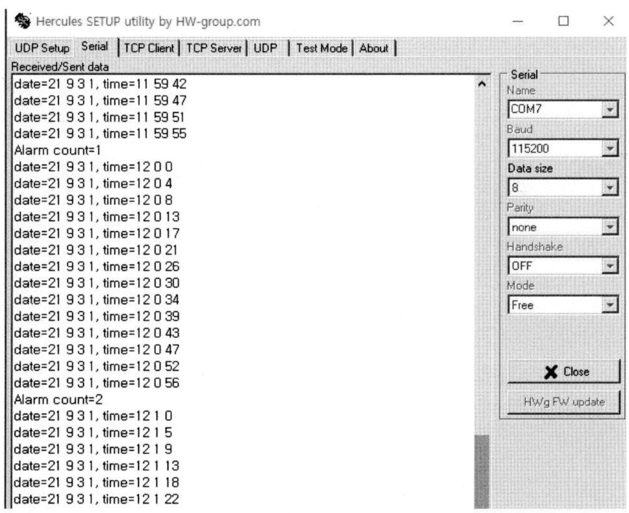
Hercules SETUP utility by HW-group.com
UDP Setup
Serial
TCP Client
TCP Server
UDP
Test Mode
About
Received/Sent data
date=21 9 3 1, time=11 59 42
date=21 9 3 1, time=11 59 47
date=21 9 3 1, time=11 59 51
date=21 9 3 1, time=11 59 55
Alarm count=1
date=21 9 3 1, time=12 0 0
date=21 9 3 1, time=12 0 4
date=21 9 3 1, time=12 0 8
date=21 9 3 1, time=12 0 13
date=21 9 3 1, time=12 0 17
date=21 9 3 1, time=12 0 21
date=21 9 3 1, time=12 0 26
date=21 9 3 1, time=12 0 30
date=21 9 3 1, time=12 0 34
date=21 9 3 1, time=12 0 39
date=21 9 3 1, time=12 0 43
date=21 9 3 1, time=12 0 47
date=21 9 3 1, time=12 0 52
date=21 9 3 1, time=12 0 56
Alarm count=2
date=21 9 3 1, time=12 1 0
date=21 9 3 1, time=12 1 5
date=21 9 3 1, time=12 1 9
date=21 9 3 1, time=12 1 13
date=21 9 3 1, time=12 1 18
date=21 9 3 1, time=12 1 22
Serial
Name
COM7
Baud
115200
Data size
8
Parity
none
Handshake
OFF
Mode
Free
Close
HWg FW update

[그림 8.44] 프로그램 결과 확인

CHAPTER 9

Timer 활용하기 2 (PWM, Compare out, Input Capture)

1. 개요
2. 실습 위한 하드웨어 설정
3. 클럭 설정
4. PWM 활용하기
5. OutPut Compare 활용하기
6. Input Capture 활용하기(내부 클럭)
7. Input Capture 활용하기(외부 클럭)

CHAPTER 09

Cortex-M4 기초 활용(ARM Core STM32)

Timer 활용하기 2 (PWM, Compare out, Input Capture)

1. 개요

타이머의 기능 중 중요하고 가장 많이 사용되는 것은 PWM, Input Capture, Output Compare 기능 등이다. 이 장에서는 이러한 기능 등을 다루도록 한다. 이러한 기능을 사용하기 위한 설정은 타이머를 기반으로 하기 때문에 타이머의 설정과 유사하다.

2. 실습위한 하드웨어 설정

[그림 9.1]과 같이 LED 제어용를 위해서 Timer3 channel 1(PA6),channel 2(PA7) Timer4 channel 1(PB6)의 출력에 맞도록 보드를 구성한다.

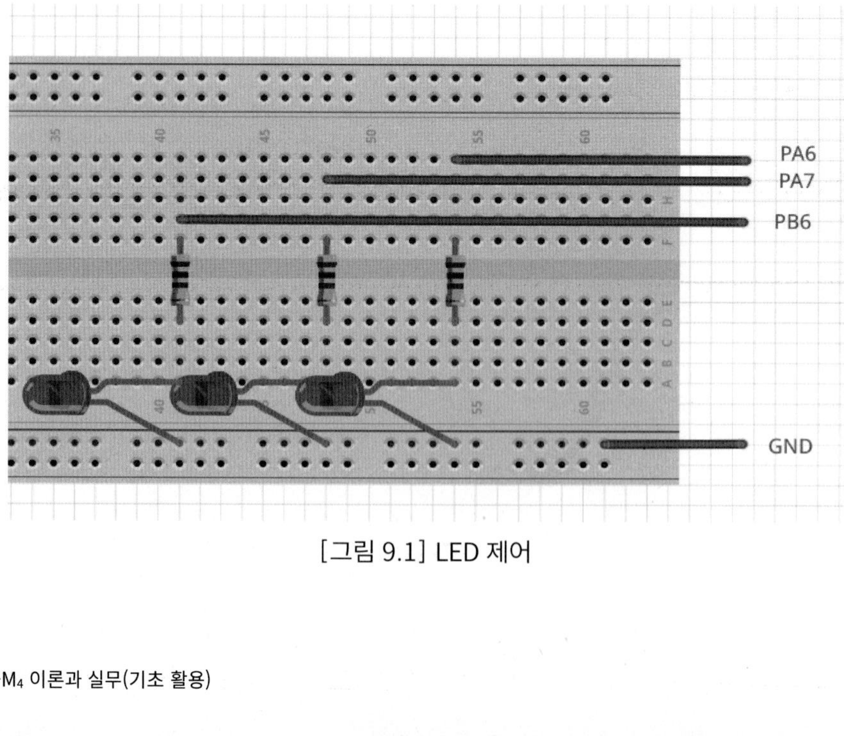

[그림 9.1] LED 제어

3. 클럭설정

RTC를 사용하기 위한 클럭 설정은 [그림 9.2]~[그림 9.3]과 같고 외부에 의해서 32.768 주파수를 설정할 수 있다.

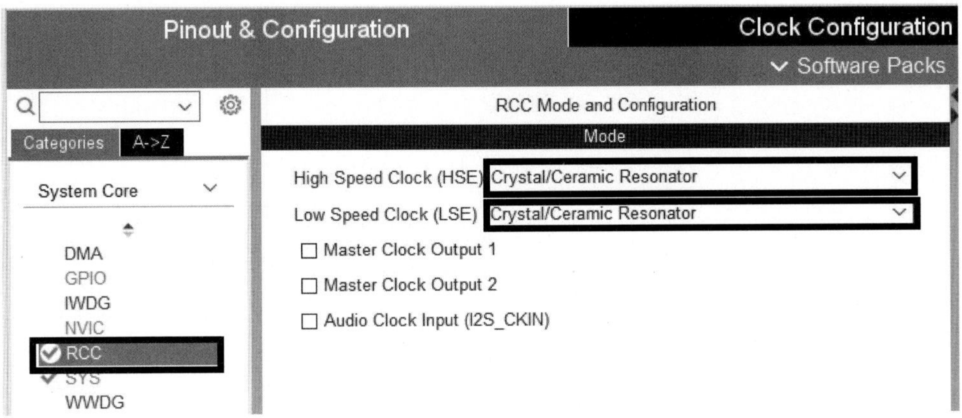

[그림 9.2] RTC 클럭 설정

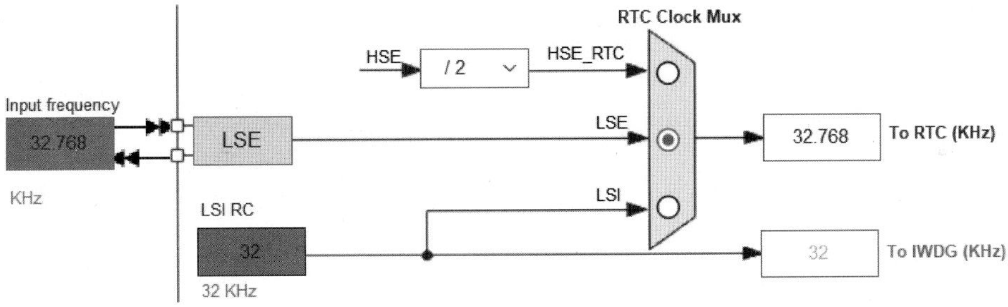

[그림 9.3] RTC 클럭 설정

4. PWM 활용하기

4.1. 개요

[그림 9.4]와 같이 PWM 파형은 주기가 일정 하고 펄스 폭을 한 주기 사이에서 조절하는 것으로 그림에서 CCR1의 파형은 OC1이고 CCR2의 펄스 폭은 OC2로 CCR 값을 조절하면 펄스 폭을 조절할 수 있다.

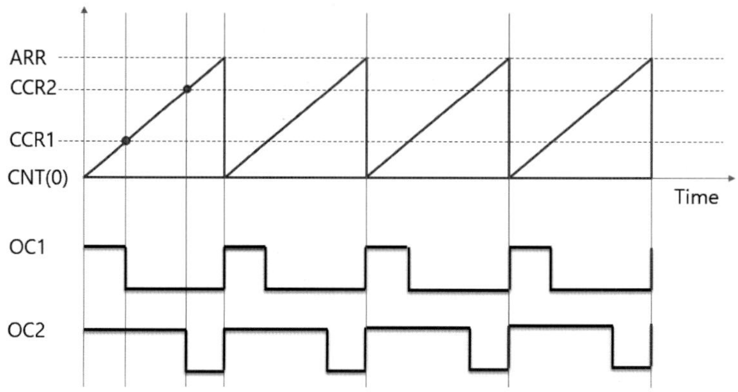

[그림 9.4] Basic PWM Mode

예를 들어 간단한 LED 디밍 애플리케이션을 위해 1000개의 서로 다른 PWM 레벨과 1000Hz의 주파수를 갖고 싶다고 가정해 본다. 먼저 카운터를 구동하는데 필요한 주파수를 계산한다.

Timer3의 클럭: 84Mhz

TIMER_Prescale =(84−1)

Count frequency = 84,000,000 / 84 =1,000,000

Count Time: T = 1/Count frequency= 1/1,000,000 // 카운터 시간 설정

ARR: PWM_Steps (1000−1) // 주기 결정함

Pulse:(0~999) / 폭 결정

4.2. PWM 설정

[그림 9.5]와 [그림 9.6]은 PWM 설정 내용이다.

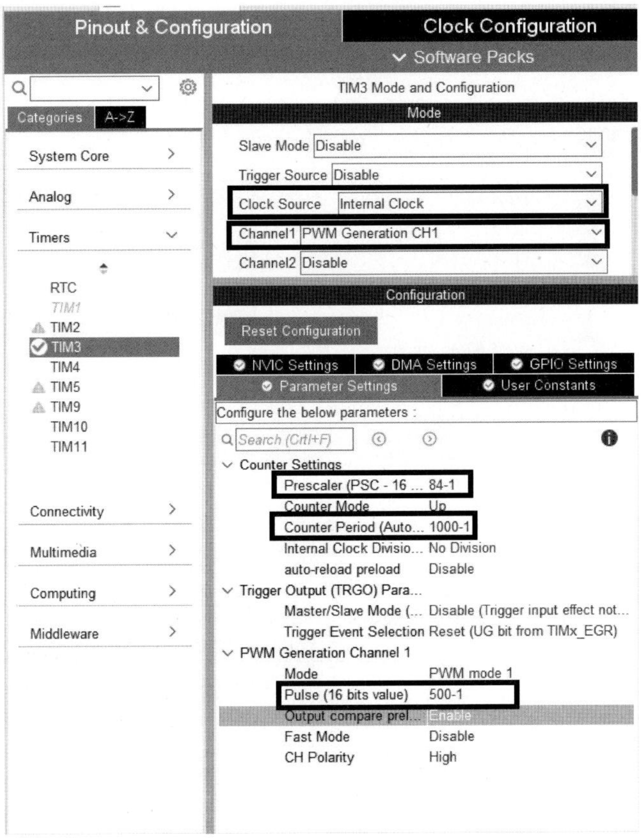

[그림 9.5] 타이머 3 PWM 설정

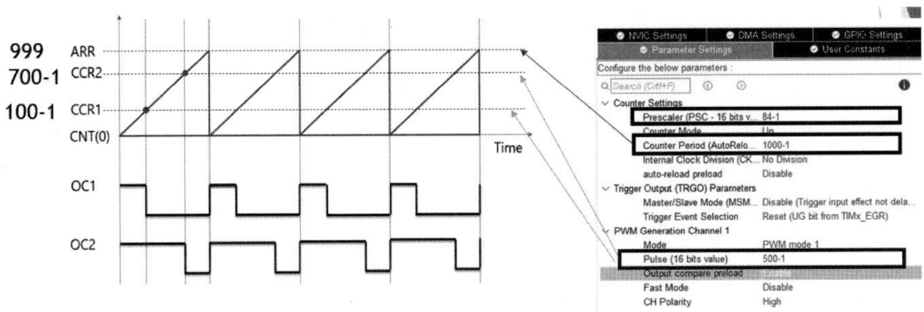

[그림 9.6] 타이머 3 PWM 설정

4.3. Generate Code 생성

[그림 9.7]은 코드를 생성하기 위한 화면이다. 그림과 같이 클릭해서 프로그램을 생성한다.

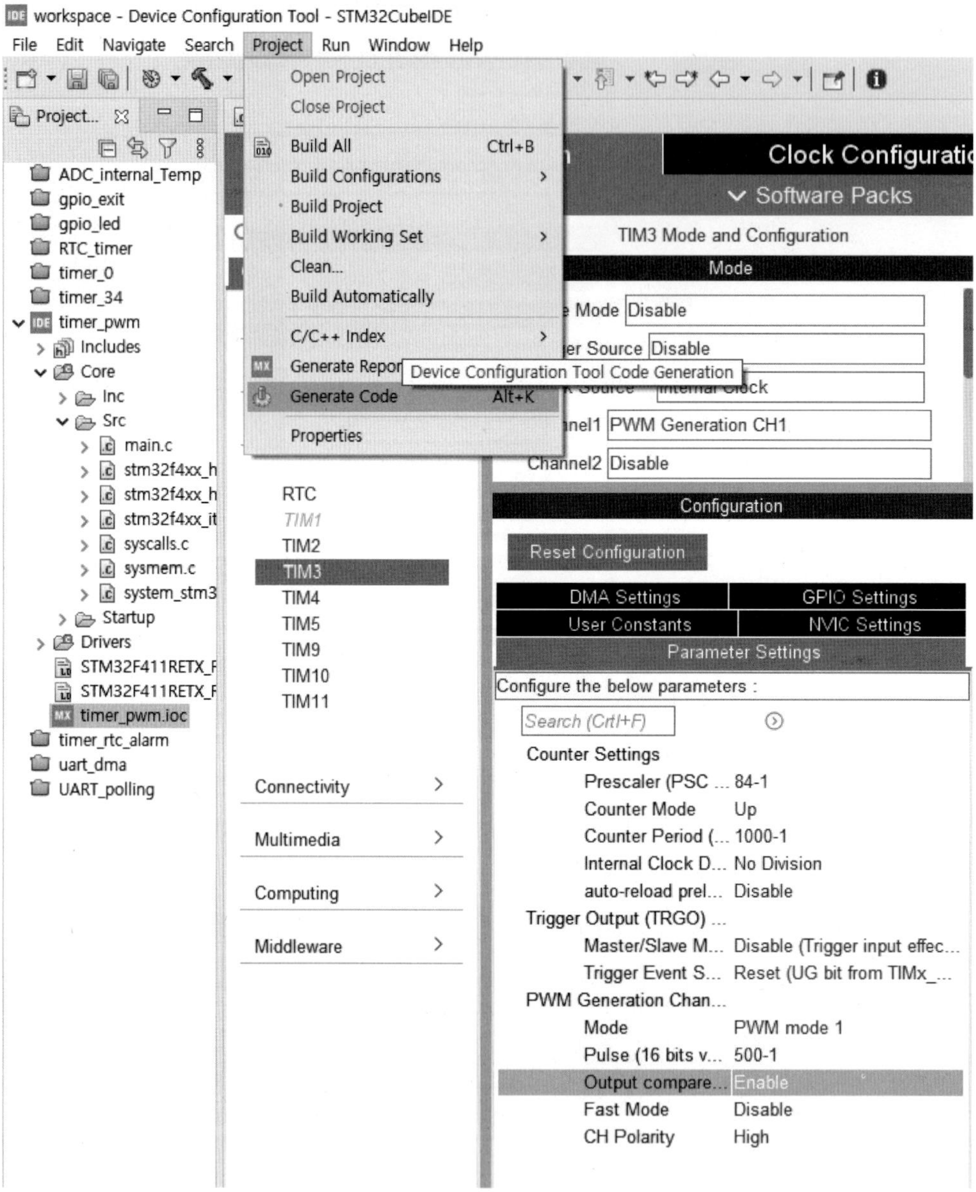

[그림 9.7] 프로그램 코드를 생성

4.4. Source Code 작성하기

LED를 1000Hz, 1000레벨로 디밍하기 위해서 타이머 3 PWM 설정은 다음과 같다. 프로그램 과정은 [그림 9.8]과 같다.

〉초기화 함수: `HAL_TIM_PWM_Start(&htim3,TIM_CHANNEL_1);`

〉듀티비(레벨)을 제어하는 방법: 0~999까지 변화, 화일 루프의 프로그램이 없으면 50% 듀티비를 갖는 파형이 출력된다.

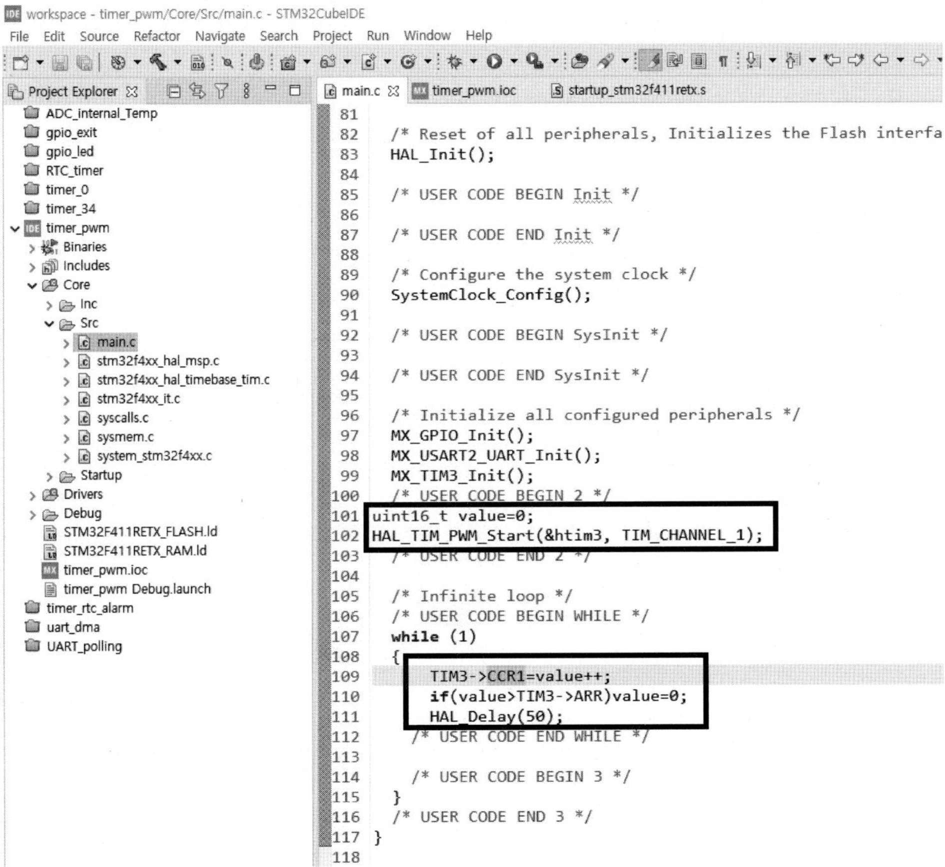

workspace - timer_pwm/Core/Src/main.c - STM32CubeIDE
File Edit Source Refactor Navigate Search Project Run Window Help
Project Explorer
main.c
timer_pwm.ioc
startup_stm32f411retx.s
ADC_internal_Temp
gpio_exit
gpio_led
RTC_timer
timer_0
timer_34
timer_pwm
Binaries
Includes
Core
Inc
Src
main.c
stm32f4xx_hal_msp.c
stm32f4xx_hal_timebase_tim.c
stm32f4xx_it.c
syscalls.c
sysmem.c
system_stm32f4xx.c
Startup
Drivers
Debug
STM32F411RETX_FLASH.ld
STM32F411RETX_RAM.ld
timer_pwm.ioc
timer_pwm Debug.launch
timer_rtc_alarm
uart_dma
UART_polling
/* Reset of all peripherals, Initializes the Flash interfa
HAL_Init();
/* USER CODE BEGIN Init */
/* USER CODE END Init */
/* Configure the system clock */
SystemClock_Config();
/* USER CODE BEGIN SysInit */
/* USER CODE END SysInit */
/* Initialize all configured peripherals */
MX_GPIO_Init();
MX_USART2_UART_Init();
MX_TIM3_Init();
/* USER CODE BEGIN 2 */
uint16_t value=0;
HAL_TIM_PWM_Start(&htim3, TIM_CHANNEL_1);
/* USER CODE END 2 */
/* Infinite loop */
/* USER CODE BEGIN WHILE */
while (1)
{
TIM3->CCR1=value++;
if(value>TIM3->ARR)value=0;
HAL_Delay(50);
/* USER CODE END WHILE */
/* USER CODE BEGIN 3 */
}
/* USER CODE END 3 */
}

[그림 9.8] 메인 코드 작성

4.5. 디버그 및 다운로드하기

메인 함수에 프로그램이 작성되었다면 다음으로 디버그는 [그림 9.9]의 를 클릭하거나 단축키 F11을 누르면 디버깅과 프로그램 다운로드가 완료된다. 그림을 보면 타이머의 스케일러 값과 주기(ARR) 값이 잘 설정되어 있는 것을 알 수 있다.

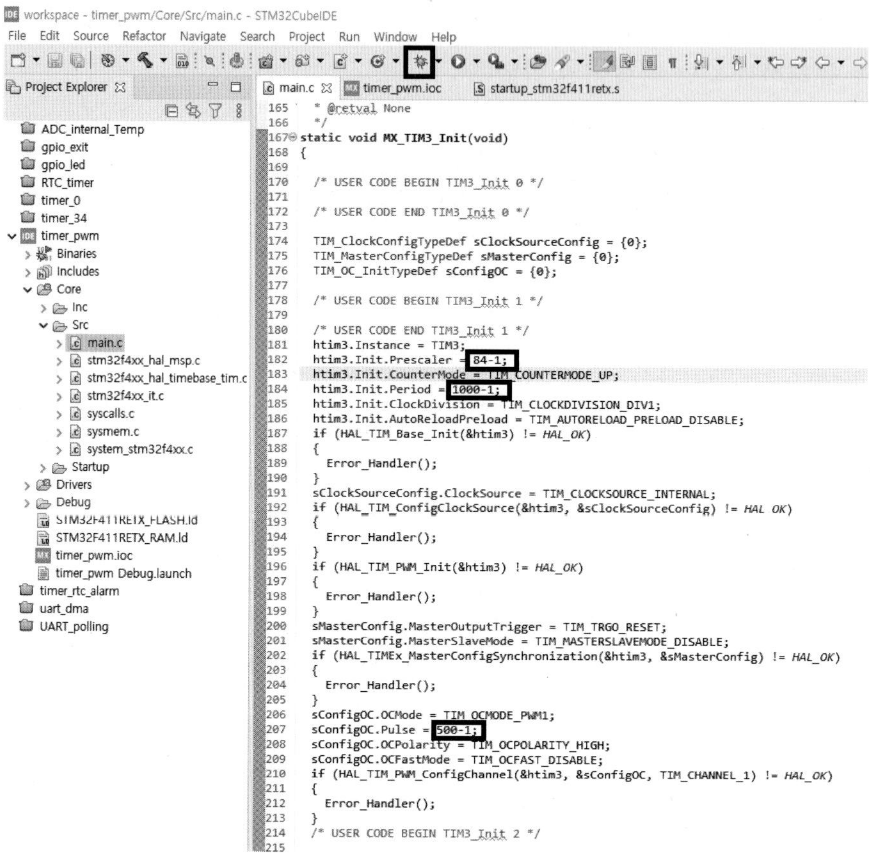

workspace - timer_pwm/Core/Src/main.c - STM32CubeIDE
File Edit Source Refactor Navigate Search Project Run Window Help
Project Explorer
main.c
timer_pwm.ioc
startup_stm32f411retx.s
ADC_internal_Temp
gpio_exit
gpio_led
RTC_timer
timer_0
timer_34
timer_pwm
Binaries
Includes
Core
Inc
Src
main.c
stm32f4xx_hal_msp.c
stm32f4xx_hal_timebase_tim.c
stm32f4xx_it.c
syscalls.c
sysmem.c
system_stm32f4xx.c
Startup
Drivers
Debug
STM32F411RETX_FLASH.ld
STM32F411RETX_RAM.ld
timer_pwm.ioc
timer_pwm Debug.launch
timer_rtc_alarm
uart_dma
UART_polling
* @retval None
*/
static void MX_TIM3_Init(void)
{
/* USER CODE BEGIN TIM3_Init 0 */
/* USER CODE END TIM3_Init 0 */
TIM_ClockConfigTypeDef sClockSourceConfig = {0};
TIM_MasterConfigTypeDef sMasterConfig = {0};
TIM_OC_InitTypeDef sConfigOC = {0};
/* USER CODE BEGIN TIM3_Init 1 */
/* USER CODE END TIM3_Init 1 */
htim3.Instance = TIM3;
htim3.Init.Prescaler = 84-1;
htim3.Init.CounterMode = TIM_COUNTERMODE_UP;
htim3.Init.Period = 1000-1;
htim3.Init.ClockDivision = TIM_CLOCKDIVISION_DIV1;
htim3.Init.AutoReloadPreload = TIM_AUTORELOAD_PRELOAD_DISABLE;
if (HAL_TIM_Base_Init(&htim3) != HAL_OK)
{
Error_Handler();
}
sClockSourceConfig.ClockSource = TIM_CLOCKSOURCE_INTERNAL;
if (HAL_TIM_ConfigClockSource(&htim3, &sClockSourceConfig) != HAL_OK)
{
Error_Handler();
}
if (HAL_TIM_PWM_Init(&htim3) != HAL_OK)
{
Error_Handler();
}
sMasterConfig.MasterOutputTrigger = TIM_TRGO_RESET;
sMasterConfig.MasterSlaveMode = TIM_MASTERSLAVEMODE_DISABLE;
if (HAL_TIMEx_MasterConfigSynchronization(&htim3, &sMasterConfig) != HAL_OK)
{
Error_Handler();
}
sConfigOC.OCMode = TIM_OCMODE_PWM1;
sConfigOC.Pulse = 500-1;
sConfigOC.OCPolarity = TIM_OCPOLARITY_HIGH;
sConfigOC.OCFastMode = TIM_OCFAST_DISABLE;
if (HAL_TIM_PWM_ConfigChannel(&htim3, &sConfigOC, TIM_CHANNEL_1) != HAL_OK)
{
Error_Handler();
}
/* USER CODE BEGIN TIM3_Init 2 */

[그림 9.9] 프로그램 디버그

4.6. 보드 프로그램 실행하기

[그림 9.10] 상단 중앙에 아이콘 ■ 클릭하면 디버그 모드가 종료되어 보드의 프로그램이 실행된다.

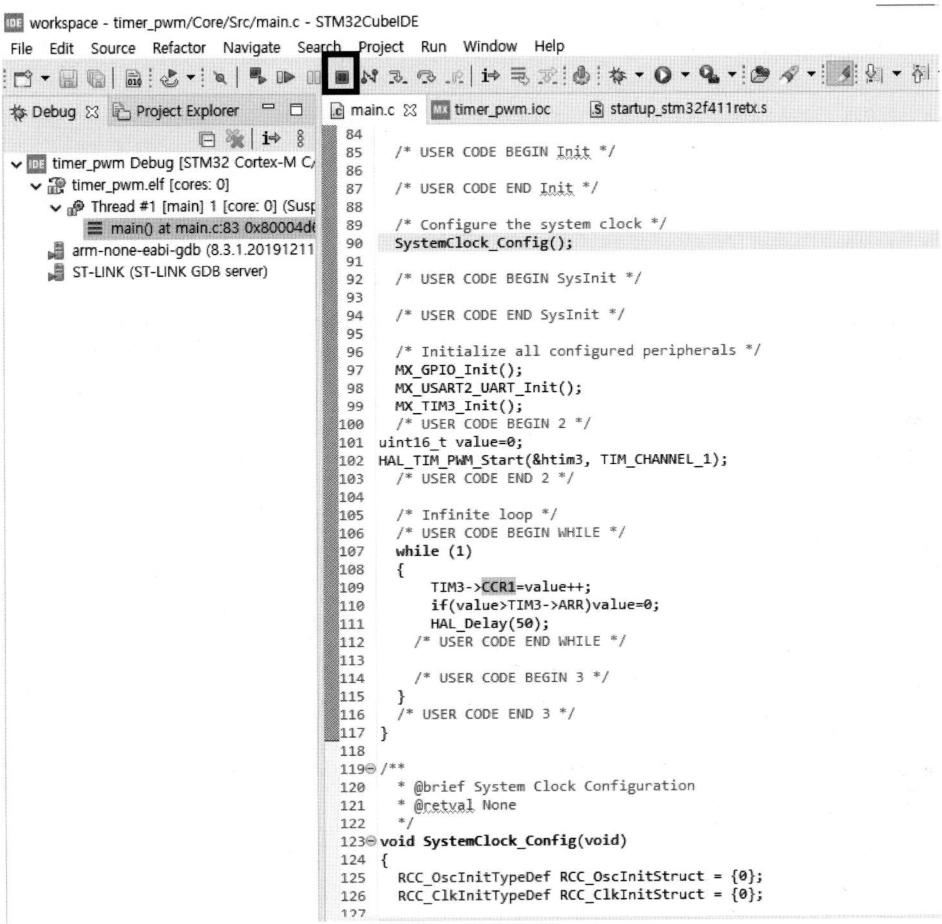
workspace - timer_pwm/Core/Src/main.c - STM32CubeIDE
File Edit Source Refactor Navigate Search Project Run Window Help
Debug
Project Explorer
main.c
timer_pwm.ioc
startup_stm32f411retx.s
timer_pwm Debug [STM32 Cortex-M C/
timer_pwm.elf [cores: 0]
Thread #1 [main] 1 [core: 0] (Susp
main() at main.c:83 0x80004d
arm-none-eabi-gdb (8.3.1.20191211
ST-LINK (ST-LINK GDB server)
/* USER CODE BEGIN Init */
/* USER CODE END Init */
/* Configure the system clock */
SystemClock_Config();
/* USER CODE BEGIN SysInit */
/* USER CODE END SysInit */
/* Initialize all configured peripherals */
MX_GPIO_Init();
MX_USART2_UART_Init();
MX_TIM3_Init();
/* USER CODE BEGIN 2 */
uint16_t value=0;
HAL_TIM_PWM_Start(&htim3, TIM_CHANNEL_1);
/* USER CODE END 2 */
/* Infinite loop */
/* USER CODE BEGIN WHILE */
while (1)
{
TIM3->CCR1=value++;
if(value>TIM3->ARR)value=0;
HAL_Delay(50);
/* USER CODE END WHILE */
/* USER CODE BEGIN 3 */
}
/* USER CODE END 3 */
}
/**
* @brief System Clock Configuration
* @retval None
*/
void SystemClock_Config(void)
{
RCC_OscInitTypeDef RCC_OscInitStruct = {0};
RCC_ClkInitTypeDef RCC_ClkInitStruct = {0};

[그림 9.10] 디버그 모드 프로그램 실행

5. Output Compare 활용하기

5.1. 개요

Timer의 Output Compare 기능은 특정 채널에 클럭을 발생하는 용도로 많이 사용된다. Prescaler에 의해서 주파수가 결정되고 주기는 Counter Period(ARR)에 의해서 결정된다. [그림 9.11]에서는 카운터 주파수가 10kHz이며 한 칸 증가하는 카운터 값 T= 1/f[sec], 1초가 되는 카운터 값은 10000이다. 출력값은 카운터 설정값(9999)이 되었을 때 토글이 된다.

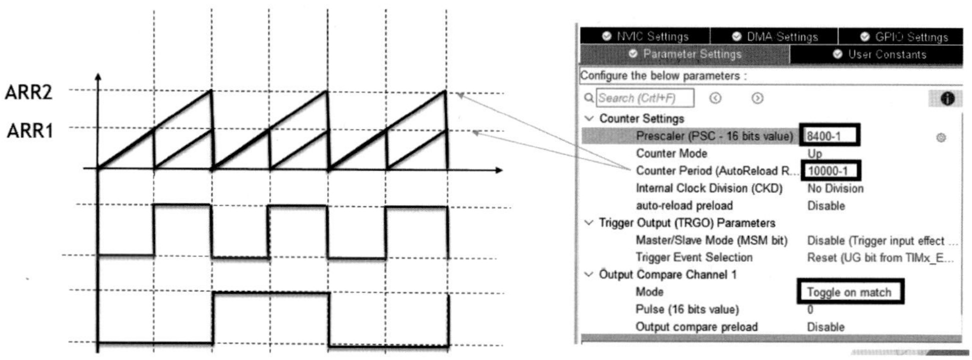

[그림 9.11] 디버그 모드 프로그램 실행

예를 들어 간단한 LED 애플리케이션을 위해 1초마다 토글이 되도록 구동하려면 아래와 같이 정리할 수 있다.

Timer3의 클럭: 84Mhz

TIMER_Prescale =(8400−1)

Count frequency = 84,000,000 /8400 =10,000

Count Time: T= 1/Count frequency= 1/10,000 // **카운터 시간 설정**

ARR: (10000−1)// **주기 결정함** 0.5초

ARR: (5000−1)// **주기 결정함** 0.25초

5.2. Output Compare 설정

[그림 9.12]는 Output Compare 설정 내용이다.

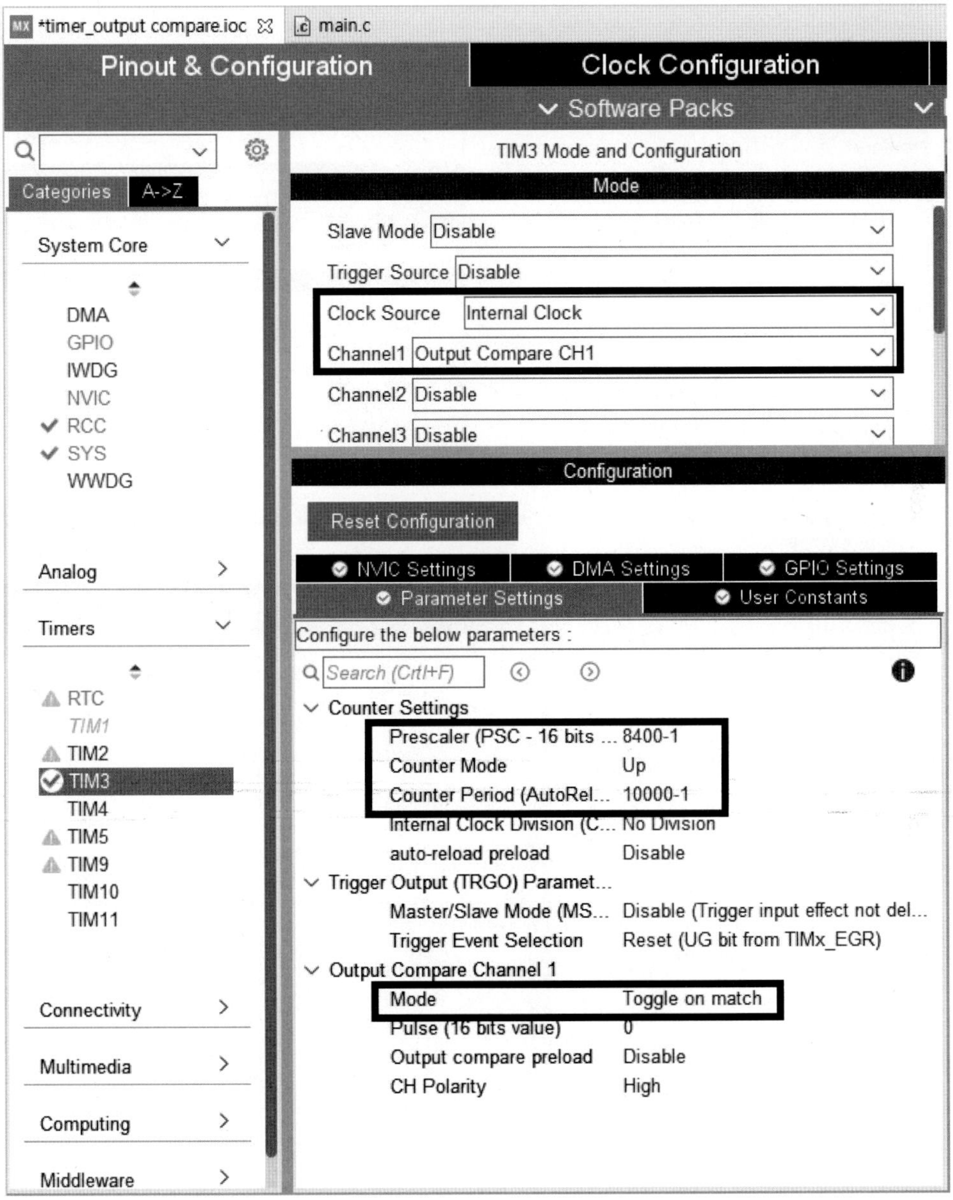

[그림 9.12] 타이머 3 Output Compare 설정

5.3. Generate Code 생성

[그림 9.13]은 코드를 생성하기 위한 화면이다. 그림과 같이 클릭해서 프로그램을 생성한다.

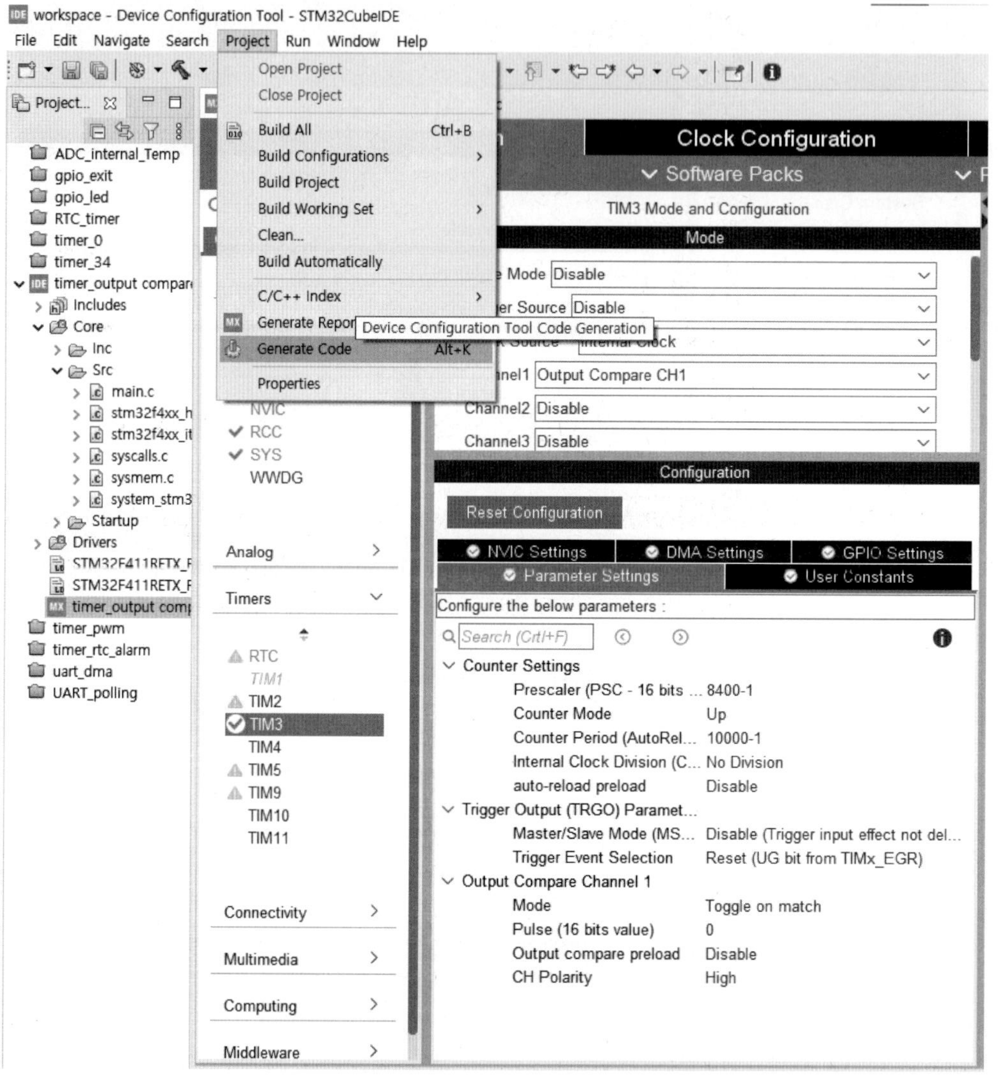

[그림 9.13] 프로그램 코드를 생성

5.4. Source Code 작성하기

LED를 1초 간격으로 토글하기 위해서 타이머 3 Output Compare 설정은 다음과 같다. 프로그램 과정은 [그림 9.14]와 같다.

> **초기화 함수**: `HAL_TIM_OC_Start(&htim3,TIM_CHANNEL_1);`

> **주기를 제어하는 방법**: ARR(0~999), 0.5Hz ~5kHz **파형이 출력된다**.

```
  * @retval int
  */
int main(void)
{
  /* USER CODE BEGIN 1 */

  /* USER CODE END 1 */

  /* MCU Configuration--------------------------------------------------------*/

  /* Reset of all peripherals, Initializes the Flash interface and the Systick. */
  HAL_Init();

  /* USER CODE BEGIN Init */

  /* USER CODE END Init */

  /* Configure the system clock */
  SystemClock_Config();

  /* USER CODE BEGIN SysInit */

  /* USER CODE END SysInit */

  /* Initialize all configured peripherals */
  MX_GPIO_Init();
  MX_USART2_UART_Init();
  MX_RTC_Init();
  MX_TIM3_Init();
  /* USER CODE BEGIN 2 */
   uint16_t value=0;
   HAL_TIM_OC_Start(&htim3, TIM_CHANNEL_1);

  /* USER CODE END 2 */

  /* Infinite loop */
  /* USER CODE BEGIN WHILE */
  while (1)
  {
      TIM3->ARR=value++;
      if(value>1000)value=0;
      HAL_Delay(100);

    /* USER CODE END WHILE */

    /* USER CODE BEGIN 3 */
  }
  /* USER CODE END 3 */
}

/**
  * @brief System Clock Configuration
```

[그림 9.14] 메인 코드 작성

5.5. 디버그 및 다운로드하기

메인 함수에 프로그램이 작성되었다면 다음으로 디버그는 [그림 9.15]의 를 클릭하거나 단축키 F11을 누르면 디버깅과 프로그램 다운로드가 완료된다. 그림을 보면 타이머의 스케일러 값과 주기(ARR) 값이 잘 설정되어 있는 것을 알 수 있다.

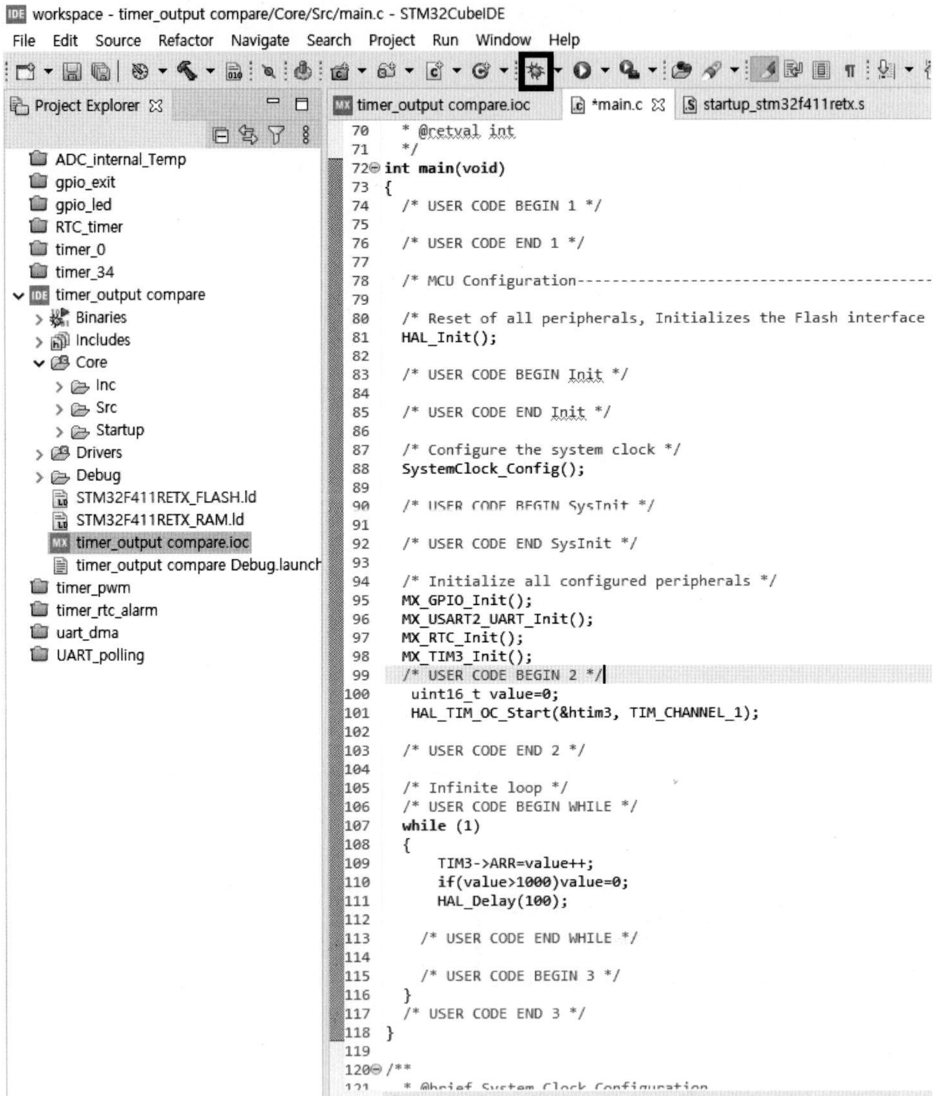
workspace - timer_output compare/Core/Src/main.c - STM32CubeIDE
File Edit Source Refactor Navigate Search Project Run Window Help
Project Explorer
timer_output compare.ioc
*main.c
startup_stm32f411retx.s
ADC_internal_Temp
gpio_exit
gpio_led
RTC_timer
timer_0
timer_34
timer_output compare
Binaries
Includes
Core
Inc
Src
Startup
Drivers
Debug
STM32F411RETX_FLASH.ld
STM32F411RETX_RAM.ld
timer_output compare.ioc
timer_output compare Debug.launch
timer_pwm
timer_rtc_alarm
uart_dma
UART_polling
* @retval int
*/
int main(void)
{
/* USER CODE BEGIN 1 */
/* USER CODE END 1 */
/* MCU Configuration--
/* Reset of all peripherals, Initializes the Flash interface
HAL_Init();
/* USER CODE BEGIN Init */
/* USER CODE END Init */
/* Configure the system clock */
SystemClock_Config();
/* USER CODE BEGIN SysInit */
/* USER CODE END SysInit */
/* Initialize all configured peripherals */
MX_GPIO_Init();
MX_USART2_UART_Init();
MX_RTC_Init();
MX_TIM3_Init();
/* USER CODE BEGIN 2 */
uint16_t value=0;
HAL_TIM_OC_Start(&htim3, TIM_CHANNEL_1);
/* USER CODE END 2 */
/* Infinite loop */
/* USER CODE BEGIN WHILE */
while (1)
{
TIM3->ARR=value++;
if(value>1000)value=0;
HAL_Delay(100);
/* USER CODE END WHILE */
/* USER CODE BEGIN 3 */
}
/* USER CODE END 3 */
}
/**

[그림 9.15] 프로그램 디버그

5.6. 보드 프로그램 실행하기

[그림 9.16] 상단 중앙에 아이콘 ■ 클릭하면 디버그 모드가 종료되어 보드의 프로그램이 실행된다.

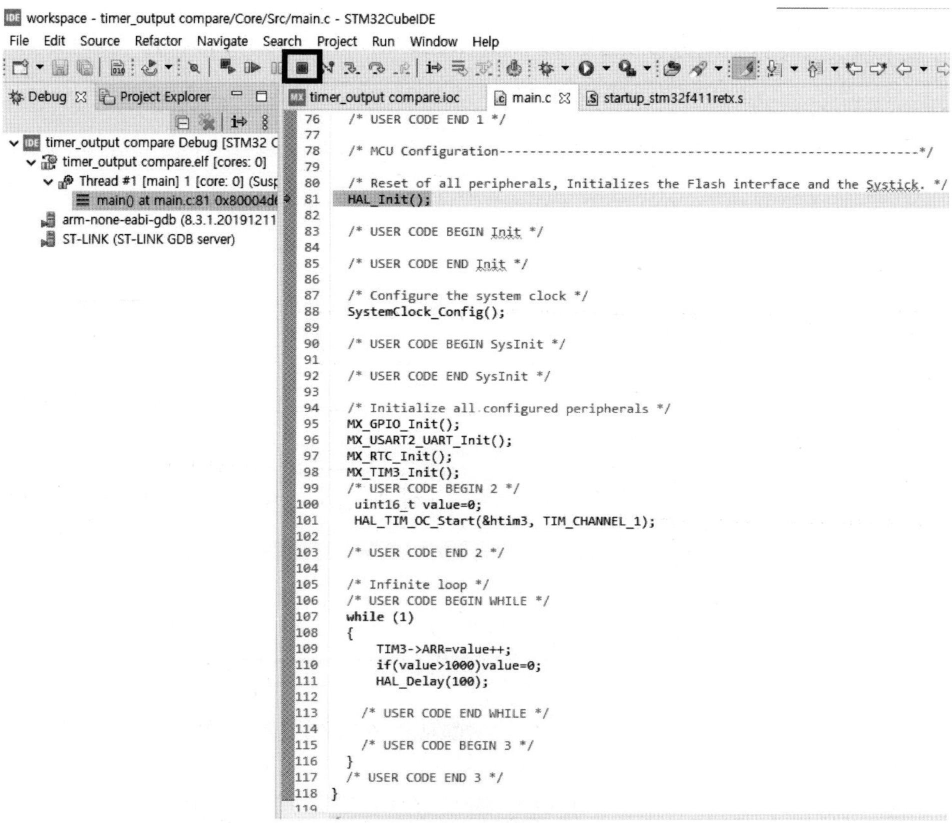
workspace - timer_output compare/Core/Src/main.c - STM32CubeIDE
File Edit Source Refactor Navigate Search Project Run Window Help
Debug
Project Explorer
timer_output compare.ioc
main.c
startup_stm32f411retx.s
timer_output compare Debug [STM32 C
timer_output compare.elf [cores: 0]
Thread #1 [main] 1 [core: 0] (Susp
main() at main.c:81 0x80004d
arm-none-eabi-gdb (8.3.1.20191211
ST-LINK (ST-LINK GDB server)
/* USER CODE END 1 */
/* MCU Configuration--*/
/* Reset of all peripherals, Initializes the Flash interface and the Systick. */
HAL_Init();
/* USER CODE BEGIN Init */
/* USER CODE END Init */
/* Configure the system clock */
SystemClock_Config();
/* USER CODE BEGIN SysInit */
/* USER CODE END SysInit */
/* Initialize all configured peripherals */
MX_GPIO_Init();
MX_USART2_UART_Init();
MX_RTC_Init();
MX_TIM3_Init();
/* USER CODE BEGIN 2 */
uint16_t value=0;
HAL_TIM_OC_Start(&htim3, TIM_CHANNEL_1);
/* USER CODE END 2 */
/* Infinite loop */
/* USER CODE BEGIN WHILE */
while (1)
{
TIM3->ARR=value++;
if(value>1000)value=0;
HAL_Delay(100);
/* USER CODE END WHILE */
/* USER CODE BEGIN 3 */
}
/* USER CODE END 3 */
}

[그림 9.16] 디버그 모드 프로그램 실행

6. Input Capture 활용하기(내부 클럭)

6.1. 개요

Timer에서 Input Capture는 MCU의 펄스 및 주어진 입력 신호의 주파수 및 펄스 폭 등을 측정하고자 할 때 사용된다. 내부 클럭과 외부 클럭을 이용하는 방법이 있다. 이 경우 카운터의 기능을 이용하는 경우에 해당한다. 주파수 발생은 Output Compare 기능에서와 같이 타이머 3의 channel1(PA6)을 이용하여 출력을 발생하고 주파수 카운터는 Input Capture의 Timer4 Channel1을 이용한다.

6.2. Input Capture 설정

[그림 9.16], [그림 9.17]은 Input Capture 및 인터럽트 설정 내용이다.

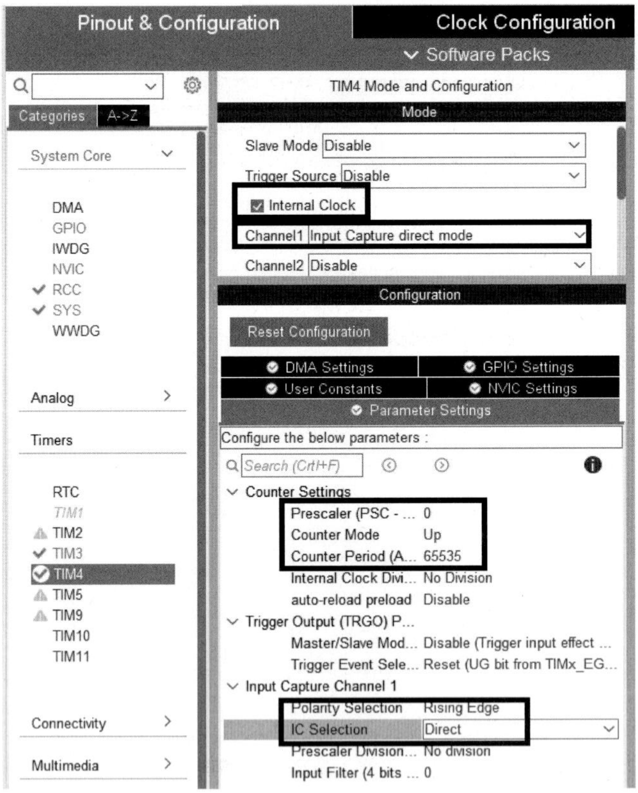

[그림 9.16] 타이머4 Input Capture 설정

6.3. Generate Code 생성

[그림 9.17]은 코드를 생성하기 위한 화면이다. 그림과 같이 클릭해서 프로그램을 생성한다.

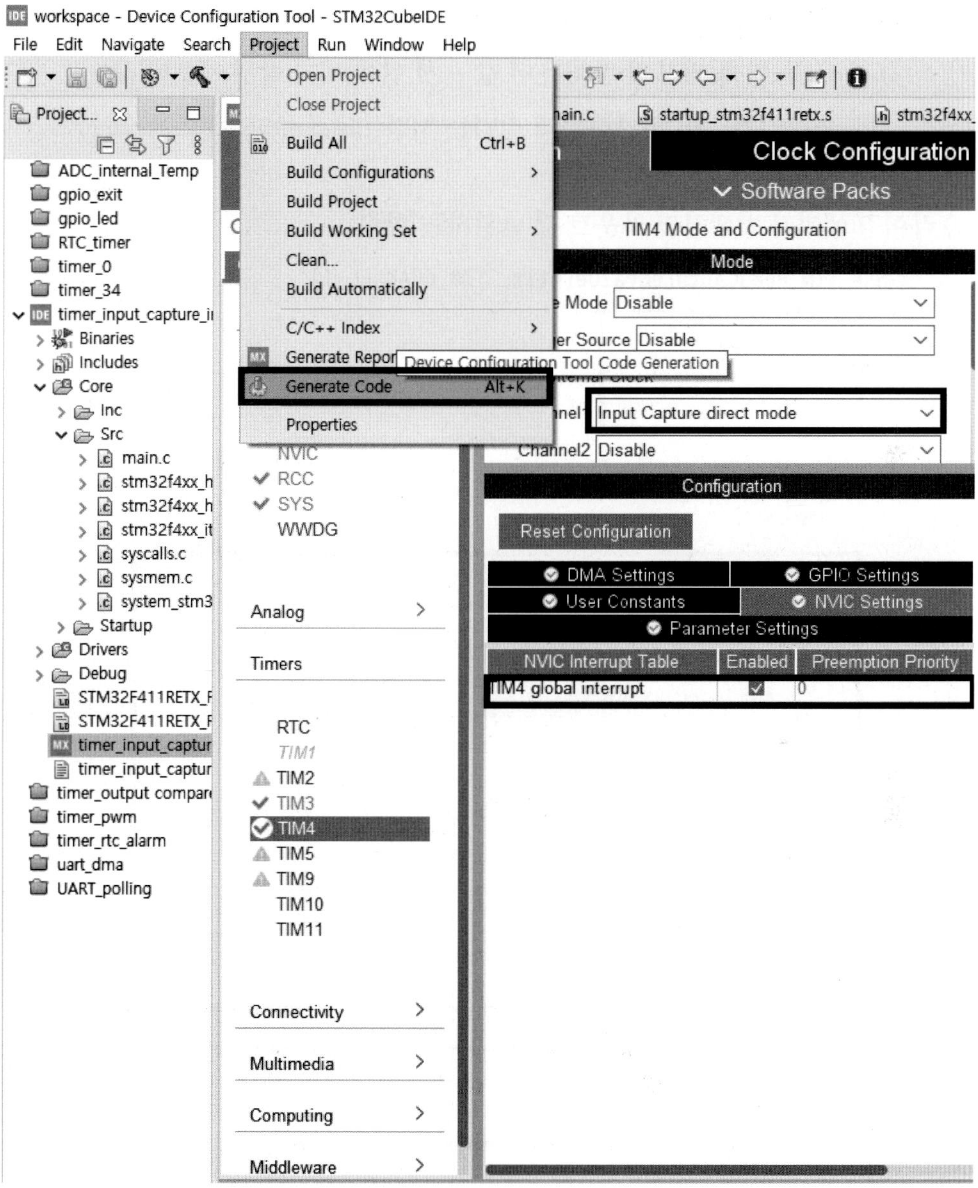

[그림 9.17] 인터럽트 설정 및 프로그램 코드를 생성

6.4. Source Code 작성하기

주파수 측정을 위한 프로그램 과정은 [그림 9.18]과 같다. 스타트 함수(`HAL_TIM_IC_Start(&htim4,TIM_CHANNEL_1)`)가 실행되면 카운트 값이 0에서 1씩 증가한다.

〉주파수 계산 방법

f=fm/(**두 번째 에지 발생 카운터 값 – 첫 번째 에지 발생 카운터 값**)

fm=**카운터가 1이 증가하는데 걸리는 시간**(timer **주파수** /Prescaler)

= `2*HAL_RCC_GetPCLK1Freq()`

〉라이징 에지가 발생하면 카운트 값 반환하는 함수:

`HAL_TIM_ReadCapturedValue(htim, TIM_CHANNEL_1):`

```c
uint32_t IC_Value1 = 0;
uint32_t IC_Value2 = 0;
uint32_t Difference = 0;
uint32_t Frequency = 0;
uint8_t First_Captured = 0;
uint32_t value=1;

int __io_putchar(int ch)
{
HAL_UART_Transmit(&huart2, (uint8_t *)&ch, 1, 0xFFFF);
 return ch;
}

void HAL_TIM_IC_CaptureCallback(TIM_HandleTypeDef *htim){

    if (htim->Channel == HAL_TIM_ACTIVE_CHANNEL_1) {
        if (First_Captured==0) {
           IC_Value1 = HAL_TIM_ReadCapturedValue(htim, TIM_CHANNEL_1);
           First_Captured =1;
        }
        else if (First_Captured){
           IC_Value2 = HAL_TIM_ReadCapturedValue(htim, TIM_CHANNEL_1);
           if (IC_Value2 > IC_Value1){
              Difference = IC_Value2-IC_Value1;
            }
           else if (IC_Value2 < IC_Value1)  {
              Difference = ((0xffff-IC_Value1)+IC_Value2) +1;
            }
           else{
              Error_Handler();
               }
           Frequency = 2*HAL_RCC_GetPCLK1Freq()/Difference;  // calculate frequency
           First_Captured = 0;  // reset the first captured
        }
    }

}
/* USER CODE BEGIN 2 */
HAL_TIM_OC_Start(&htim3, TIM_CHANNEL_1);
HAL_TIM_IC_Start_IT(&htim4, TIM_CHANNEL_1);
/* USER CODE END 2 */

/* Infinite loop */
/* USER CODE BEGIN WHILE */
while (1)
{
   TIM3->ARR=value++;
   if(value>200)value=0;
   printf("ARR value=%ld, frequency=%ldHz\r\n",value,Frequency);
   HAL_Delay(100);
  /* USER CODE END WHILE */

  /* USER CODE BEGIN 3 */
}
/* USER CODE END 3 */
```

[그림 9.18] 메인 코드 작성

6.5. 디버그 및 다운로드하기

메인 함수에 프로그램이 작성되었다면 다음으로 디버그는 [그림 9.19]의 를 클릭하거나 단축키 F11을 누르면 디버깅과 프로그램 다운로드가 완료된다. 그림을 보면 타이머의 스케일러 값과 주기(ARR) 값이 잘 설정되어 있는 것을 알 수 있다.

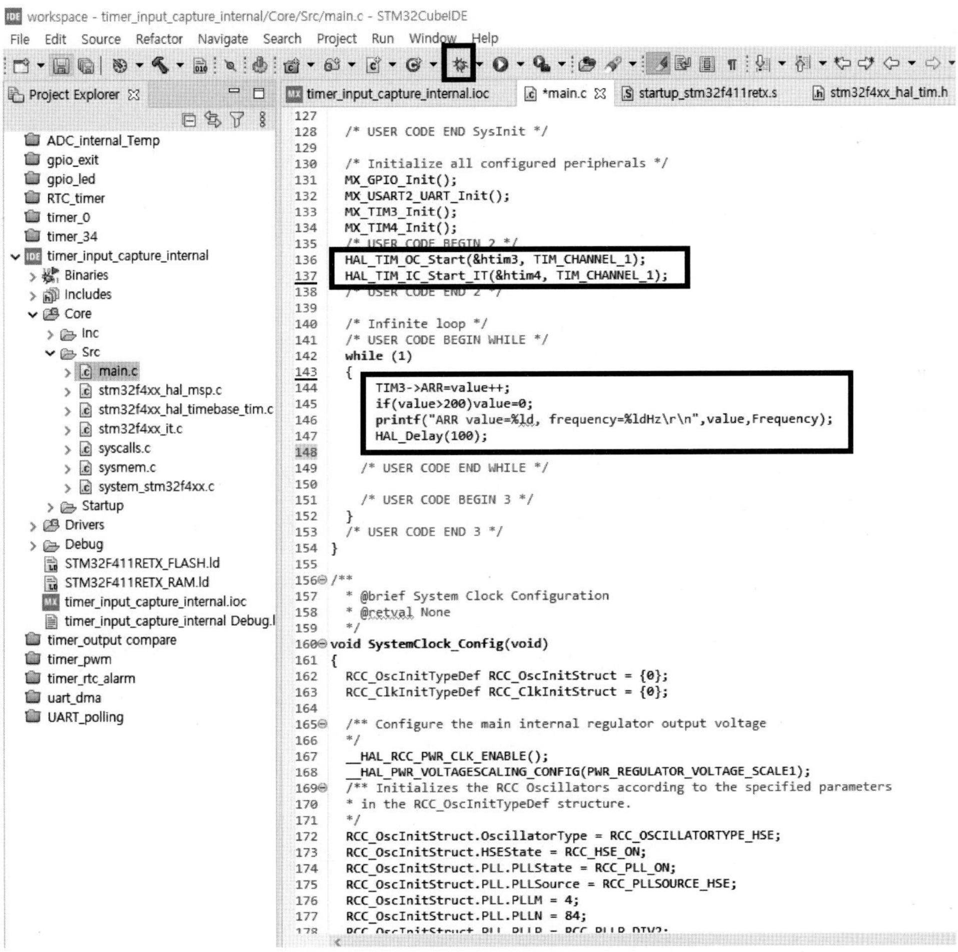
workspace - timer_input_capture_internal/Core/Src/main.c - STM32CubeIDE
File Edit Source Refactor Navigate Search Project Run Window Help
Project Explorer
timer_input_capture_internal.ioc
*main.c
startup_stm32f411retx.s
stm32f4xx_hal_tim.h
ADC_internal_Temp
gpio_exit
gpio_led
RTC_timer
timer_0
timer_34
timer_input_capture_internal
Binaries
Includes
Core
Inc
Src
main.c
stm32f4xx_hal_msp.c
stm32f4xx_hal_timebase_tim.c
stm32f4xx_it.c
syscalls.c
sysmem.c
system_stm32f4xx.c
Startup
Drivers
Debug
STM32F411RETX_FLASH.ld
STM32F411RETX_RAM.ld
timer_input_capture_internal.ioc
timer_input_capture_internal Debug.l
timer_output compare
timer_pwm
timer_rtc_alarm
uart_dma
UART_polling
/* USER CODE END SysInit */
/* Initialize all configured peripherals */
MX_GPIO_Init();
MX_USART2_UART_Init();
MX_TIM3_Init();
MX_TIM4_Init();
/* USER CODE BEGIN 2 */
HAL_TIM_OC_Start(&htim3, TIM_CHANNEL_1);
HAL_TIM_IC_Start_IT(&htim4, TIM_CHANNEL_1);
/* USER CODE END 2 */
/* Infinite loop */
/* USER CODE BEGIN WHILE */
while (1)
{
TIM3->ARR=value++;
if(value>200)value=0;
printf("ARR value=%ld, frequency=%ldHz\r\n",value,Frequency);
HAL_Delay(100);
/* USER CODE END WHILE */
/* USER CODE BEGIN 3 */
}
/* USER CODE END 3 */
}
/**
* @brief System Clock Configuration
* @retval None
*/
void SystemClock_Config(void)
{
RCC_OscInitTypeDef RCC_OscInitStruct = {0};
RCC_ClkInitTypeDef RCC_ClkInitStruct = {0};
/** Configure the main internal regulator output voltage
*/
__HAL_RCC_PWR_CLK_ENABLE();
__HAL_PWR_VOLTAGESCALING_CONFIG(PWR_REGULATOR_VOLTAGE_SCALE1);
/** Initializes the RCC Oscillators according to the specified parameters
* in the RCC_OscInitTypeDef structure.
*/
RCC_OscInitStruct.OscillatorType = RCC_OSCILLATORTYPE_HSE;
RCC_OscInitStruct.HSEState = RCC_HSE_ON;
RCC_OscInitStruct.PLL.PLLState = RCC_PLL_ON;
RCC_OscInitStruct.PLL.PLLSource = RCC_PLLSOURCE_HSE;
RCC_OscInitStruct.PLL.PLLM = 4;
RCC_OscInitStruct.PLL.PLLN = 84;

[그림 9.19] 프로그램 디버그

6.6. 보드 프로그램 실행하기

[그림 9.20] 상단 중앙에 아이콘 ■ 클릭하면 디버그 모드가 종료되어 보드의 프로그램이 실행된다.

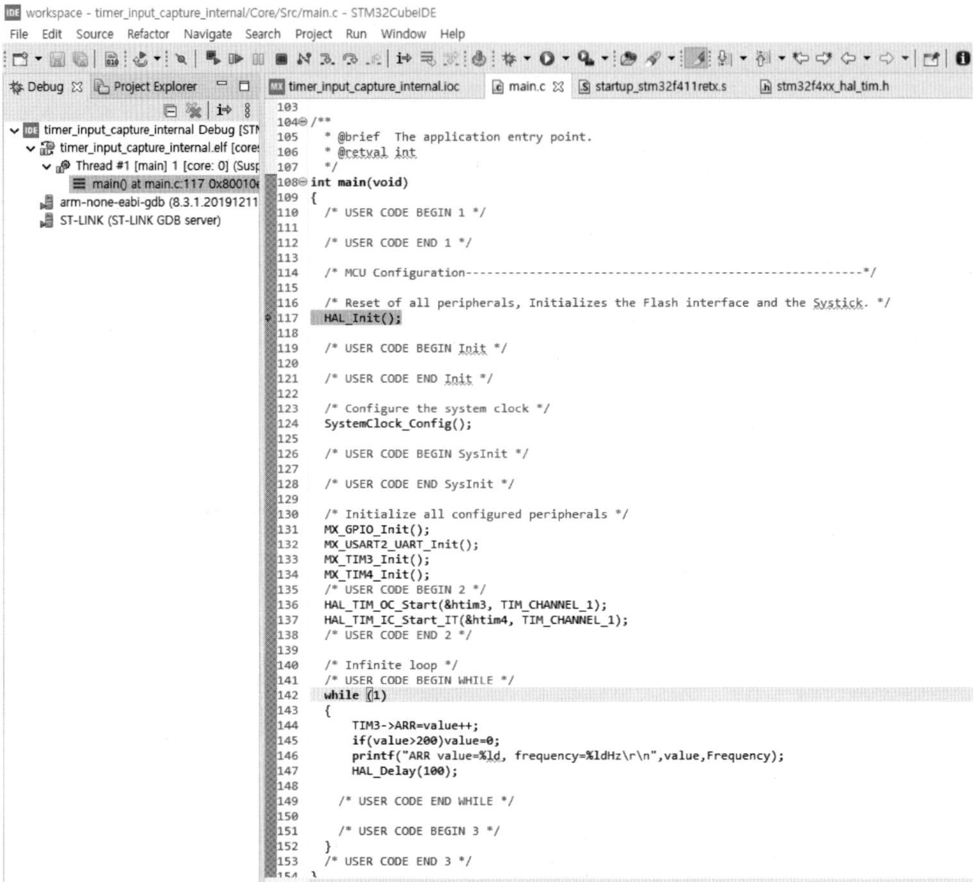
workspace - timer_input_capture_internal/Core/Src/main.c - STM32CubeIDE
File Edit Source Refactor Navigate Search Project Run Window Help
Debug
Project Explorer
timer_input_capture_internal.ioc
main.c
startup_stm32f411retx.s
stm32f4xx_hal_tim.h
timer_input_capture_internal Debug [STM
timer_input_capture_internal.elf [cores
Thread #1 [main] 1 [core: 0] (Susp
main() at main.c:117 0x80010
arm-none-eabi-gdb (8.3.1.20191211
ST-LINK (ST-LINK GDB server)
/**
* @brief The application entry point.
* @retval int
*/
int main(void)
{
/* USER CODE BEGIN 1 */
/* USER CODE END 1 */
/* MCU Configuration--*/
/* Reset of all peripherals, Initializes the Flash interface and the Systick. */
HAL_Init();
/* USER CODE BEGIN Init */
/* USER CODE END Init */
/* Configure the system clock */
SystemClock_Config();
/* USER CODE BEGIN SysInit */
/* USER CODE END SysInit */
/* Initialize all configured peripherals */
MX_GPIO_Init();
MX_USART2_UART_Init();
MX_TIM3_Init();
MX_TIM4_Init();
/* USER CODE BEGIN 2 */
HAL_TIM_OC_Start(&htim3, TIM_CHANNEL_1);
HAL_TIM_IC_Start_IT(&htim4, TIM_CHANNEL_1);
/* USER CODE END 2 */
/* Infinite loop */
/* USER CODE BEGIN WHILE */
while (1)
{
TIM3->ARR=value++;
if(value>200)value=0;
printf("ARR value=%ld, frequency=%ldHz\r\n",value,Frequency);
HAL_Delay(100);
/* USER CODE END WHILE */
/* USER CODE BEGIN 3 */
}
/* USER CODE END 3 */

[그림 9.20] 디버그 모드 프로그램 실행

6.7. 시리얼 포트로 출력 확인하기

[그림 9.21]에 실행 결과가 나타고 있다.

```
ARR=1000: f=  2,264Hz
ARR= 100: f=  4,950Hz
ARR=  10: f= 45,454Hz
ARR=   1: f=250,000Hz
```

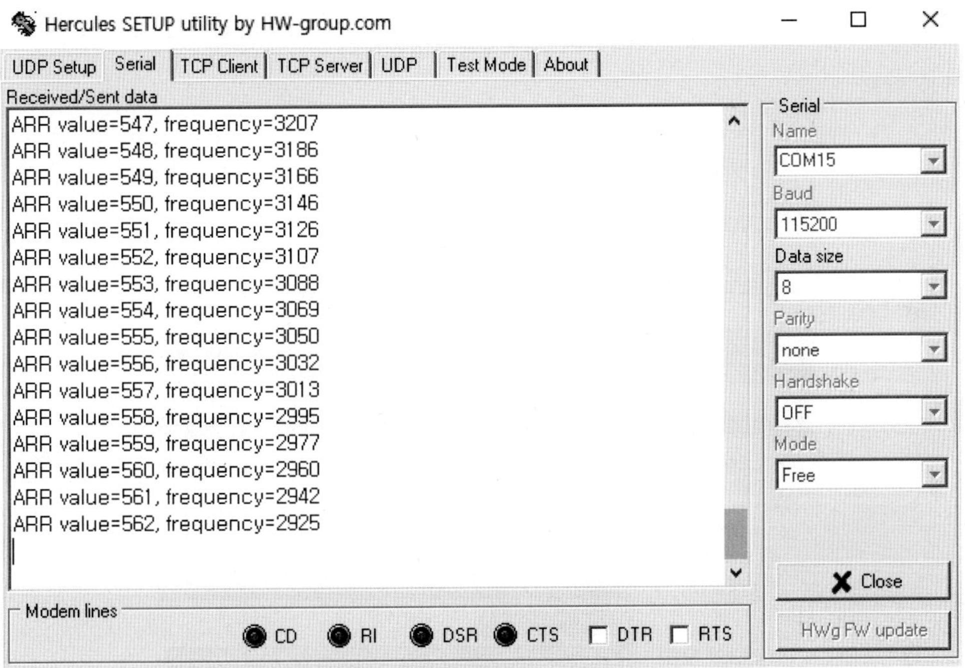

[그림 9.21] 주파수 측정 결과

7. Input Capture 활용하기(외부 클럭)

7.1. 개요

Input Capture 기능 중에서 Trigger Source와 channel 2개를 이용하여 Rising Edge 및 Falling Edge를 읽어 주파수와 펄스 폭을 알아내 활용할 수가 있다. 주파수 및 펄스 폭 변조를 위해서는 PWM 기능에서와 같이 타이머 3의 channel1(PA6)을 이용하여 출력을 발생하고 주파수 카운터는 Input Capture의 Timer4 Channel1,2을 이용 한다.

7.2. Input Capture 설정

[그림 9.22], [그림 9.23]은 Input Capture 및 인터럽트 설정 내용이다.

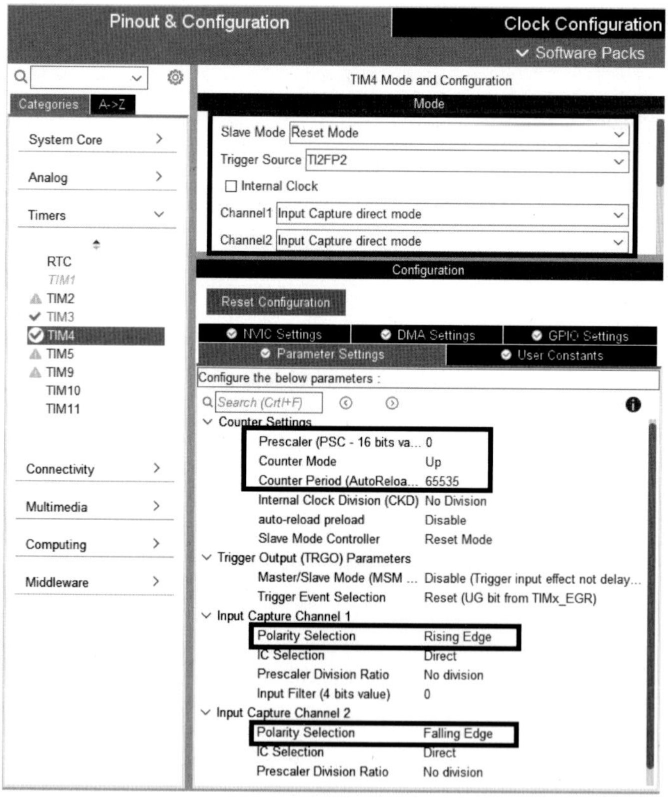

[그림 9.22] 타이머4 Input Capture 설정

7.3. Generate Code 생성

[그림 9.23]은 코드를 생성하기 위한 화면이다. 그림과 같이 클릭해서 프로그램을 생성한다.

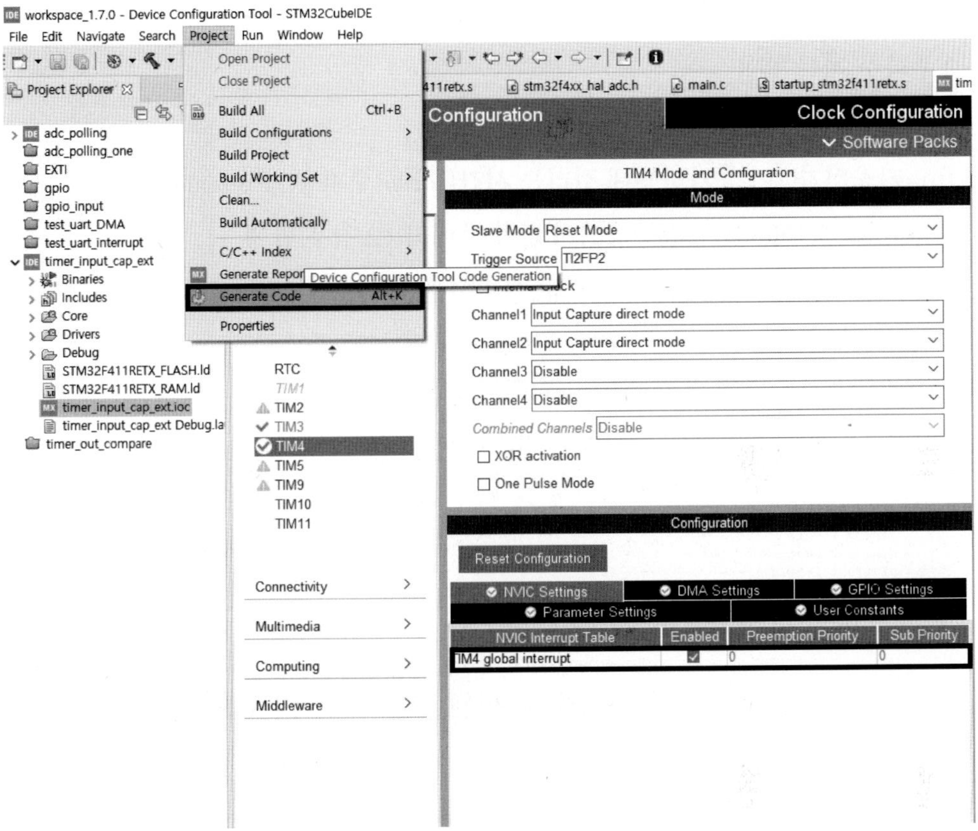

[그림 9.23] 인터럽트 설정 및 프로그램 코드를 생성

7.4. Source Code 작성하기

주파수 측정을 위한 프로그램 과정은 [그림 9.18]과 같다. 스타트 함수(HAL_TIM_IC_Start(&htim4,TIM_CHANNEL_1))가 실행되면 카운트 값이 0에서 1씩 증가한다.

〉주파수 계산 방법

f=fm/(라이징(폴링)에지 발생 경우 카운터 값)

fm=카운터가 1이 증가하는데 걸리는 시간(timer 주파수 /Prescaler)

= HAL_RCC_GetPCLK1Freq() //타이머 종류에 따라 *2 경우 발생

〉라이징 에지가 발생하면 카운트 값 반환하는 함수:

HAL_TIM_ReadCapturedValue(htim, TIM_CHANNEL_1):

〉폴링 에지가 발생하면 카운트 값 반환하는 함수:

HAL_TIM_ReadCapturedValue(htim, TIM_CHANNEL_1):

```
uint32_t IC_Value1 = 0;
uint32_t IC_Value2 = 0;
uint32_t Duty_Cycle = 0;
uint32_t Frequency = 0;
uint16_t i=100;
void HAL_TIM_IC_CaptureCallback(TIM_HandleTypeDef *htim){
  if (htim->Channel == HAL_TIM_ACTIVE_CHANNEL_1){
      IC_Value1 = HAL_TIM_ReadCapturedValue(htim, TIM_CHANNEL_1);  // capture the first value
      if(IC_Value1 !=0){
           IC_Value2 = HAL_TIM_ReadCapturedValue(htim, TIM_CHANNEL_2);  // capture second value
           if(IC_Value1>IC_Value2){
               Duty_Cycle=(IC_Value2*100/IC_Value1);
               Frequency = HAL_RCC_GetPCLK1Freq()/IC_Value1;  // calculate frequency
                   }
           else{
               Duty_Cycle=(IC_Value1*100/IC_Value2);
               Duty_Cycle=100-Duty_Cycle;
               Frequency = HAL_RCC_GetPCLK1Freq()/IC_Value2;  // calculate frequency
           }
      }
      else{
           Duty_Cycle=0;
           Frequency=0;
      }
  }
}

  /* USER CODE BEGIN 2 */
    HAL_TIM_IC_Start_IT(&htim4, TIM_CHANNEL_1);
    HAL_TIM_IC_Start_IT(&htim4, TIM_CHANNEL_2);
    HAL_TIM_PWM_Start(&htim3, TIM_CHANNEL_1);

  /* USER CODE END 2 */

  /* Infinite loop */
  /* USER CODE BEGIN WHILE */
  while (1)
  {
      TIM3->CCR1=10-1;
      printf("frequency=%ld, Duty cycle=%ld\r\n",Frequency,Duty_Cycle);
      HAL_Delay(100);
    /* USER CODE END WHILE */

    /* USER CODE BEGIN 3 */
  }
  /* USER CODE END 3 */
```

[그림 9.24] 메인 코드 작성

7.5. 디버그 및 다운로드하기

메인 함수에 프로그램이 작성되었다면 다음으로 디버그는 [그림 9.25]의 ⚙를 클릭하거나 단축키 F11을 누르면 디버깅과 프로그램 다운로드가 완료된다. 그림을 보면 타이머의 스케일러 값과 주기(ARR) 값이 잘 설정되어 있는 것을 알 수 있다.

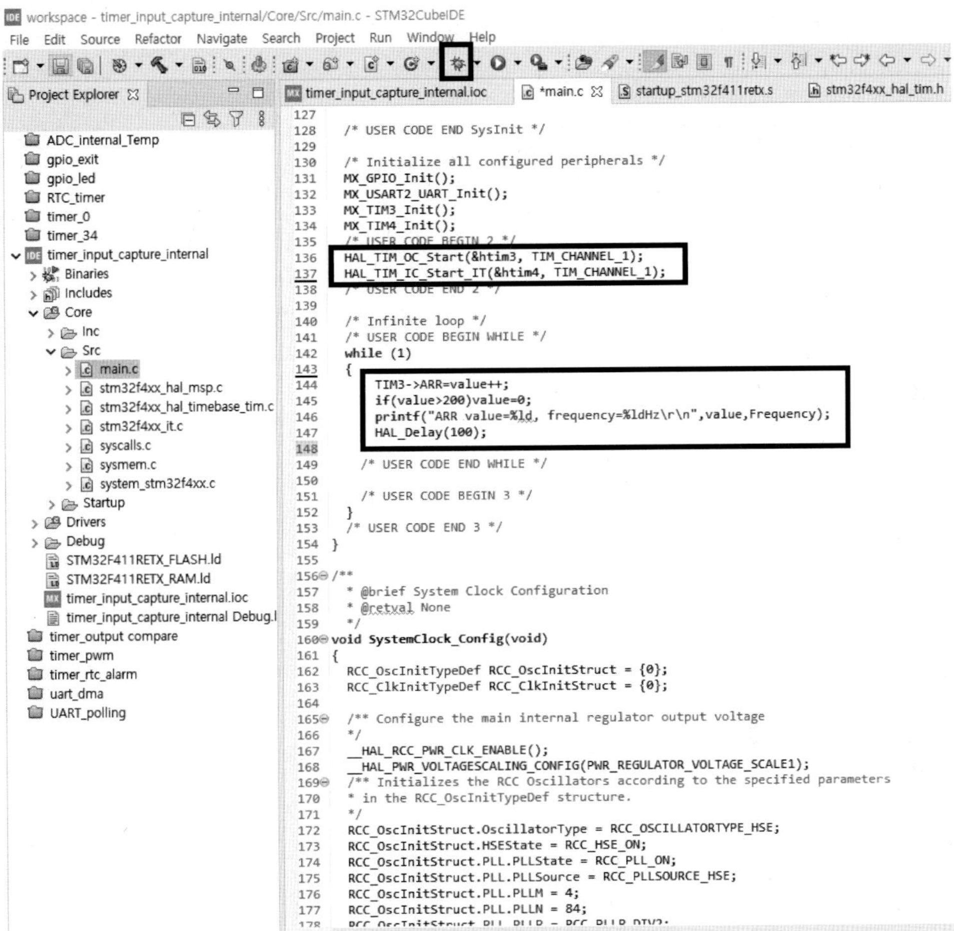
workspace - timer_input_capture_internal/Core/Src/main.c - STM32CubeIDE
File Edit Source Refactor Navigate Search Project Run Window Help
Project Explorer
ADC_internal_Temp
gpio_exit
gpio_led
RTC_timer
timer_0
timer_34
timer_input_capture_internal
Binaries
Includes
Core
Inc
Src
main.c
stm32f4xx_hal_msp.c
stm32f4xx_hal_timebase_tim.c
stm32f4xx_it.c
syscalls.c
sysmem.c
system_stm32f4xx.c
Startup
Drivers
Debug
STM32F411RETX_FLASH.ld
STM32F411RETX_RAM.ld
timer_input_capture_internal.ioc
timer_input_capture_internal Debug.l
timer_output compare
timer_pwm
timer_rtc_alarm
uart_dma
UART_polling
timer_input_capture_internal.ioc
*main.c
startup_stm32f411retx.s
stm32f4xx_hal_tim.h
/* USER CODE END SysInit */
/* Initialize all configured peripherals */
MX_GPIO_Init();
MX_USART2_UART_Init();
MX_TIM3_Init();
MX_TIM4_Init();
HAL_TIM_OC_Start(&htim3, TIM_CHANNEL_1);
HAL_TIM_IC_Start_IT(&htim4, TIM_CHANNEL_1);
/* Infinite loop */
/* USER CODE BEGIN WHILE */
while (1)
{
TIM3->ARR=value++;
if(value>200)value=0;
printf("ARR value=%ld, frequency=%ldHz\r\n",value,Frequency);
HAL_Delay(100);
/* USER CODE END WHILE */
/* USER CODE BEGIN 3 */
}
/* USER CODE END 3 */
}
/**
* @brief System Clock Configuration
* @retval None
*/
void SystemClock_Config(void)
{
RCC_OscInitTypeDef RCC_OscInitStruct = {0};
RCC_ClkInitTypeDef RCC_ClkInitStruct = {0};
/** Configure the main internal regulator output voltage
*/
__HAL_RCC_PWR_CLK_ENABLE();
__HAL_PWR_VOLTAGESCALING_CONFIG(PWR_REGULATOR_VOLTAGE_SCALE1);
/** Initializes the RCC Oscillators according to the specified parameters
* in the RCC_OscInitTypeDef structure.
*/
RCC_OscInitStruct.OscillatorType = RCC_OSCILLATORTYPE_HSE;
RCC_OscInitStruct.HSEState = RCC_HSE_ON;
RCC_OscInitStruct.PLL.PLLState = RCC_PLL_ON;
RCC_OscInitStruct.PLL.PLLSource = RCC_PLLSOURCE_HSE;
RCC_OscInitStruct.PLL.PLLM = 4;
RCC_OscInitStruct.PLL.PLLN = 84;

[그림 9.25] 프로그램 디버그

7.6. 보드 프로그램 실행하기

[그림 9.26] 상단 중앙에 아이콘 클릭하면 디버그 모드가 종료되어 보드의 프로그램이 실행된다.

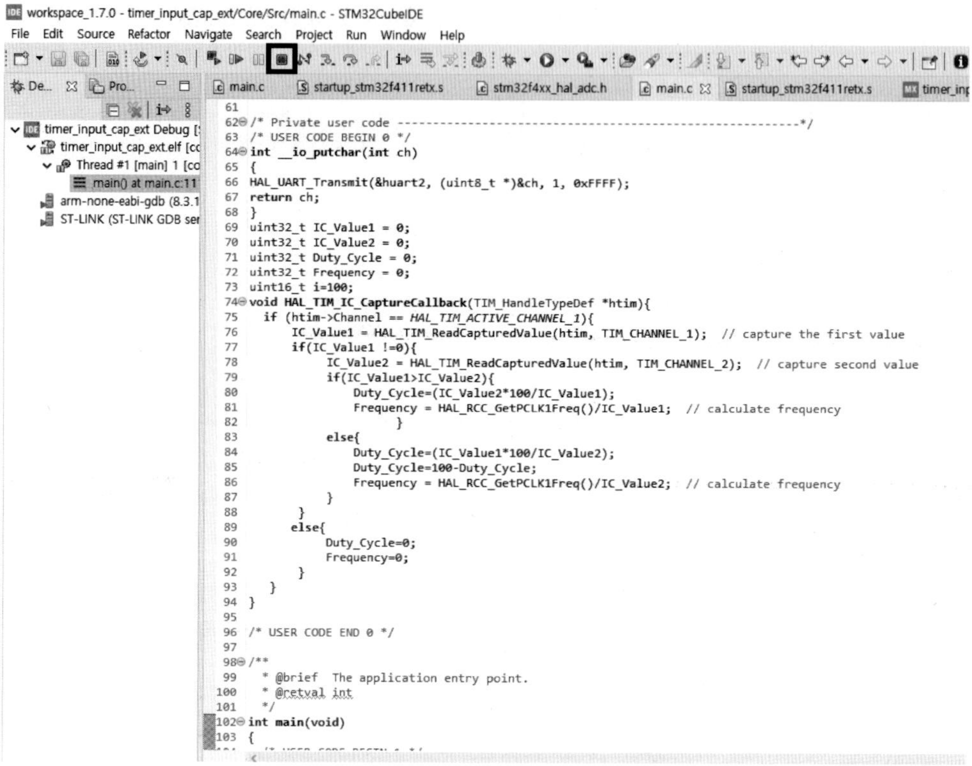
workspace_1.7.0 - timer_input_cap_ext/Core/Src/main.c - STM32CubeIDE
File Edit Source Refactor Navigate Search Project Run Window Help
main.c
startup_stm32f411retx.s
stm32f4xx_hal_adc.h
timer_input_cap_ext Debug
timer_input_cap_ext.elf
Thread #1 [main] 1
main() at main.c:11
arm-none-eabi-gdb (8.3.1
ST-LINK (ST-LINK GDB ser
/* Private user code ---*/
/* USER CODE BEGIN 0 */
int __io_putchar(int ch)
{
HAL_UART_Transmit(&huart2, (uint8_t *)&ch, 1, 0xFFFF);
return ch;
}
uint32_t IC_Value1 = 0;
uint32_t IC_Value2 = 0;
uint32_t Duty_Cycle = 0;
uint32_t Frequency = 0;
uint16_t i=100;
void HAL_TIM_IC_CaptureCallback(TIM_HandleTypeDef *htim){
if (htim->Channel == HAL_TIM_ACTIVE_CHANNEL_1){
IC_Value1 = HAL_TIM_ReadCapturedValue(htim, TIM_CHANNEL_1); // capture the first value
if(IC_Value1 !=0){
IC_Value2 = HAL_TIM_ReadCapturedValue(htim, TIM_CHANNEL_2); // capture second value
if(IC_Value1>IC_Value2){
Duty_Cycle=(IC_Value2*100/IC_Value1);
Frequency = HAL_RCC_GetPCLK1Freq()/IC_Value1; // calculate frequency
}
else{
Duty_Cycle=(IC_Value1*100/IC_Value2);
Duty_Cycle=100-Duty_Cycle;
Frequency = HAL_RCC_GetPCLK1Freq()/IC_Value2; // calculate frequency
}
}
else{
Duty_Cycle=0;
Frequency=0;
}
}
}
/* USER CODE END 0 */
/**
* @brief The application entry point.
* @retval int
*/
int main(void)
{

[그림 9.26] 디버그 모드 프로그램 실행

7.7. 시리얼 포트로 출력확인 하기

[그림 9.27]에 실행 결과가 나타고 있다.

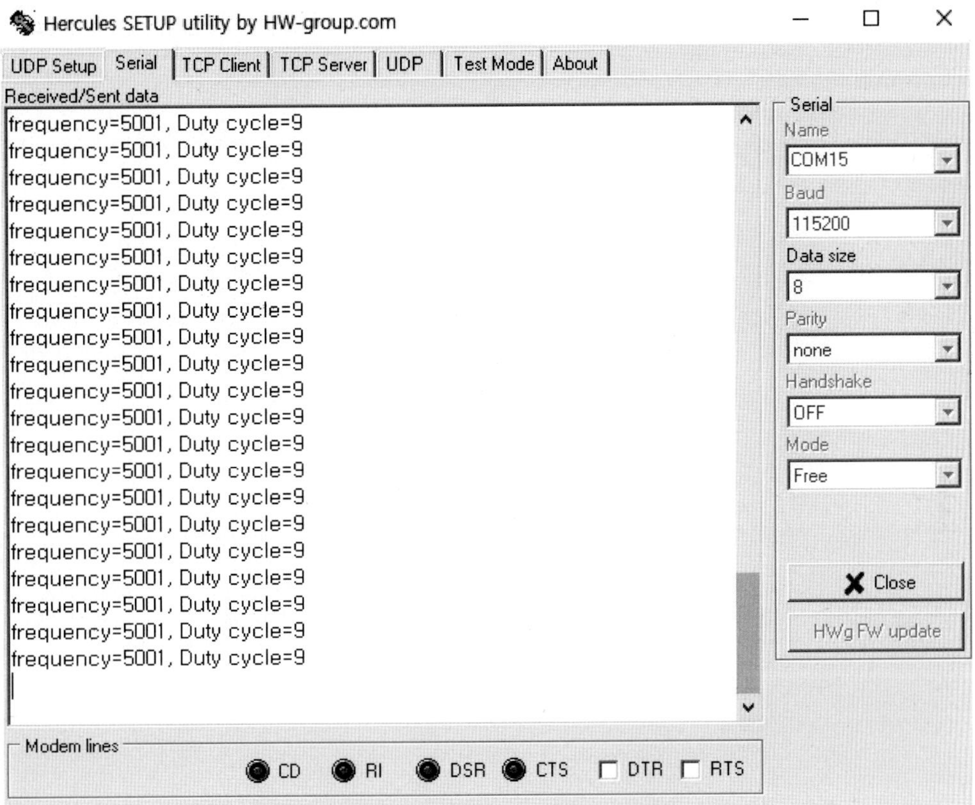

[그림 9.27] 주파수 및 펄스 폭 측정 결과

CHAPTER 10

디스플레이 장치 활용하기 (CLCD, OLED, GLCD)

CHAPTER 10

Cortex-M_4 기초 활용(ARM Core STM32)

디스플레이 장치 활용하기 (CLCD, OLED, GLCD)

1. 개요

본 장에서는 디스플레이 장치로 Character LCD(CLCD, I2C 방식, 16*2), OLED(I2C, 128*64), GLCD(Serial 방식, 128*64)를 활용하는 방법을 소개한다. 위 장치들을 활용하기 위한 소스 코드를 작성하는 것은 프로그램 개발자들의 몫으로 남겨 두고 여기서는 인터넷 등에서 제공하는 자료를 활용한다. 따라서 STM32F에서는 I2C를 활용할 수 있도록 STM32CubeIDE 설정이 필요하고 해당 구동 드라이버 파일(i2c-lcd.h ,i2c-lcd.c)이 필요하며 헤더 파일에 있는 명령어를 이용하면 된다.

2. CLCD 활용하기

2.1. 하드웨어 설정

I2C BUS I/O 확장을 지원하는 디바이스 중 많이 사용되는 것은 PCF8574이다. LCD 장치와 모듈을 연결해서 사용하면 된다. I2C 모듈은 4천 원대 후반 정도하며 인터넷에서 쉽게 구매할 수 있다. 일반적으로 I2C 모듈이 장착된 LCD를 구매하는 것이 편리하다. 하드웨어 결선은 다음과 같다(i2c1 경우).

〉VCC=3.3V, GND=0V,

〉SCL(LCD장치)-〉STM32F4 (PB6)

〉SDA(LCD장치)-〉STM32F4(PB7)

〉Anode(LCD장치)-〉3.3V

〉Kathode(LCD장치)-〉GND , 가변저항: 백라이트 조절

[그림 10.1] I2C 모듈 및 LCD 장치

2.2. I2C 설정

클럭 및 시스템 설정은 이전과 동일하며 [그림 10.2]와 같이 I2C만 선택하면 된다.

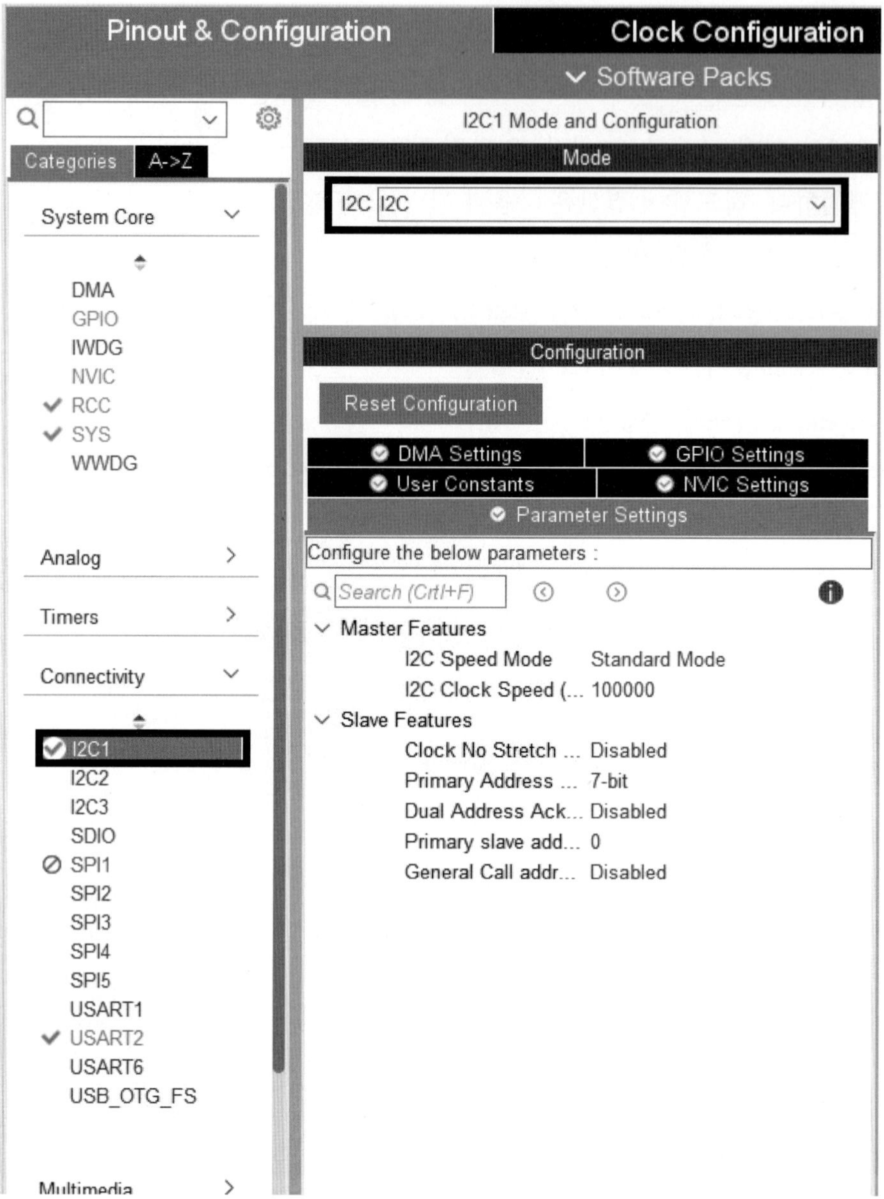

[그림 10.2] I2C 설정

2.3. Generate Code 생성

[그림 10.3]은 코드를 생성하기 위한 화면이다. 그림과 같이 클릭해서 프로그램을 생성한다.

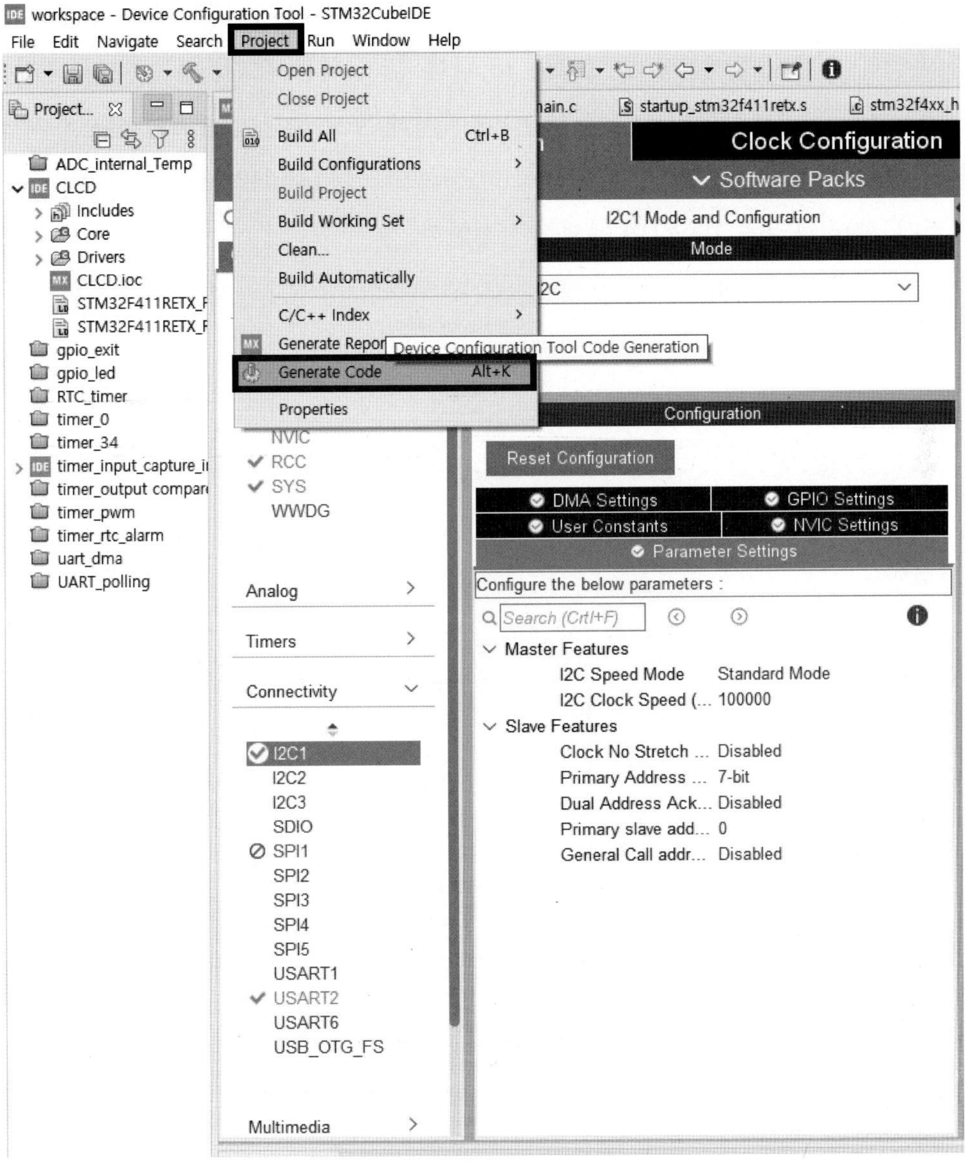

[그림 10.3] 프로그램 코드를 생성

2.4. Source Code 작성하기

Library를 활용하려면 인터넷에서 stm32 CLCD library 검색하고 그중에서 본인의 LCD에 맞는 Library를 활용하거나 수정(stm32f1xx_hal.h->stm32f4xx_hal.h)하여 사용하면 된다. 보통 파일이 2(소스 파일, 헤더 파일)개 존재하는데 이를 프로그램 파일에 포함시켜 사용한다. [그림 10.4]와 같이 사용하고자 하는 해당 파일을 Core->Inc(헤더 파일)와 Src(소스 파일)에 복사해 두고 해당 명령어를 사용하여 코딩한다. 여기서는 https://controllerstech.com/i2c-lcd-in-stm32cm32/에서 제공하는 소스 및 헤더 파일을 이용하여 예제를 작성한다.

```
/* Private includes ----------------------------------------------------------
/* USER CODE BEGIN Includes */
#include "stdio.h"
#include "i2c-lcd.h"
/* USER CODE END Includes */
```

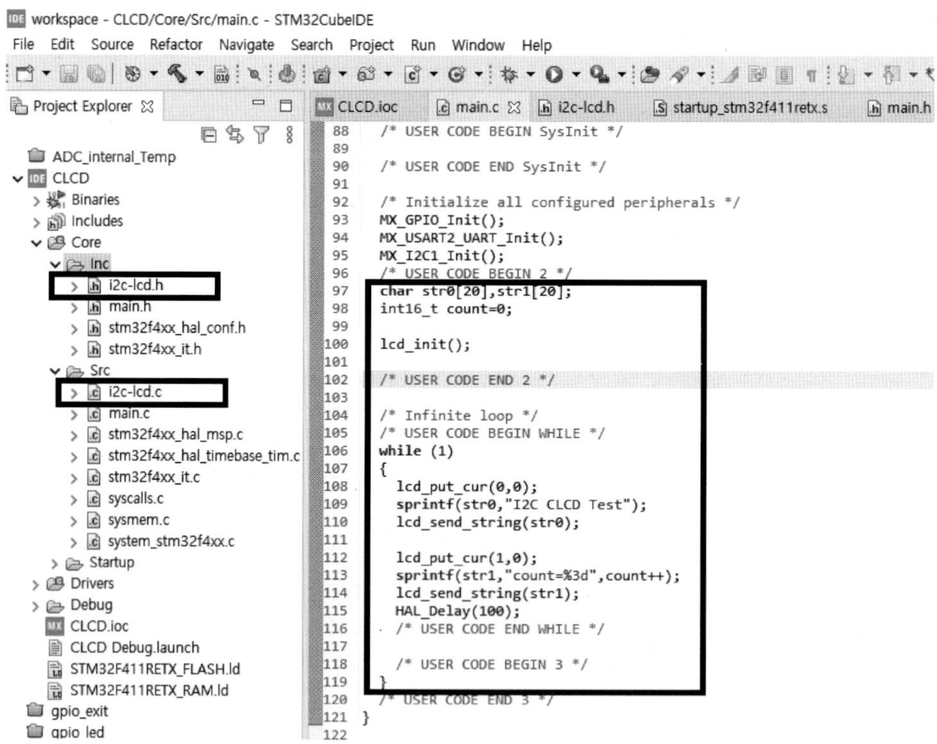

workspace - CLCD/Core/Src/main.c - STM32CubeIDE
File Edit Source Refactor Navigate Search Project Run Window Help
Project Explorer
CLCD.ioc
main.c
i2c-lcd.h
startup_stm32f411retx.s
main.h
ADC_internal_Temp
CLCD
Binaries
Includes
Core
Inc
i2c-lcd.h
main.h
stm32f4xx_hal_conf.h
stm32f4xx_it.h
Src
i2c-lcd.c
main.c
stm32f4xx_hal_msp.c
stm32f4xx_hal_timebase_tim.c
stm32f4xx_it.c
syscalls.c
sysmem.c
system_stm32f4xx.c
Startup
Drivers
Debug
CLCD.ioc
CLCD Debug.launch
STM32F411RETX_FLASH.ld
STM32F411RETX_RAM.ld
gpio_exit
gpio led
/* USER CODE BEGIN SysInit */
/* USER CODE END SysInit */
/* Initialize all configured peripherals */
MX_GPIO_Init();
MX_USART2_UART_Init();
MX_I2C1_Init();
/* USER CODE BEGIN 2 */
char str0[20],str1[20];
int16_t count=0;
lcd_init();
/* USER CODE END 2 */
/* Infinite loop */
/* USER CODE BEGIN WHILE */
while (1)
{
lcd_put_cur(0,0);
sprintf(str0,"I2C CLCD Test");
lcd_send_string(str0);
lcd_put_cur(1,0);
sprintf(str1,"count=%3d",count++);
lcd_send_string(str1);
HAL_Delay(100);
/* USER CODE END WHILE */
/* USER CODE BEGIN 3 */
}
/* USER CODE END 3 */
}

[그림 10.4] 메인 코드 작성

2.5. 디버그 및 다운로드하기

메인 함수에 프로그램이 작성되었다면 다음으로 디버그는 [그림 10.5]의 를 클릭하거나 단축키 F11을 누르면 디버깅과 프로그램 다운로드가 완료된다.

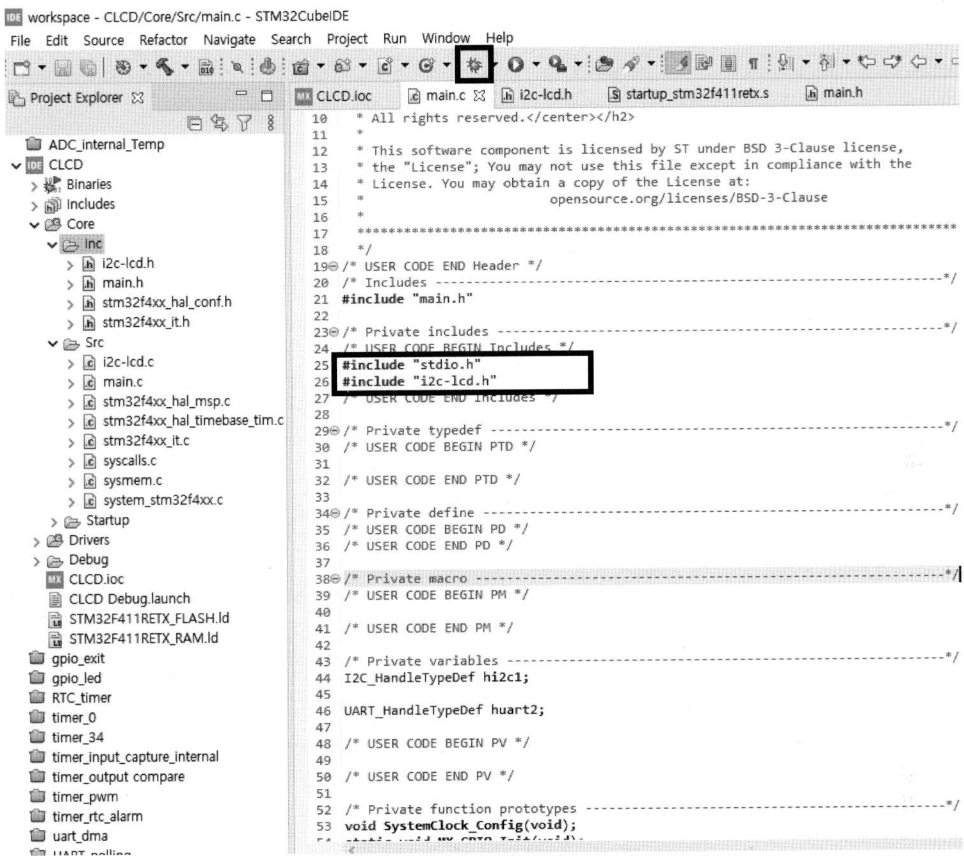
workspace - CLCD/Core/Src/main.c - STM32CubeIDE
File Edit Source Refactor Navigate Search Project Run Window Help
Project Explorer
CLCD.ioc
main.c
i2c-lcd.h
startup_stm32f411retx.s
main.h
ADC_internal_Temp
CLCD
Binaries
Includes
Core
Inc
i2c-lcd.h
main.h
stm32f4xx_hal_conf.h
stm32f4xx_it.h
Src
i2c-lcd.c
main.c
stm32f4xx_hal_msp.c
stm32f4xx_hal_timebase_tim.c
stm32f4xx_it.c
syscalls.c
sysmem.c
system_stm32f4xx.c
Startup
Drivers
Debug
CLCD.ioc
CLCD Debug.launch
STM32F411RETX_FLASH.ld
STM32F411RETX_RAM.ld
gpio_exit
gpio_led
RTC_timer
timer_0
timer_34
timer_input_capture_internal
timer_output compare
timer_pwm
timer_rtc_alarm
uart_dma
* All rights reserved.</center></h2>
* This software component is licensed by ST under BSD 3-Clause license,
* the "License"; You may not use this file except in compliance with the
* License. You may obtain a copy of the License at:
opensource.org/licenses/BSD-3-Clause
/* USER CODE END Header */
/* Includes */
#include "main.h"
/* Private includes */
#include "stdio.h"
#include "i2c-lcd.h"
/* Private typedef */
/* USER CODE BEGIN PTD */
/* USER CODE END PTD */
/* Private define */
/* USER CODE BEGIN PD */
/* USER CODE END PD */
/* Private macro */
/* USER CODE BEGIN PM */
/* USER CODE END PM */
/* Private variables */
I2C_HandleTypeDef hi2c1;
UART_HandleTypeDef huart2;
/* USER CODE BEGIN PV */
/* USER CODE END PV */
/* Private function prototypes */
void SystemClock_Config(void);

[그림 10.5] 프로그램 디버그

2.6. 보드 프로그램 실행하기

[그림 10.6] 상단 중앙에 아이콘 ■ 클릭하면 디버그 모드가 종료되어 보드의 프로그램이 실행된다.

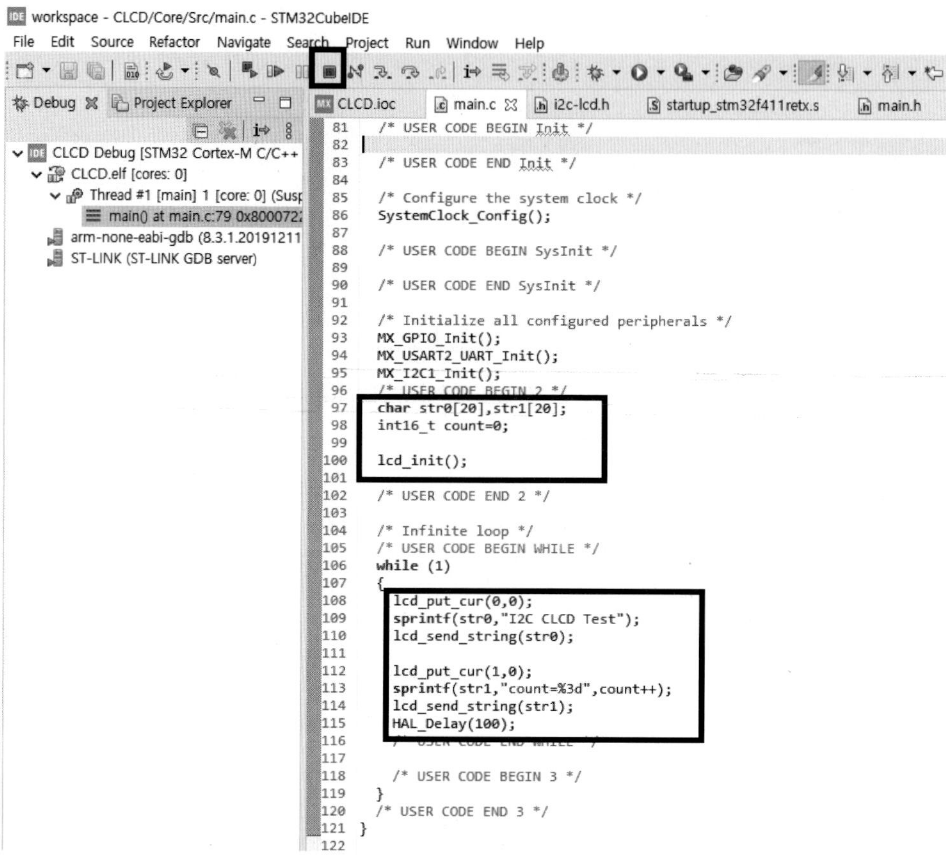
workspace - CLCD/Core/Src/main.c - STM32CubeIDE
File Edit Source Refactor Navigate Search Project Run Window Help
Debug
Project Explorer
CLCD.ioc
main.c
i2c-lcd.h
startup_stm32f411retx.s
main.h
CLCD Debug [STM32 Cortex-M C/C++
CLCD.elf [cores: 0]
Thread #1 [main] 1 [core: 0] (Susp
main() at main.c:79 0x800072
arm-none-eabi-gdb (8.3.1.20191211
ST-LINK (ST-LINK GDB server)
/* USER CODE BEGIN Init */
/* USER CODE END Init */
/* Configure the system clock */
SystemClock_Config();
/* USER CODE BEGIN SysInit */
/* USER CODE END SysInit */
/* Initialize all configured peripherals */
MX_GPIO_Init();
MX_USART2_UART_Init();
MX_I2C1_Init();
char str0[20],str1[20];
int16_t count=0;
lcd_init();
/* USER CODE END 2 */
/* Infinite loop */
/* USER CODE BEGIN WHILE */
while (1)
{
lcd_put_cur(0,0);
sprintf(str0,"I2C CLCD Test");
lcd_send_string(str0);
lcd_put_cur(1,0);
sprintf(str1,"count=%3d",count++);
lcd_send_string(str1);
HAL_Delay(100);
/* USER CODE BEGIN 3 */
}
/* USER CODE END 3 */
}

[그림 10.6] 디버그 모드 프로그램 실행

3. OLED 활용하기

3.1. 하드웨어 설정

여기서는 I2C를 이용하여 OLED(0.96", 128*64) 사용 방법을 다룬다. 하드웨어 결선은 다음과 같다(i2c1 경우)[그림 10.7].

〉VCC=3.3V, GND=0V,

〉SCL(LCD장치)-〉STM32F4 (PB6)

〉SDA(LCD장치)-〉STM32F4(PB7)

[그림 10.7] I2C모듈 및 OLED 장치

3.2. I2C 설정

클럭 및 시스템 설정은 이전과 동일하며 [그림 10.8]과 같이 I2C만 선택하면 된다.

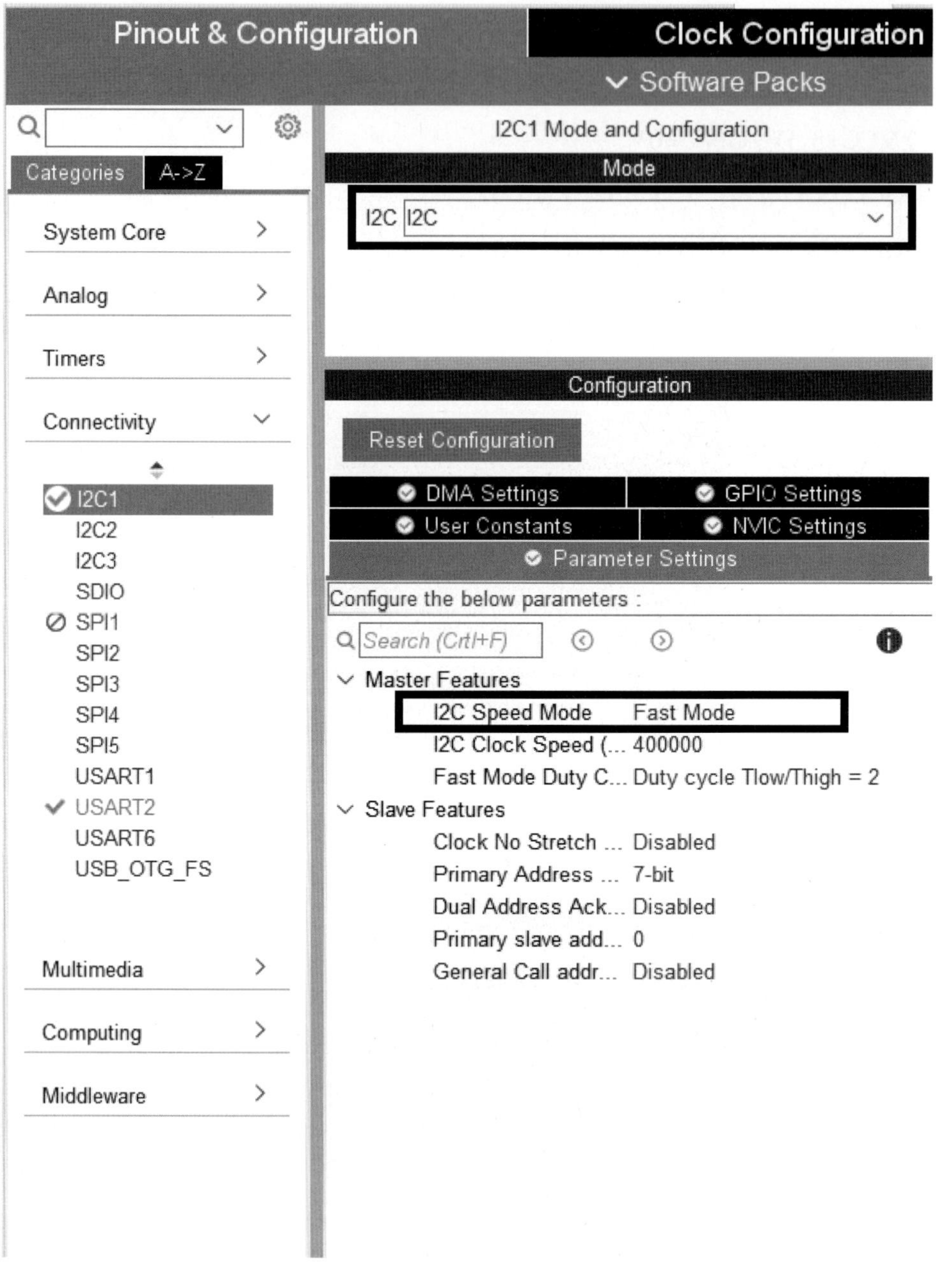

[그림 10.8] I2C 설정

3.3. Generate Code 생성

[그림 10.9]는 코드를 생성하기 위한 화면이다. 그림과 같이 클릭해서 프로그램을 생성한다.

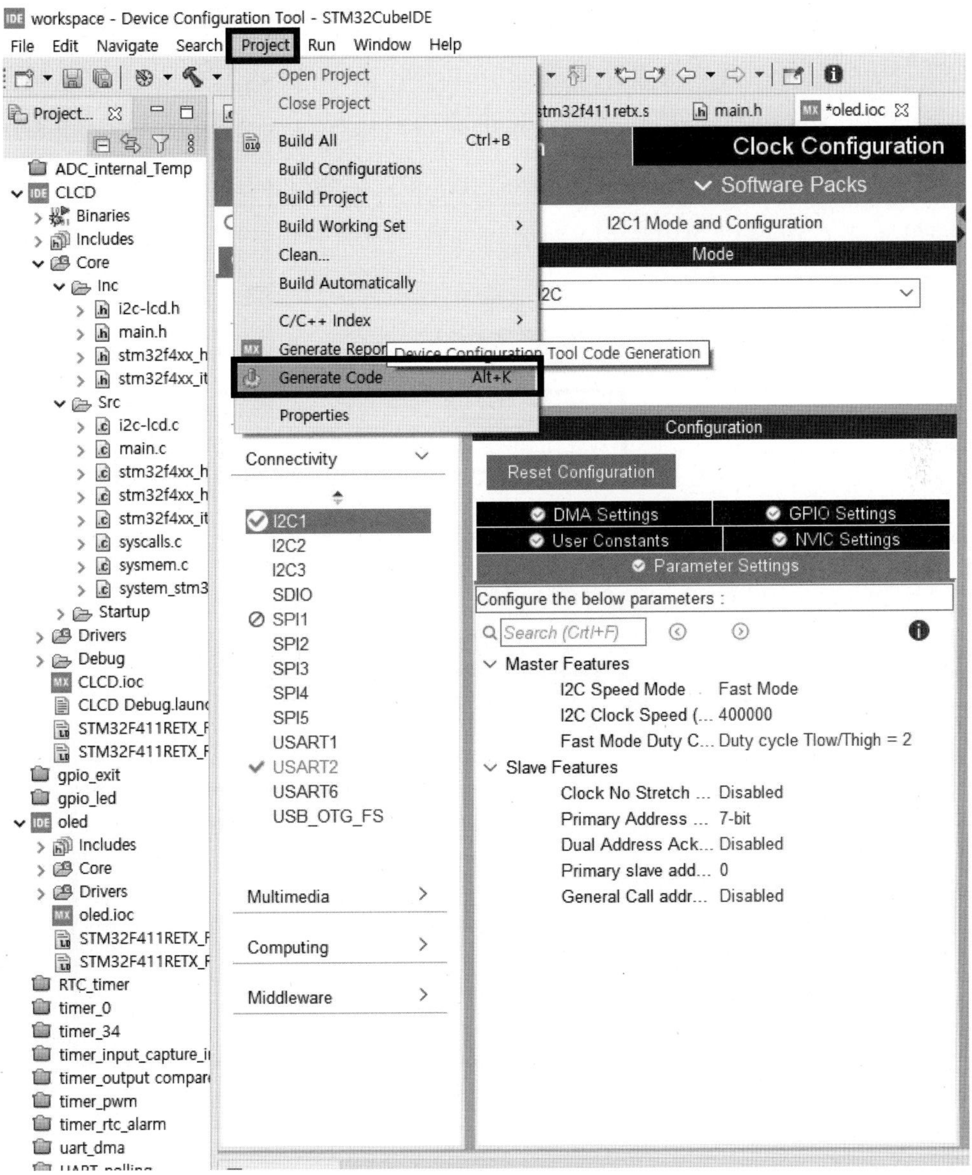

[그림 10.9] 프로그램 코드를 생성

3.4. Source Code 작성하기

Library를 활용하려면 인터넷에서 stm32 OLED library 검색하고 그중에서 본인의 OLED에 맞는 Library를 활용하거나 수정(stm32f1xx_hal.h->stm32f4xx_hal.h)하여 사용하면 된다. 파일이 3, 4개씩(소스 파일, 헤더 파일)이 존재하는데 이를 프로그램 파일에 포함시켜 사용한다. [그림 10.10]과 같이 사용하고자 하는 해당 파일을 Core->Inc(헤더 파일)와 Src(소스 파일)에 복사해 두고 해당 명령어를 사용하여 코딩한다. 여기서는 https://controllerstech.com/oled-display-using-i2c-stm32/ 에서 제공하는 소스 및 헤더 파일을 이용하여 예제를 작성해 본다.

#include "main.h"
/* Private includes --
/* USER CODE BEGIN Includes */
#include "fonts.h"
#include "ssd1306.h"
#include "test.h"
#include "bitmap.h"
/* USER CODE END Includes */
ADC_internal_Temp
CLCD
gpio_exit
gpio_led
oled
Binaries
Includes
Core
Inc
bitmap.h
fonts.h
main.h
ssd1306.h
stm32f4xx_hal_conf.h
stm32f4xx_it.h
test.h
Src
fonts.c
main.c
ssd1306.c
stm32f4xx_hal_msp.c
stm32f4xx_hal_timebase_tim.c
stm32f4xx_it.c
syscalls.c
sysmem.c
system_stm32f4xx.c
test.c
Startup
Drivers
Debug
/* USER CODE BEGIN SysInit */
/* USER CODE END SysInit */
/* Initialize all configured peripherals */
MX_GPIO_Init();
MX_USART2_UART_Init();
MX_I2C1_Init();
SSD1306_Init(); // initialise
/* Infinite loop */
/* USER CODE BEGIN WHILE */
while (1)
{
SSD1306_GotoXY (0,0);
SSD1306_Puts ("OLED Test ^*", &Font_7x10, 1);
SSD1306_GotoXY (10, 30);
SSD1306_Puts ("STM32F4", &Font_16x26, 1);
SSD1306_UpdateScreen(); //display
HAL_Delay (2000);
SSD1306_ScrollRight(0,7); // scroll entire screen
HAL_Delay(2000); // 2 sec
SSD1306_ScrollLeft(0,7); // scroll entire screen
HAL_Delay(2000); // 2 sec
SSD1306_Stopscroll();
SSD1306_Clear();
/* USER CODE END WHILE */
/* USER CODE BEGIN 3 */
}
/* USER CODE END 3 */

[그림 10.10] 메인 코드 작성

3.5. 디버그 및 다운로드하기

메인 함수에 프로그램이 작성되었다면 다음으로 디버그는 [그림 10.11]의 를 클릭하거나 단축키 F11을 누르면 디버깅과 프로그램 다운로드가 완료된다.

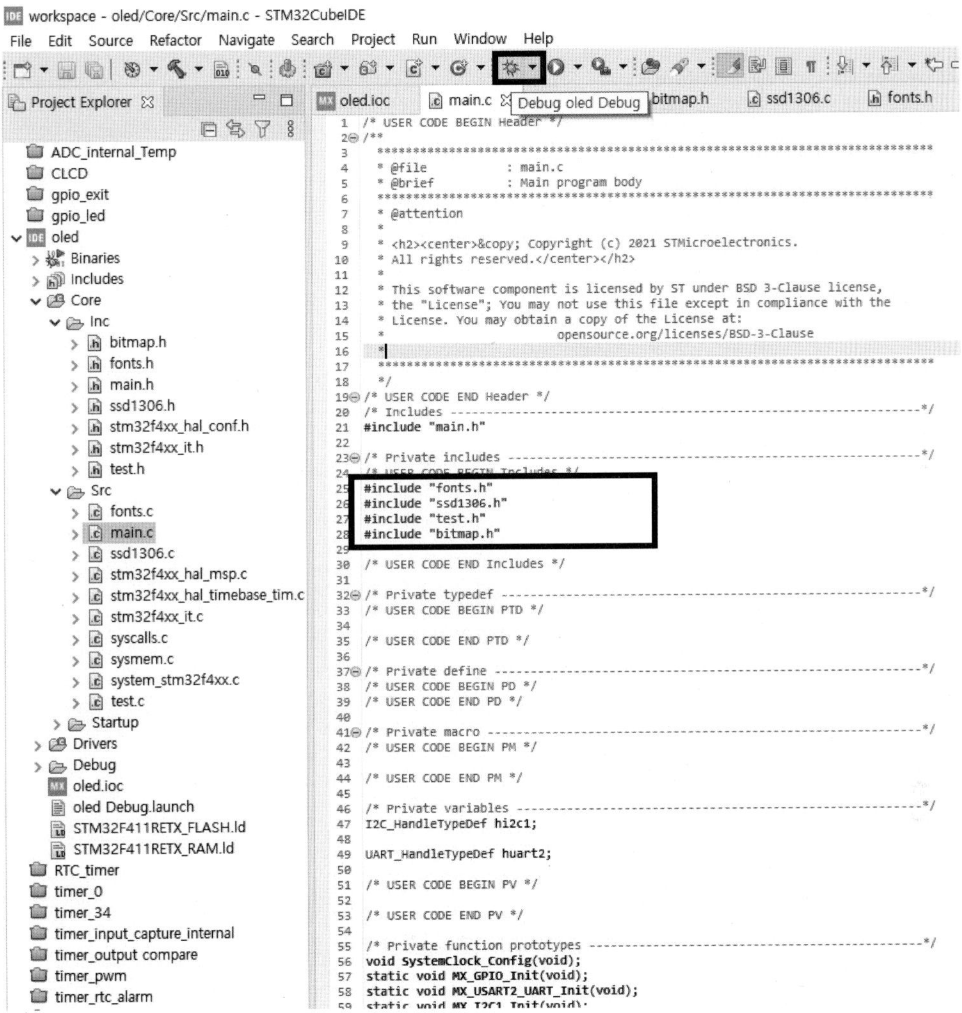

[그림 10.11] 프로그램 디버그

3.6. 보드 프로그램 실행하기

[그림 10.12] 상단 중앙에 아이콘 ■ 클릭하면 디버그 모드가 종료되어 보드의 프로그램이 실행된다. 폰트도 조정해 보고 예제를 보면 애니메이션도 있으니 실행해 보길 바란다.

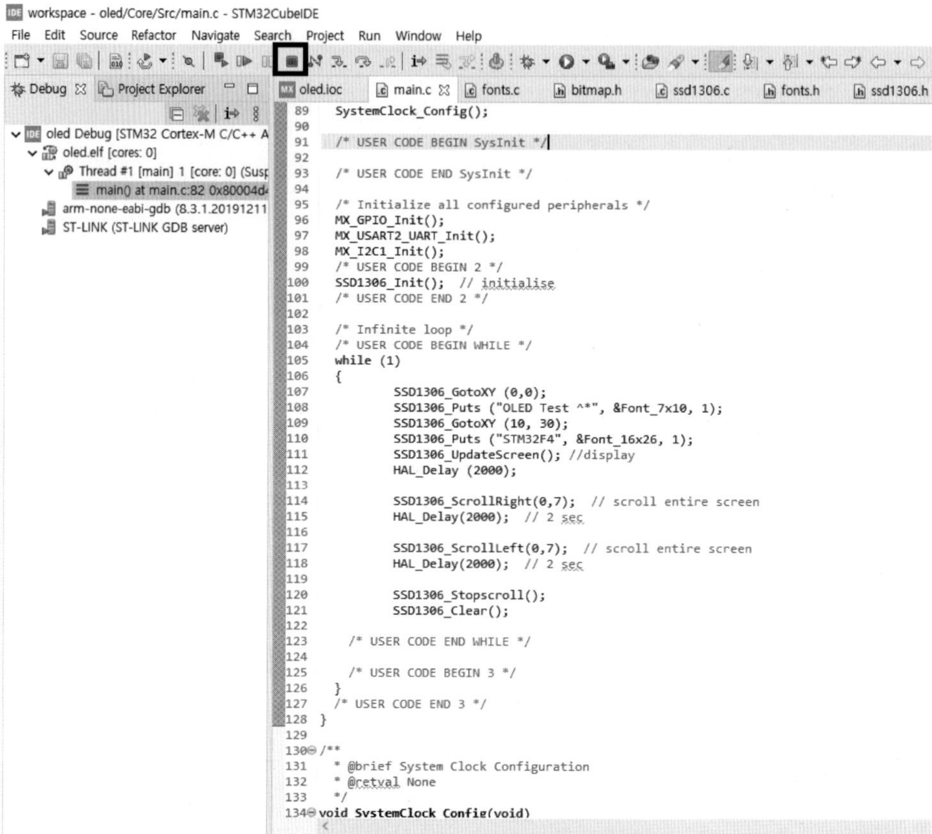
workspace - oled/Core/Src/main.c - STM32CubeIDE
File Edit Source Refactor Navigate Search Project Run Window Help
Debug
Project Explorer
oled.ioc
main.c
fonts.c
bitmap.h
ssd1306.c
fonts.h
ssd1306.h
oled Debug [STM32 Cortex-M C/C++ A
oled.elf [cores: 0]
Thread #1 [main] 1 [core: 0] (Susp
main() at main.c:82 0x80004d
arm-none-eabi-gdb (8.3.1.20191211
ST-LINK (ST-LINK GDB server)
SystemClock_Config();
/* USER CODE BEGIN SysInit */
/* USER CODE END SysInit */
/* Initialize all configured peripherals */
MX_GPIO_Init();
MX_USART2_UART_Init();
MX_I2C1_Init();
/* USER CODE BEGIN 2 */
SSD1306_Init(); // initialise
/* USER CODE END 2 */
/* Infinite loop */
/* USER CODE BEGIN WHILE */
while (1)
{
SSD1306_GotoXY (0,0);
SSD1306_Puts ("OLED Test ^*", &Font_7x10, 1);
SSD1306_GotoXY (10, 30);
SSD1306_Puts ("STM32F4", &Font_16x26, 1);
SSD1306_UpdateScreen(); //display
HAL_Delay (2000);
SSD1306_ScrollRight(0,7); // scroll entire screen
HAL_Delay(2000); // 2 sec
SSD1306_ScrollLeft(0,7); // scroll entire screen
HAL_Delay(2000); // 2 sec
SSD1306_Stopscroll();
SSD1306_Clear();
/* USER CODE END WHILE */
/* USER CODE BEGIN 3 */
}
/* USER CODE END 3 */
}
/**
* @brief System Clock Configuration
* @retval None
*/
void SystemClock_Config(void)

[그림 10.12] 디버그 모드 프로그램 실행

4. GLCD 활용하기

4.1. 하드웨어 설정

여기서는 시리얼(SPI)을 이용하여 GLCD(ST7920, 168*64) 사용 방법을 다룬다. 하드웨어 결선은 다음과 같다(SPI). [그림10.13]은 하드웨어 인터페이스를 나타낸다.

〉VCC=3.3V, GND=0V,

〉RST -〉STM32F4 (PB12)

〉SCLK-〉STM32F4 (PB13)

〉SID -〉STM32F4 (PB14)

〉CS -〉STM32F4 (PB15)

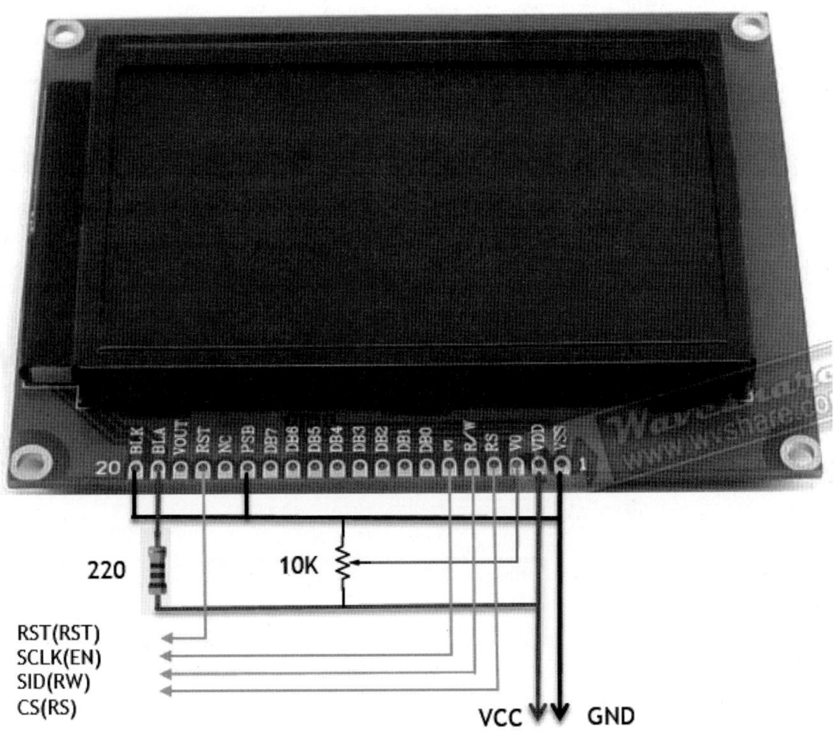

[그림 10.13] GLCD 인터페이스

4.2. GLCD 설정

클럭 및 시스템 설정은 이전과 동일하며 [그림 10.14]와 같이 GPIO 및 타이머를 설정한다.

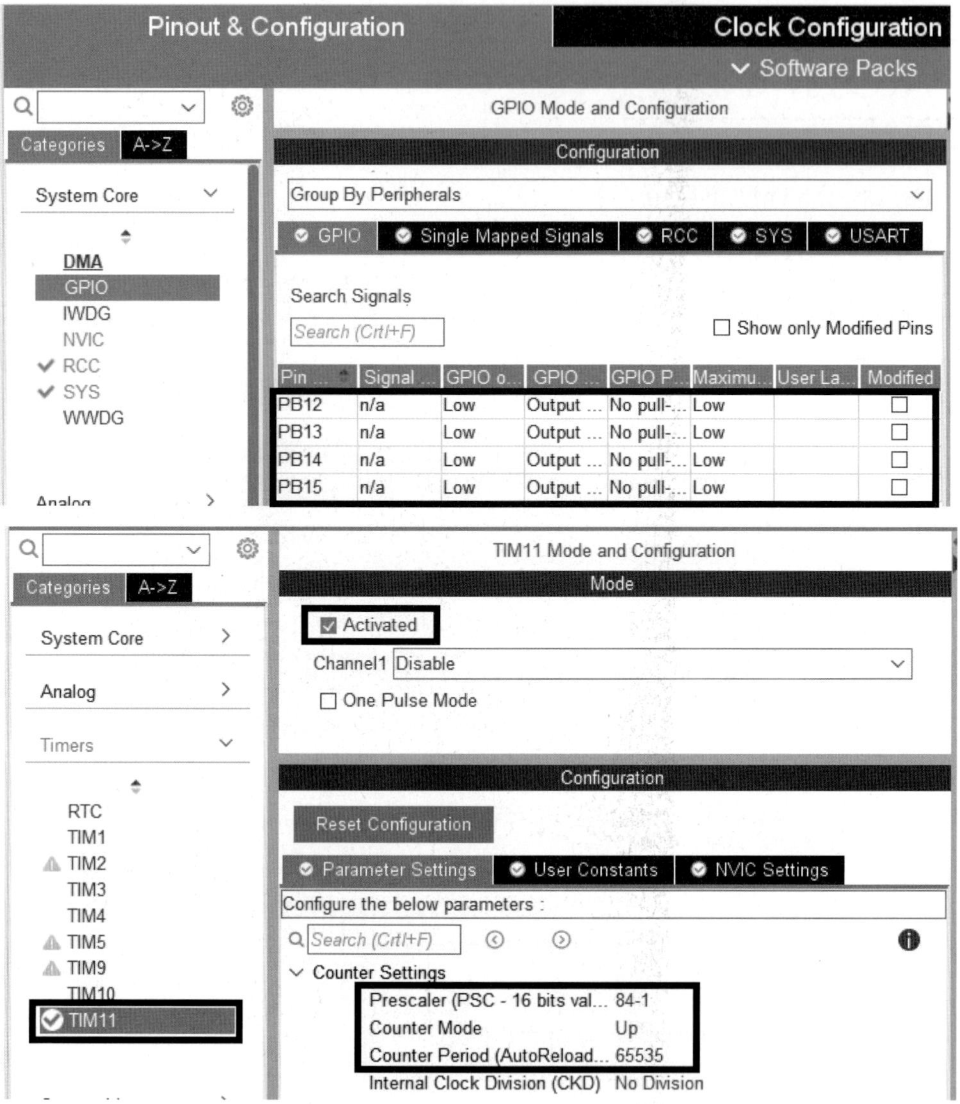

[그림 10.14] GLCD GPIO 및 타이머 설정

4.3. Generate Code 생성

[그림 10.15]는 코드를 생성하기 위한 화면이다. 그림과 같이 클릭해서 프로그램을 생성한다.

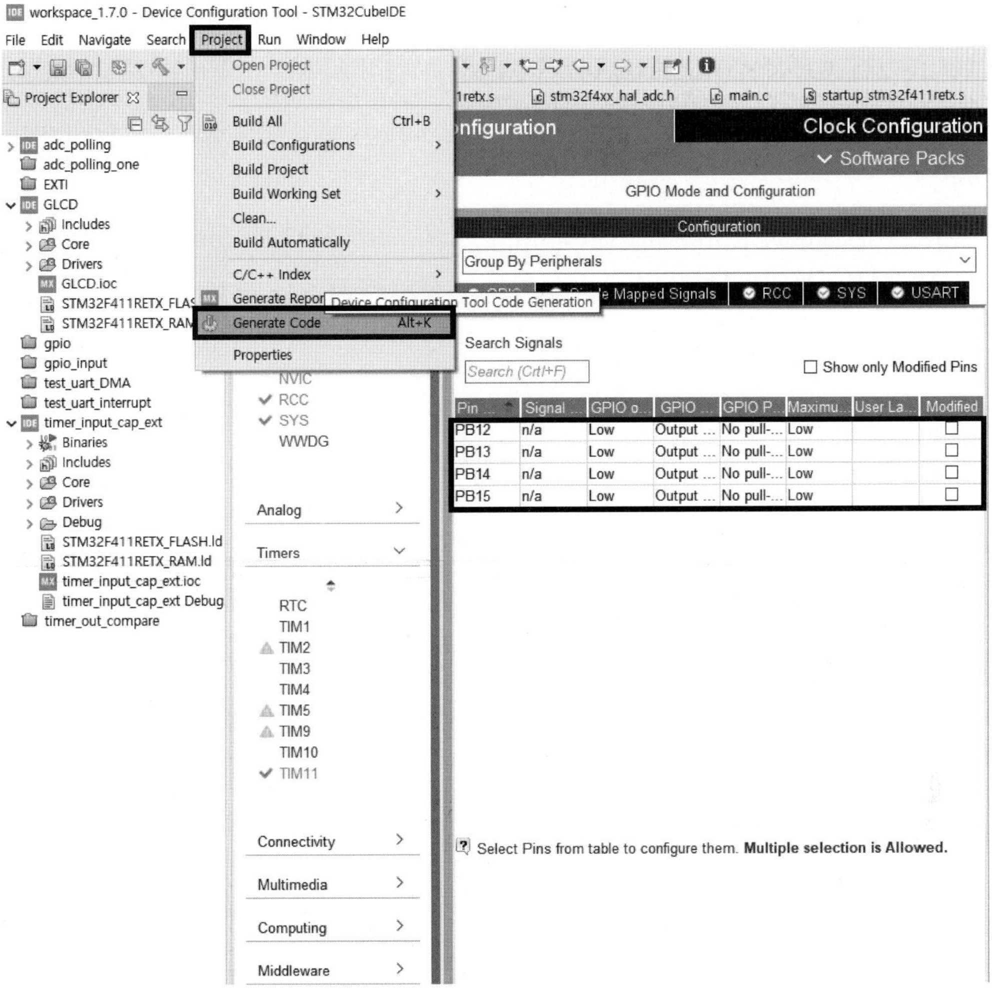

[그림 10.15] 프로그램 코드를 생성

4.4. Source Code 작성하기

Library를 활용하려면 인터넷에서 stm32 GLCD library 검색하고 그중에서 본인의 OLED에 맞는 Library를 활용하거나 수정(stm32f1xx_hal.h-〉stm32f4xx_hal.h)하여 사용하면 된다. 파일이 2, 3개씩(소스 파일, 헤더 파일)이 존재하는데 이를 프로그램 파일에 포함시켜 사용한다. [그림 10.16]과 같이 사용하고자 하는 해당 파일을 Core-〉Inc(헤더 파일)와 Src(소스 파일)에 복사해 두고 해당 명령어를 사용하여 코딩한다.

여기서는 https://controllerstech.com/glcd-128x64-st7920 -interfacing-with-stm32/에서 제공하는 소스 및 헤더 파일을 이용하여 예제를 작성해 본다.

[그림 10.16] 추가 파일 코드 수정(수정 내용)

```
#include "ST7920_SERIAL.h"
#include "delay.h"
#include "bitmap.h"
```

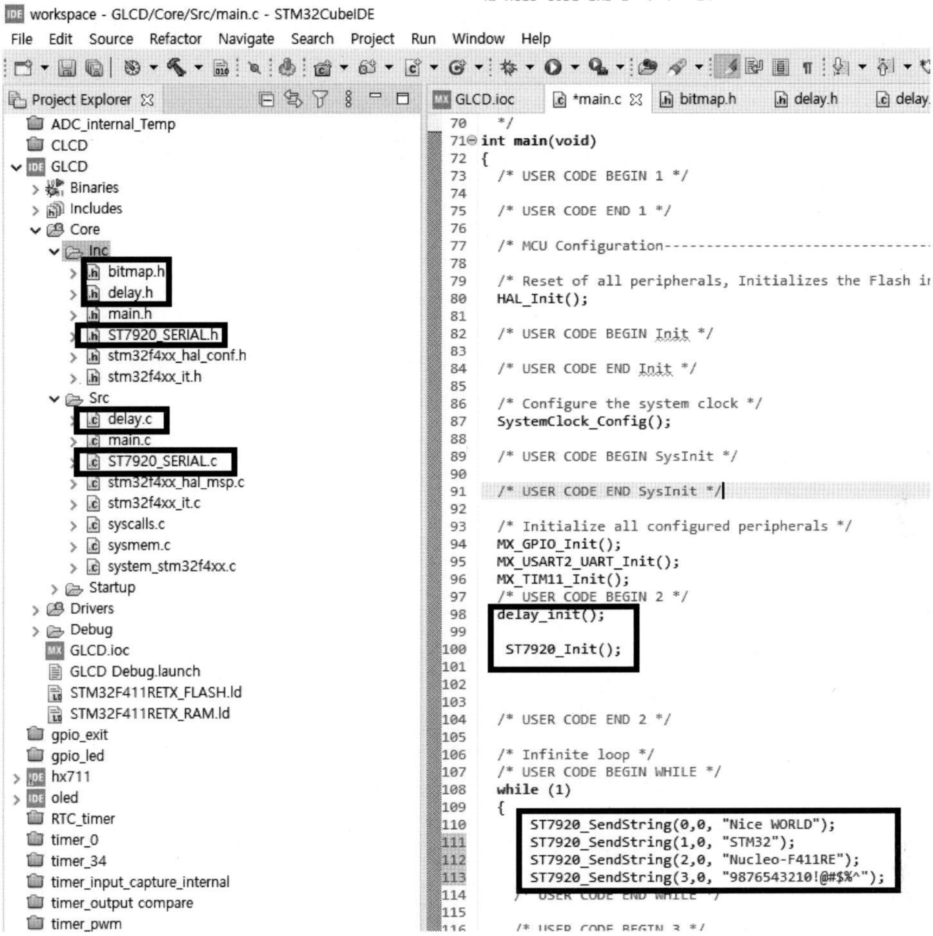

workspace - GLCD/Core/Src/main.c - STM32CubeIDE
File Edit Source Refactor Navigate Search Project Run Window Help
Project Explorer
ADC_internal_Temp
CLCD
GLCD
Binaries
Includes
Core
Inc
bitmap.h
delay.h
main.h
ST7920_SERIAL.h
stm32f4xx_hal_conf.h
stm32f4xx_it.h
Src
delay.c
main.c
ST7920_SERIAL.c
stm32f4xx_hal_msp.c
stm32f4xx_it.c
syscalls.c
sysmem.c
system_stm32f4xx.c
Startup
Drivers
Debug
GLCD.ioc
GLCD Debug.launch
STM32F411RETX_FLASH.ld
STM32F411RETX_RAM.ld
gpio_exit
gpio_led
hx711
oled
RTC_timer
timer_0
timer_34
timer_input_capture_internal
timer_output compare
timer_pwm
GLCD.ioc
*main.c
bitmap.h
delay.h
int main(void)
{
/* USER CODE BEGIN 1 */
/* USER CODE END 1 */
/* MCU Configuration--------------------------------
/* Reset of all peripherals, Initializes the Flash i
HAL_Init();
/* USER CODE BEGIN Init */
/* USER CODE END Init */
/* Configure the system clock */
SystemClock_Config();
/* USER CODE BEGIN SysInit */
/* USER CODE END SysInit */
/* Initialize all configured peripherals */
MX_GPIO_Init();
MX_USART2_UART_Init();
MX_TIM11_Init();
/* USER CODE BEGIN 2 */
delay_init();
ST7920_Init();
/* USER CODE END 2 */
/* Infinite loop */
/* USER CODE BEGIN WHILE */
while (1)
{
ST7920_SendString(0,0, "Nice WORLD");
ST7920_SendString(1,0, "STM32");
ST7920_SendString(2,0, "Nucleo-F411RE");
ST7920_SendString(3,0, "9876543210!@#$%^");

[그림 10.16_a] 메인 코드 작성(본문 내용)

4.5. 디버그 및 다운로드하기

메인 함수에 프로그램이 작성되었다면 다음으로 디버그는 [그림10.17]의 를 클릭하거나 단축키 F11을 누르면 디버깅과 프로그램 다운로드가 완료된다.

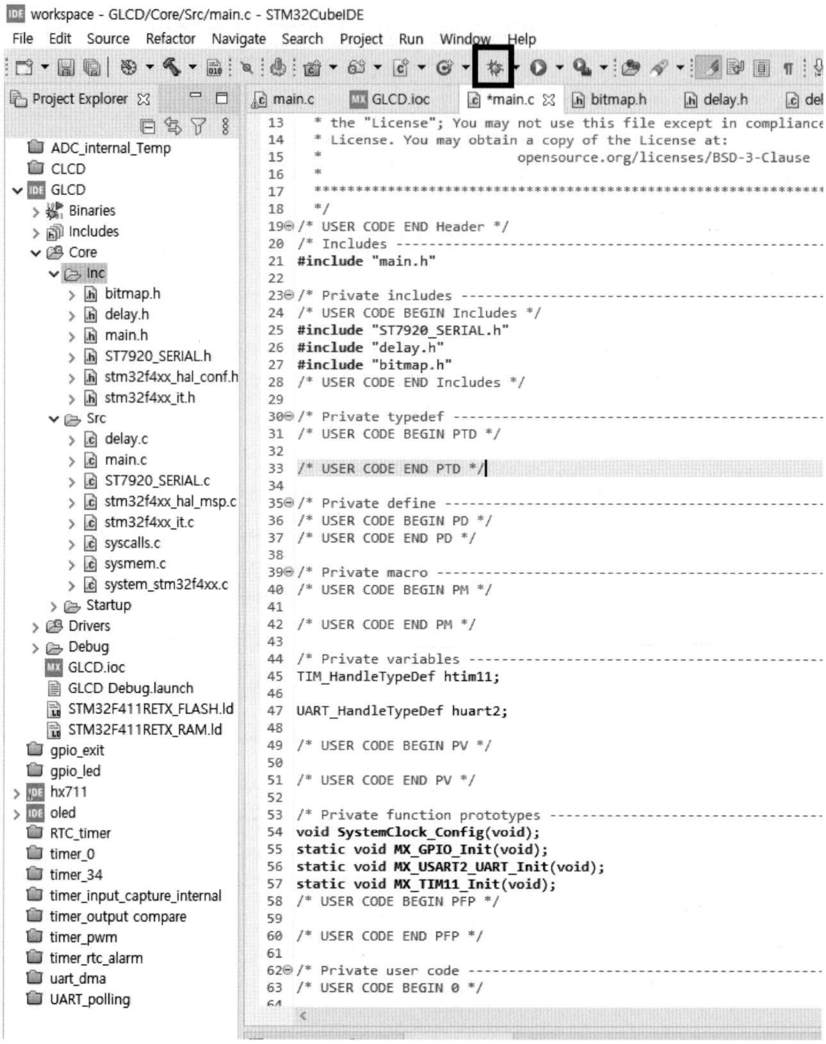

[그림 10.17] 프로그램 디버그

4.6. 보드 프로그램 실행하기

[그림 10.18] 상단 중앙에 아이콘 클릭하면 디버그 모드가 종료되어 보드의 프로그램이 실행된다. 폰트도 조정해 보고 예제를 보면 애니메이션도 있으니 실행해 보길 바란다.

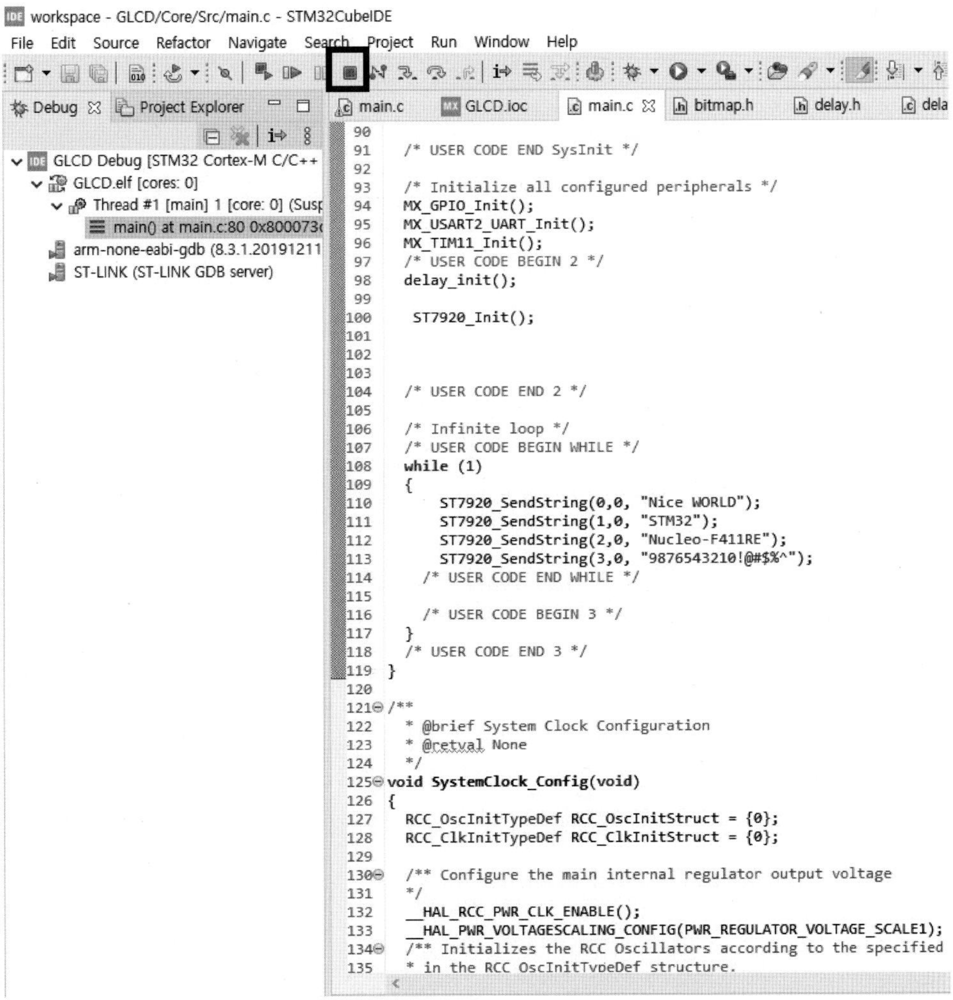
workspace - GLCD/Core/Src/main.c - STM32CubeIDE
File Edit Source Refactor Navigate Search Project Run Window Help
Debug
Project Explorer
main.c
GLCD.ioc
main.c
bitmap.h
delay.h
GLCD Debug [STM32 Cortex-M C/C++
GLCD.elf [cores: 0]
Thread #1 [main] 1 [core: 0] (Susp
main() at main.c:80 0x800073
arm-none-eabi-gdb (8.3.1.20191211
ST-LINK (ST-LINK GDB server)
/* USER CODE END SysInit */
/* Initialize all configured peripherals */
MX_GPIO_Init();
MX_USART2_UART_Init();
MX_TIM11_Init();
/* USER CODE BEGIN 2 */
delay_init();
ST7920_Init();
/* USER CODE END 2 */
/* Infinite loop */
/* USER CODE BEGIN WHILE */
while (1)
{
ST7920_SendString(0,0, "Nice WORLD");
ST7920_SendString(1,0, "STM32");
ST7920_SendString(2,0, "Nucleo-F411RE");
ST7920_SendString(3,0, "9876543210!@#$%^");
/* USER CODE END WHILE */
/* USER CODE BEGIN 3 */
}
/* USER CODE END 3 */
}
/**
* @brief System Clock Configuration
* @retval None
*/
void SystemClock_Config(void)
{
RCC_OscInitTypeDef RCC_OscInitStruct = {0};
RCC_ClkInitTypeDef RCC_ClkInitStruct = {0};
/** Configure the main internal regulator output voltage
*/
__HAL_RCC_PWR_CLK_ENABLE();
__HAL_PWR_VOLTAGESCALING_CONFIG(PWR_REGULATOR_VOLTAGE_SCALE1);
/** Initializes the RCC Oscillators according to the specified
* in the RCC_OscInitTypeDef structure.

[그림 10.18] 디버그 모드 프로그램 실행

CHAPTER 11

FreeRTOS 활용하기

CHAPTER 11

Cortex-M4 기초 활용(ARM Core STM32)

FreeRTOS 활용하기

1. 개요

프로그램 2개 이상을 서로 간섭을 받지 않고 독립적으로 동작시키고자 할 때 FreeRTOS의 도움을 받으면 쉽게 해결할 수 있다. 본서에서는 RTOS의 개념보다는 활용 측면을 다룬다. 나머지는 독자의 영역으로 남긴다.

2. FreeRTOS 실습 위한 하드웨어 설정

내부 클럭을 사용하고 별도의 외부 트리거 신호를 사용하지 않으면 하드웨어 없이 동작시킬 수 있다. 따라서 [그림 11.1]과 같이 LED 제어 용도로 사용한다.

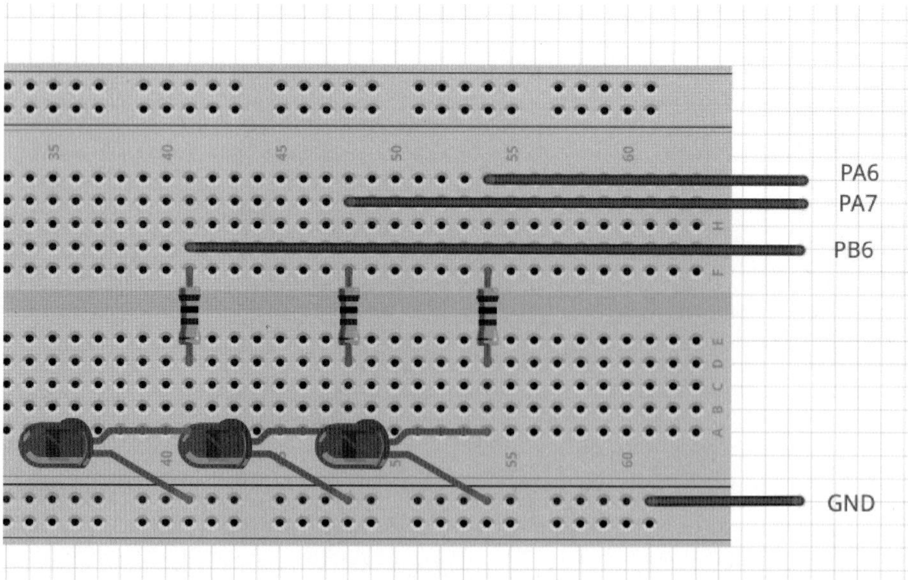

[그림 11.1] LED 제어

3. 클럭 및 SYS 설정

[그림 11.2]~[그림 11.3]은 클럭이 설정된 경우를 나타낸다. 메인 클럭은 외부 클럭(Crystal/Ceramic Resonator)과 내부 클럭(HSI RC(BYPASS Clock Source) 둘 중의 하나를 이용하여야 한다. 내부 클럭보다는 외부클럭이 정확하며, 본 보드에서는 Crystal/Ceramic Resonator는 연결되어 있지 않아 사용하지 않으며 ST-LINK MCO 출력 8MHz를 사용한다.

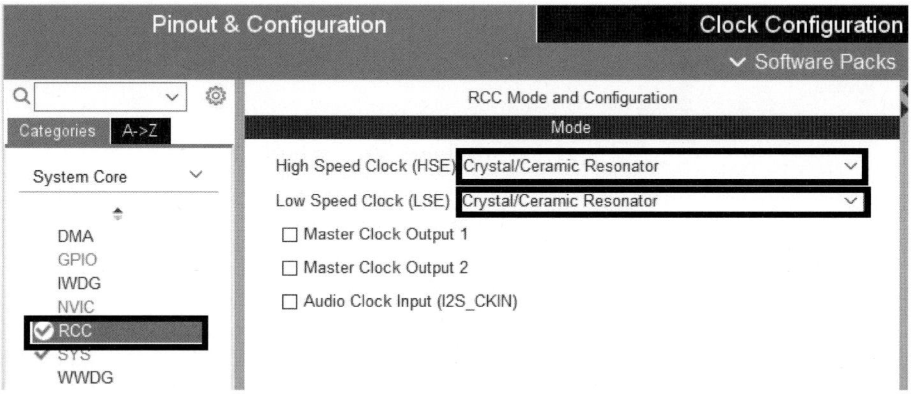

[그림 11.2] NUCLEO-F411RE 클럭 설정

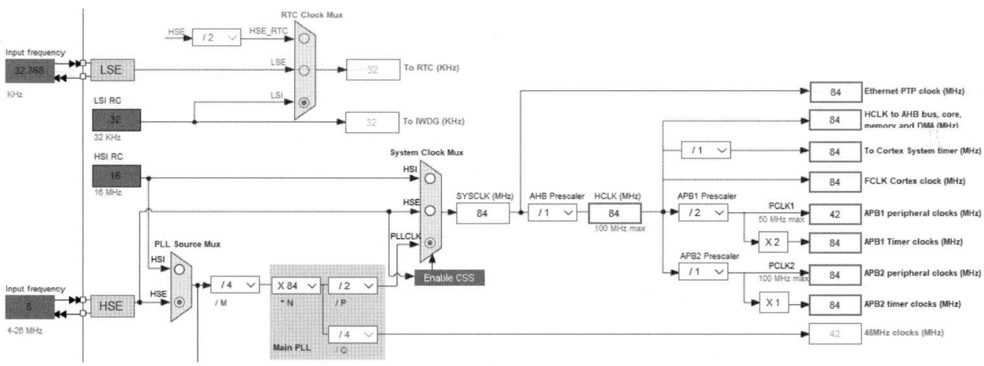

[그림 11.3] NUCLEO-F411RE 클럭 설정

[그림 11.4]는 Timebase 소스를 타이머 1로 설정한다. 일반적으로 Sys tick보다는 정확하다.

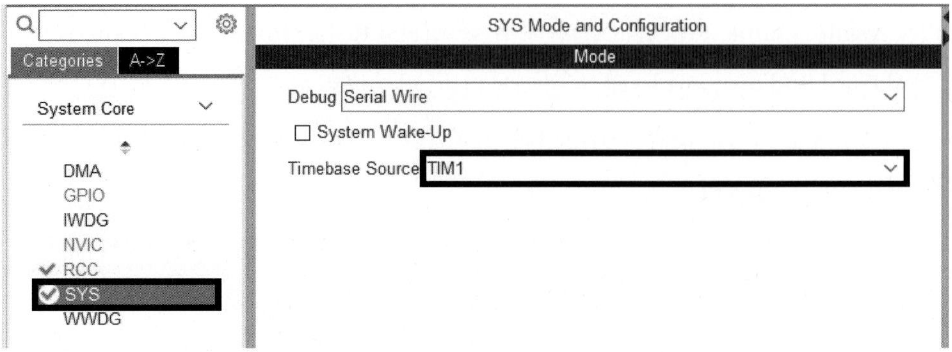

[그림 11.4] Timebase Source 설정

4. FreeRTOS 활용하기

4.1. 범용 FreeRTOS 설정

Middleware-〉 Interface를 선정하고, Configuration-〉Tasks and Queues로 들어간다. DefaultTask의 이름을 LD1Task로 수정하고 Priority를 osPriorityNormal로 설정한다.

그다음 하단에 Add를 클릭하여 Task를 2개 더 생성하고 Task 이름과 Priority를 [그림 11.5]~[그림 11.7]과 같이 설정한다.

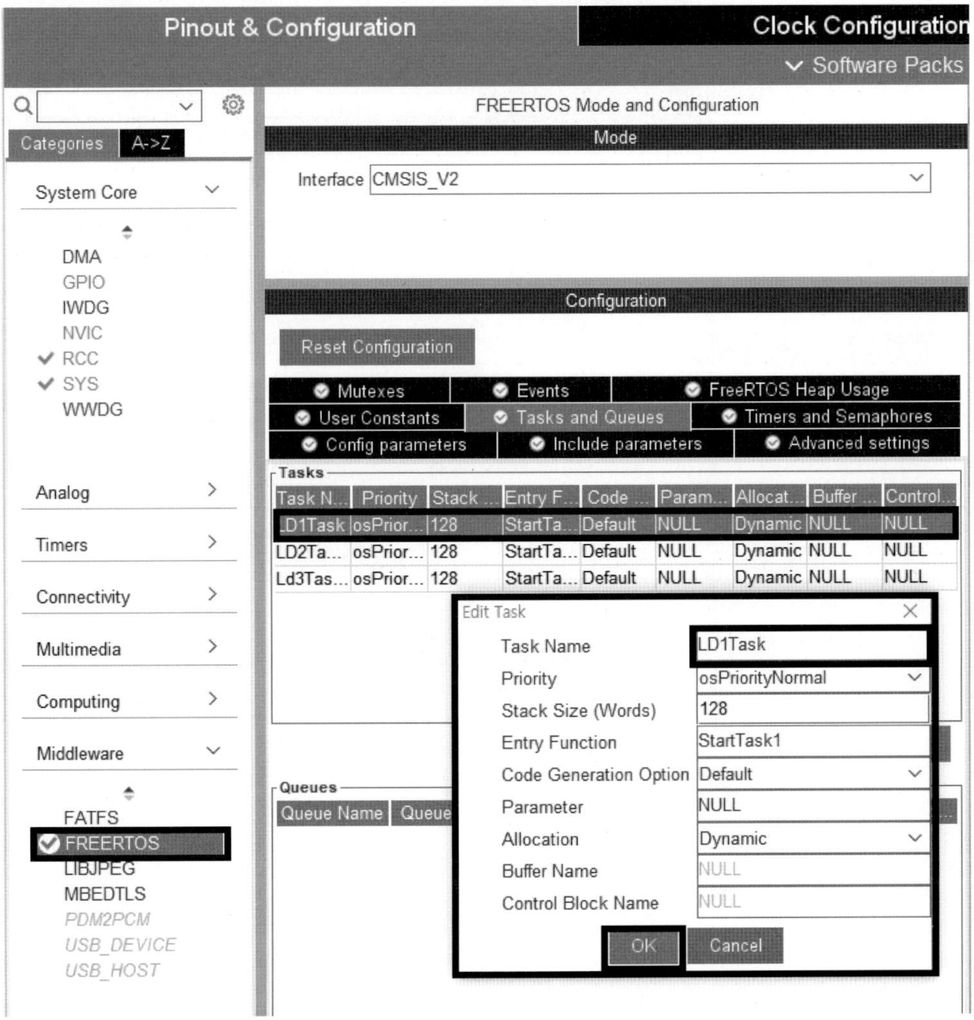

[그림 11.5] LD1Task 생성

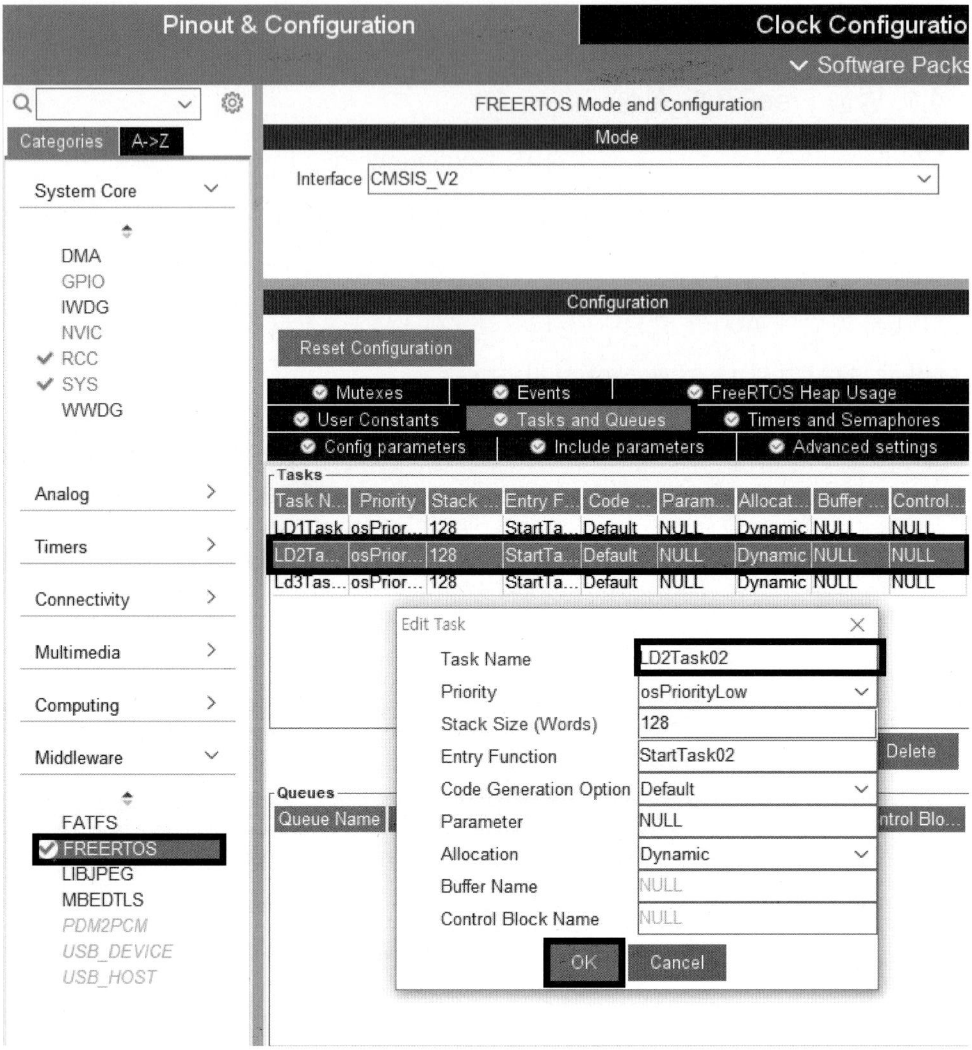

[그림 11.6] LD3Task02 생성

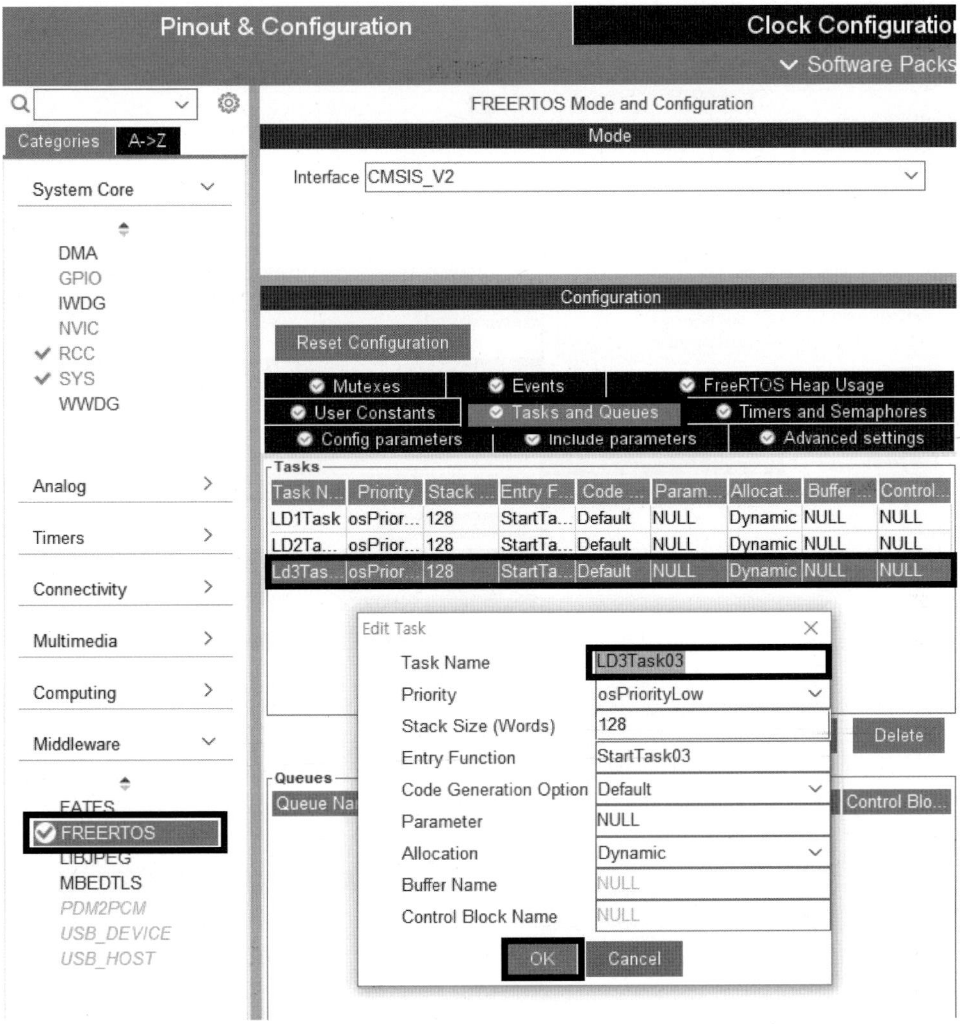

[그림 11.7] LD3Task03 생성

4.2. Generate Code 생성

[그림 11.8]은 코드를 생성하기 위한 화면이다. 그림과 같이 클릭해서 프로그램을 생성한다.

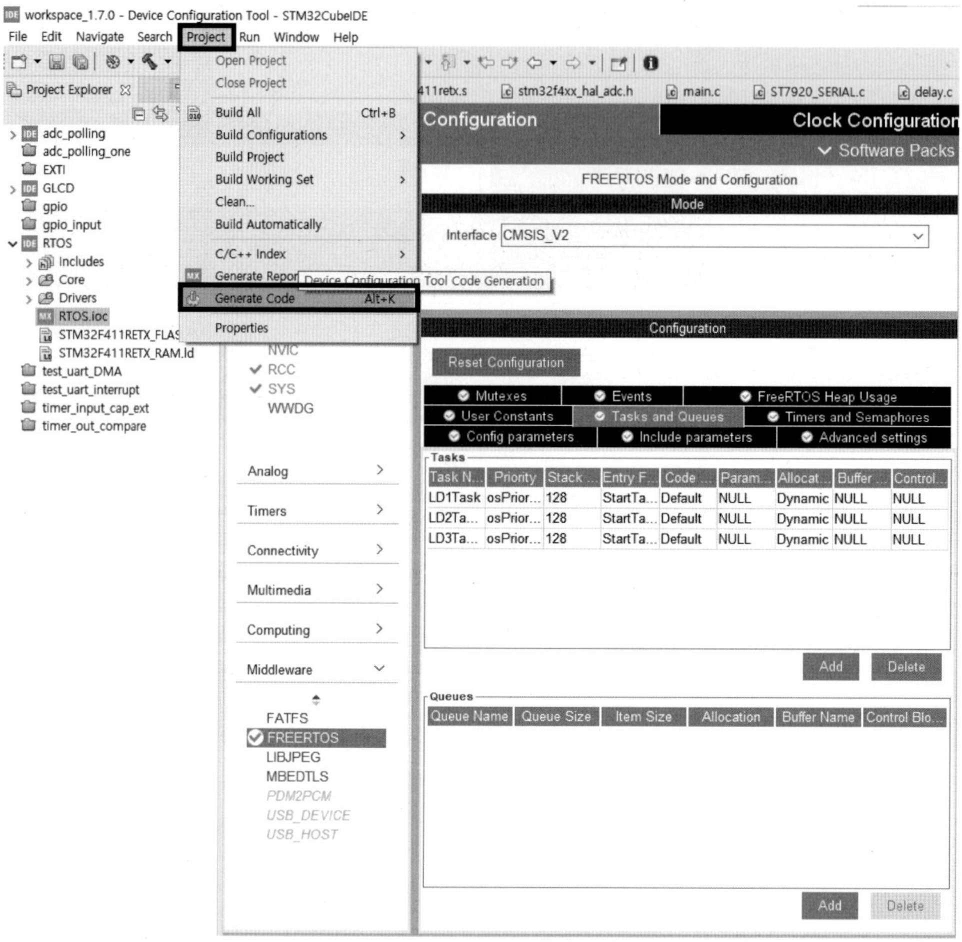

[그림 11.8] 프로그램 코드를 생성

4.3. Source Code 작성하기

이번 장에서는 LED 3개를 독립적으로 동작시키는 프로그램을 FreeRTOS를 통해서 구현한다. main.c 파일에는 task 핸들러들과 스레드, os 커널들이 설정되어 있으며 RTOS 프로그램을 작성할 수 있는 Task 함수들이 선언되어 있다. 이 함수들에 프로그램을 하면 된다. LED가 다음 시간 간격으로 토글 되도록 프로그램을 작성한다 [그림 11.9].

LD1=0.1s, LD2=0.2s, LD3=0.3s

```
void StartTask1(void *argument);
void StartTask02(void *argument);
void StartTask03(void *argument);
```

```
void StartTask1(void *argument)
{
  /* USER CODE BEGIN 5 */
  /* Infinite loop */
  for(;;)
  {
    HAL_GPIO_TogglePin(GPIOA, GPIO_PIN_6);
    printf("TASK1=0.1\r\n");
    osDelay(100);
  }
  /* USER CODE END 5 */
}

void StartTask02(void *argument)
{
  /* USER CODE BEGIN StartTask02 */
  /* Infinite loop */
  for(;;)
  {
    HAL_GPIO_TogglePin(GPIOA, GPIO_PIN_7);
    printf("TASK2=0.2\r\n");
    osDelay(200);
  }
  /* USER CODE END StartTask02 */
}

void StartTask03(void *argument)
{
  /* USER CODE BEGIN StartTask03 */
  /* Infinite loop */
  for(;;)
  {
    HAL_GPIO_TogglePin(GPIOB, GPIO_PIN_6);
    printf("TASK3=0.3\r\n");
    osDelay(300);
  }
  /* USER CODE END StartTask03 */
}
```

[그림 11.9] 주요 메인 코드 내용

4.4. 디버그 및 다운로드하기

메인 함수에 프로그램이 작성되었다면 다음으로 디버그는 [그림 11.10]의 를 클릭하거나 단축키 F11을 누르면 디버깅과 프로그램 다운로드가 완료된다.

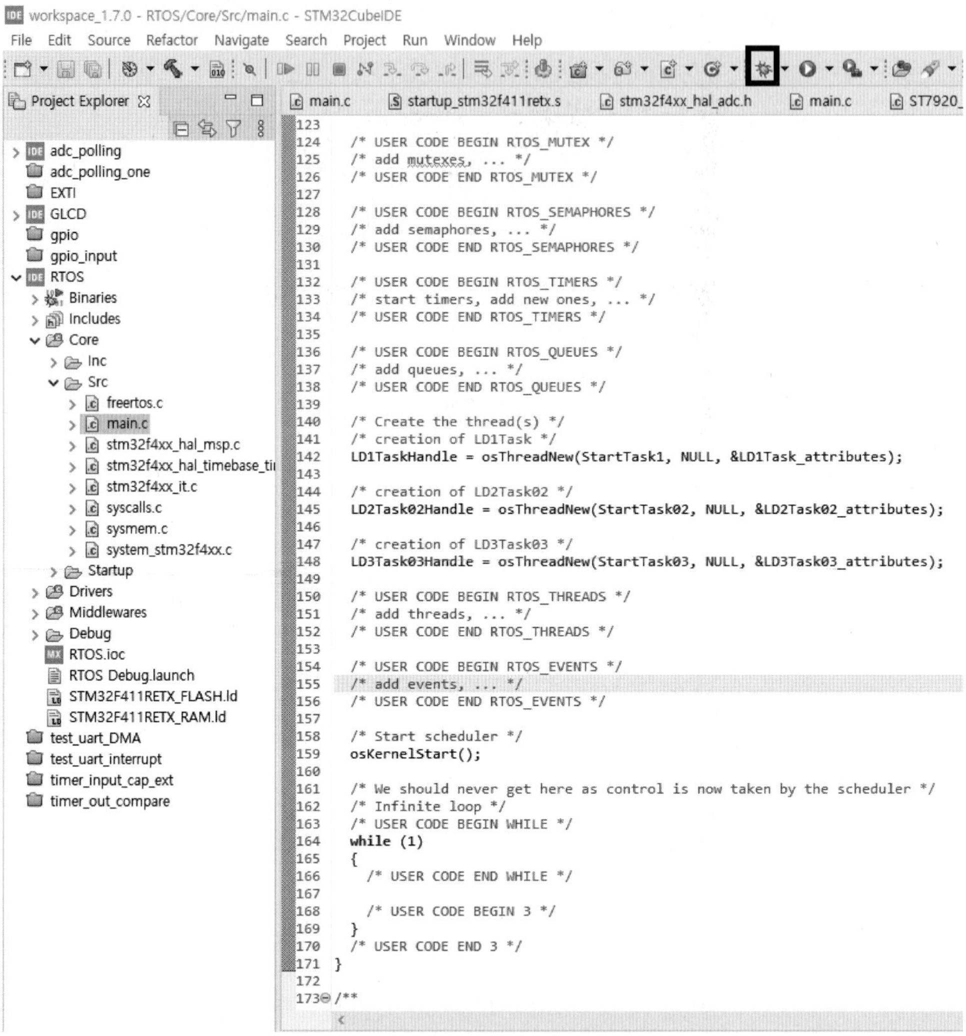

workspace_1.7.0 - RTOS/Core/Src/main.c - STM32CubeIDE
File Edit Source Refactor Navigate Search Project Run Window Help
Project Explorer
adc_polling
adc_polling_one
EXTI
GLCD
gpio
gpio_input
RTOS
Binaries
Includes
Core
Inc
Src
freertos.c
main.c
stm32f4xx_hal_msp.c
stm32f4xx_hal_timebase_ti
stm32f4xx_it.c
syscalls.c
sysmem.c
system_stm32f4xx.c
Startup
Drivers
Middlewares
Debug
RTOS.ioc
RTOS Debug.launch
STM32F411RETX_FLASH.ld
STM32F411RETX_RAM.ld
test_uart_DMA
test_uart_interrupt
timer_input_cap_ext
timer_out_compare
main.c
startup_stm32f411retx.s
stm32f4xx_hal_adc.h
main.c
ST7920_
/* USER CODE BEGIN RTOS_MUTEX */
/* add mutexes, ... */
/* USER CODE END RTOS_MUTEX */
/* USER CODE BEGIN RTOS_SEMAPHORES */
/* add semaphores, ... */
/* USER CODE END RTOS_SEMAPHORES */
/* USER CODE BEGIN RTOS_TIMERS */
/* start timers, add new ones, ... */
/* USER CODE END RTOS_TIMERS */
/* USER CODE BEGIN RTOS_QUEUES */
/* add queues, ... */
/* USER CODE END RTOS_QUEUES */
/* Create the thread(s) */
/* creation of LD1Task */
LD1TaskHandle = osThreadNew(StartTask1, NULL, &LD1Task_attributes);
/* creation of LD2Task02 */
LD2Task02Handle = osThreadNew(StartTask02, NULL, &LD2Task02_attributes);
/* creation of LD3Task03 */
LD3Task03Handle = osThreadNew(StartTask03, NULL, &LD3Task03_attributes);
/* USER CODE BEGIN RTOS_THREADS */
/* add threads, ... */
/* USER CODE END RTOS_THREADS */
/* USER CODE BEGIN RTOS_EVENTS */
/* add events, ... */
/* USER CODE END RTOS_EVENTS */
/* Start scheduler */
osKernelStart();
/* We should never get here as control is now taken by the scheduler */
/* Infinite loop */
/* USER CODE BEGIN WHILE */
while (1)
{
/* USER CODE END WHILE */
/* USER CODE BEGIN 3 */
}
/* USER CODE END 3 */
}
/**

[그림 11.10] 프로그램 디버그

4.5. 보드 프로그램 실행하기

[그림 11.11] 상단 중앙에 아이콘 ■ 클릭하면 디버그 모드가 종료되어 보드의 프로그램이 실행된다.

workspace_1.7.0 - RTOS/Core/Src/main.c - STM32CubeIDE
File Edit Source Refactor Navigate Search Project Run Window Help
De... Pro...
main.c ST7920_SERIAL.c delay.c delay.h main.h
RTOS Debug [STM32 Cortex-
RTOS.elf [cores: 0]
Thread #1 [main] 1 [co
main() at main.c:10
arm-none-eabi-gdb (8.3.1
ST-LINK (ST-LINK GDB ser
303 * @param argument: Not used
304 * @retval None
305 */
306 /* USER CODE END Header_StartTask1 */
307 void StartTask1(void *argument)
308 {
309 /* USER CODE BEGIN 5 */
310 /* Infinite loop */
311 for(;;)
312 {
313 HAL_GPIO_TogglePin(GPIOA, GPIO_PIN_6);
314 printf("TASK1=0.1\r\n");
315 osDelay(100);
316 }
317 /* USER CODE END 5 */
318 }
319
320 /* USER CODE BEGIN Header_StartTask02 */
321 /**
322 * @brief Function implementing the LD2Task02 thread.
323 * @param argument: Not used
324 * @retval None
325 */
326 /* USER CODE END Header_StartTask02 */
327 void StartTask02(void *argument)
328 {
329 /* USER CODE BEGIN StartTask02 */
330 /* Infinite loop */
331 for(;;)
332 {
333 HAL_GPIO_TogglePin(GPIOA, GPIO_PIN_7);
334 printf("TASK2=0.2\r\n");
335 osDelay(200);
336 }
337 /* USER CODE END StartTask02 */
338 }
339
340 /* USER CODE BEGIN Header_StartTask03 */
341 /**
342 * @brief Function implementing the LD3Task03 thread.
343 * @param argument: Not used
344 * @retval None
345 */
346 /* USER CODE END Header_StartTask03 */
347 void StartTask03(void *argument)
348 {
349 /* USER CODE BEGIN StartTask03 */
350 /* Infinite loop */
351 for(;;)
352 {
353 HAL_GPIO_TogglePin(GPIOB, GPIO_PIN_6);

[그림 11.11] 디버그 모드 프로그램 실행

4.6. 프로그램 실행 확인

LED 확인 및 시리얼 통신 [그림 11.12] 데이터를 확인할 수 있다.

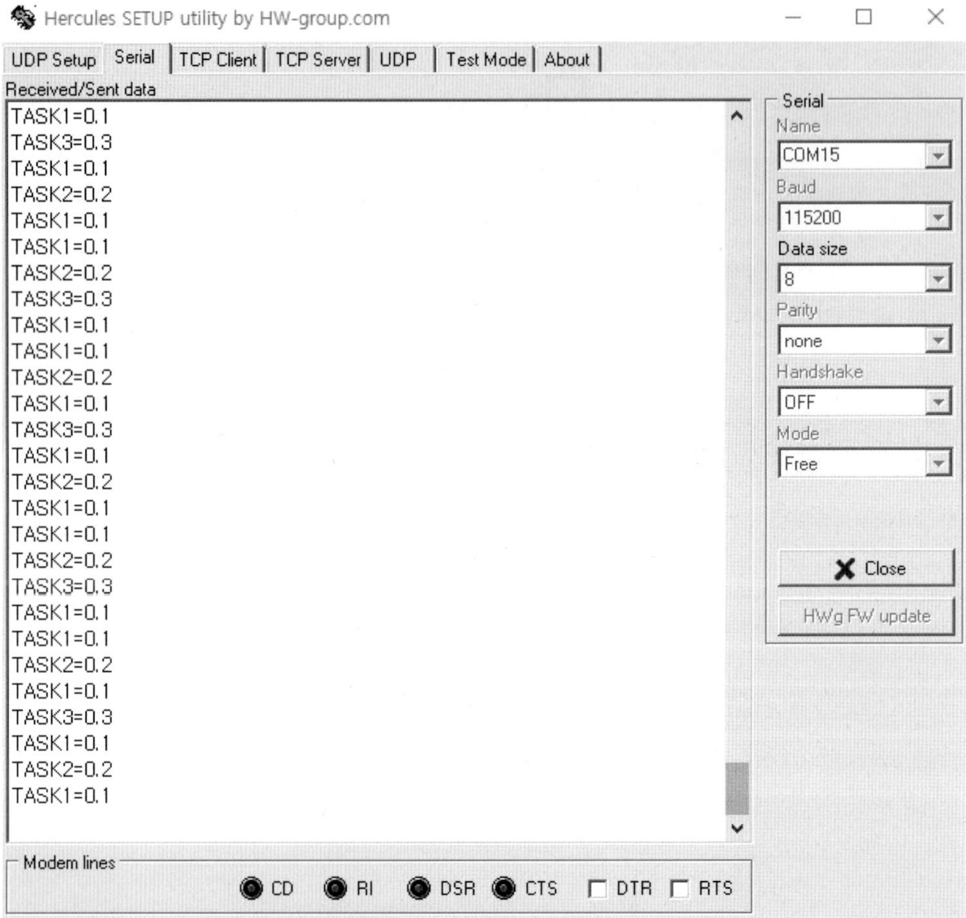

[그림 11.12] LED 3개 독립 동작

5. Binary Semaphore 활용하기

이번 장에서는 Task 3개를 우선순위에 의해서 프로그램이 돌아가도록 작성해 본다.

5.1. Binary Semaphore 설정

기본 설정은 범용 FreeRTOS와 동일하며 Middleware-> Interface를 선정하고, Configuration->Tasks and Queues로 들어간다. DefaultTask의 이름을 HighTask로 수정하고 Priority를 osPriorityHigh로 설정한다. 그다음 하단에 Add를 클릭하여 Task를 2개 더 생성하고 Task 이름과 Priority를 [그림 11.13]~[그림 11.14]와 같이 설정한다.

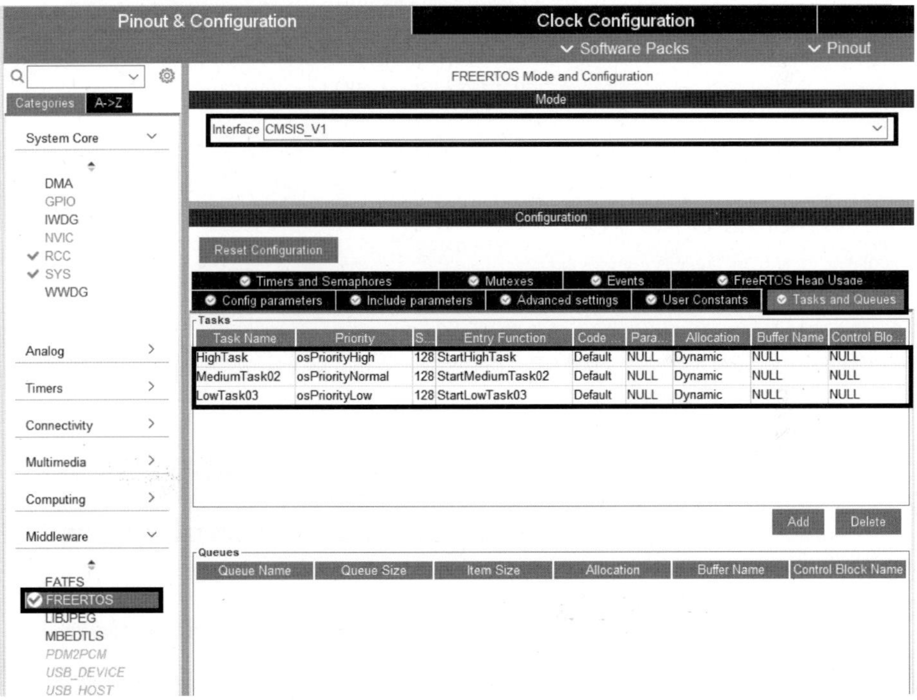

[그림 11.13] Task 3개 생성

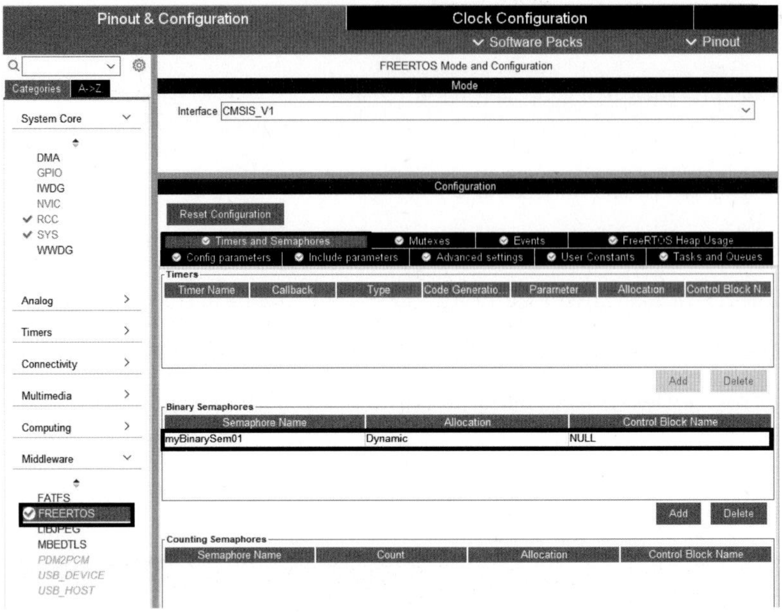

[그림 11.14] Task 생성

5.2. Generate Code 생성

[그림 11.15]는 코드를 생성하기 위한 화면이다. 그림과 같이 클릭해서 프로그램을 생성한다.

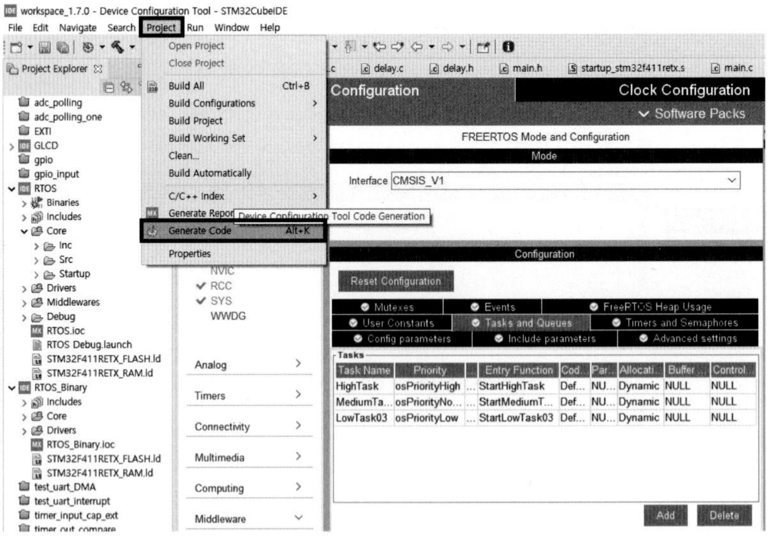

[그림 11.15] 프로그램 코드를 생성

5.3. Source Code 작성하기

FreeRTOS의 Task 우선순위는 [그림11.16]과 같다. [그림 11.17]은 Task 프로그램이다.

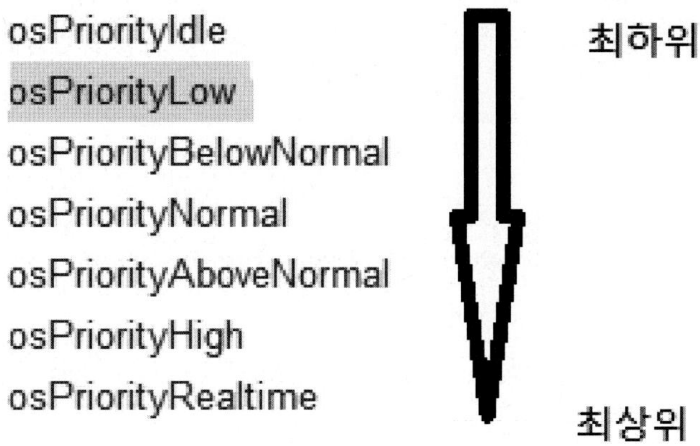

[그림 11.16] Task 우선순위

```
void StartHighTask(void const * argument)
{
  /* USER CODE BEGIN 5 */
  /* Infinite loop */
  for(;;)
  {
    printf(" Entered High Task\r\n");
    printf(" Leaving High Task\r\n");
    osDelay(200);
  }
  /* USER CODE END 5 */
}
```

```
void StartLowTask03(void const * argument)
{
  /* USER CODE BEGIN StartLowTask03 */
  /* Infinite loop */
  for(;;)
  {
      printf(" Entered Low Task\r\n");
      printf(" Leaving Low Task\r\n");
      osDelay(200);
  }
  /* USER CODE END StartLowTask03 */
}
```

```
void StartMediumTask02(void const * argument)
{
  /* USER CODE BEGIN StartMediumTask02 */
  /* Infinite loop */
  for(;;)
  {
      printf(" Entered Medium Task\r\n");
      printf(" Leaving Medium Task\r\n");
      osDelay(200);
  }
  /* USER CODE END StartMediumTask02 */
}
```

[그림 11.17] 주요 메인 코드 내용

5.4. 디버그 및 다운로드하기

메인 함수에 프로그램이 작성되었다면 다음으로 디버그는 [그림 11.18]의 를 클릭하거나 단축키 F11을 누르면 디버깅과 프로그램 다운로드가 완료된다.

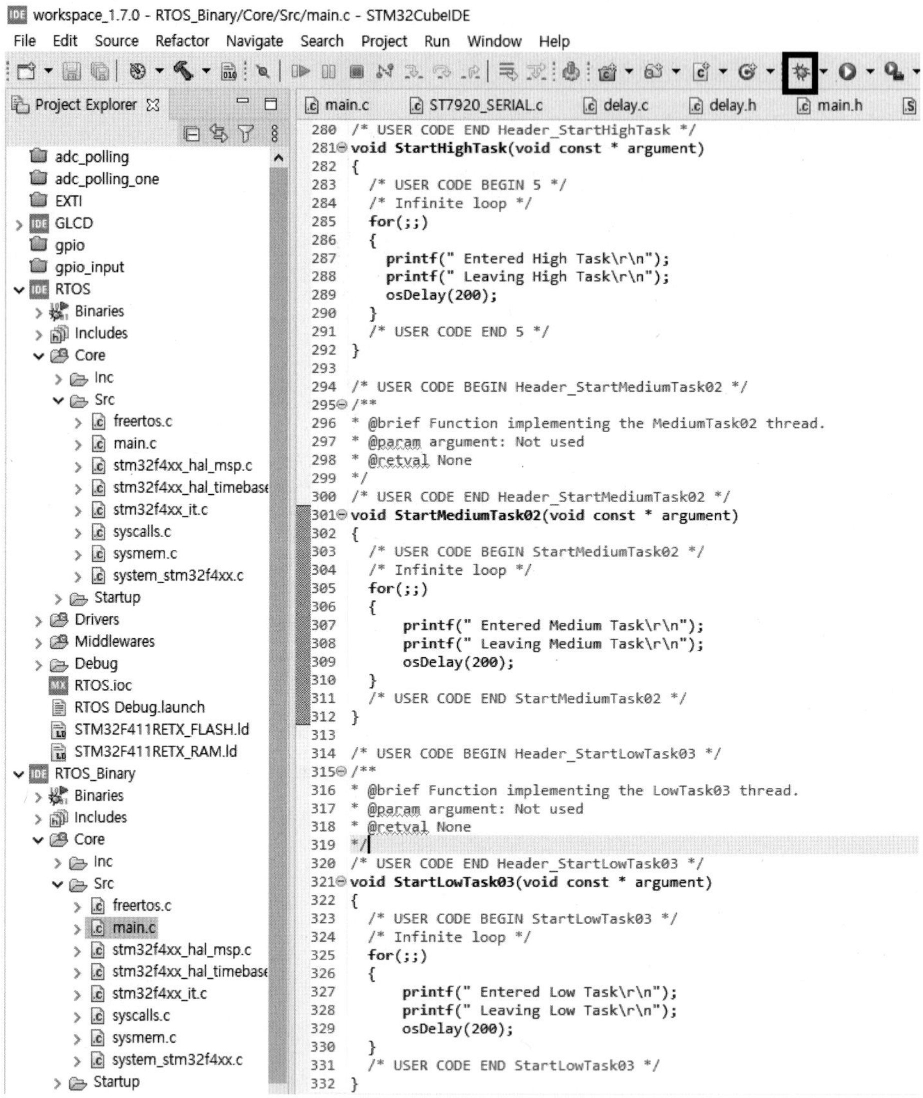
workspace_1.7.0 - RTOS_Binary/Core/Src/main.c - STM32CubeIDE
File Edit Source Refactor Navigate Search Project Run Window Help
Project Explorer
main.c
ST7920_SERIAL.c
delay.c
delay.h
main.h
adc_polling
adc_polling_one
EXTI
GLCD
gpio
gpio_input
RTOS
Binaries
Includes
Core
Inc
Src
freertos.c
main.c
stm32f4xx_hal_msp.c
stm32f4xx_hal_timebase
stm32f4xx_it.c
syscalls.c
sysmem.c
system_stm32f4xx.c
Startup
Drivers
Middlewares
Debug
RTOS.ioc
RTOS Debug.launch
STM32F411RETX_FLASH.ld
STM32F411RETX_RAM.ld
RTOS_Binary
Binaries
Includes
Core
Inc
Src
freertos.c
main.c
stm32f4xx_hal_msp.c
stm32f4xx_hal_timebase
stm32f4xx_it.c
syscalls.c
sysmem.c
system_stm32f4xx.c
Startup
/* USER CODE END Header_StartHighTask */
void StartHighTask(void const * argument)
{
/* USER CODE BEGIN 5 */
/* Infinite loop */
for(;;)
{
printf(" Entered High Task\r\n");
printf(" Leaving High Task\r\n");
osDelay(200);
}
/* USER CODE END 5 */
}
/* USER CODE BEGIN Header_StartMediumTask02 */
/**
* @brief Function implementing the MediumTask02 thread.
* @param argument: Not used
* @retval None
*/
/* USER CODE END Header_StartMediumTask02 */
void StartMediumTask02(void const * argument)
{
/* USER CODE BEGIN StartMediumTask02 */
/* Infinite loop */
for(;;)
{
printf(" Entered Medium Task\r\n");
printf(" Leaving Medium Task\r\n");
osDelay(200);
}
/* USER CODE END StartMediumTask02 */
}
/* USER CODE BEGIN Header_StartLowTask03 */
/**
* @brief Function implementing the LowTask03 thread.
* @param argument: Not used
* @retval None
*/
/* USER CODE END Header_StartLowTask03 */
void StartLowTask03(void const * argument)
{
/* USER CODE BEGIN StartLowTask03 */
/* Infinite loop */
for(;;)
{
printf(" Entered Low Task\r\n");
printf(" Leaving Low Task\r\n");
osDelay(200);
}
/* USER CODE END StartLowTask03 */
}

[그림 11.18] 프로그램 디버그

5.5. 보드 프로그램 실행하기

[그림 11.19] 상단 중앙에 아이콘 클릭하면 디버그 모드가 종료되어 보드의 프로그램이 실행된다.

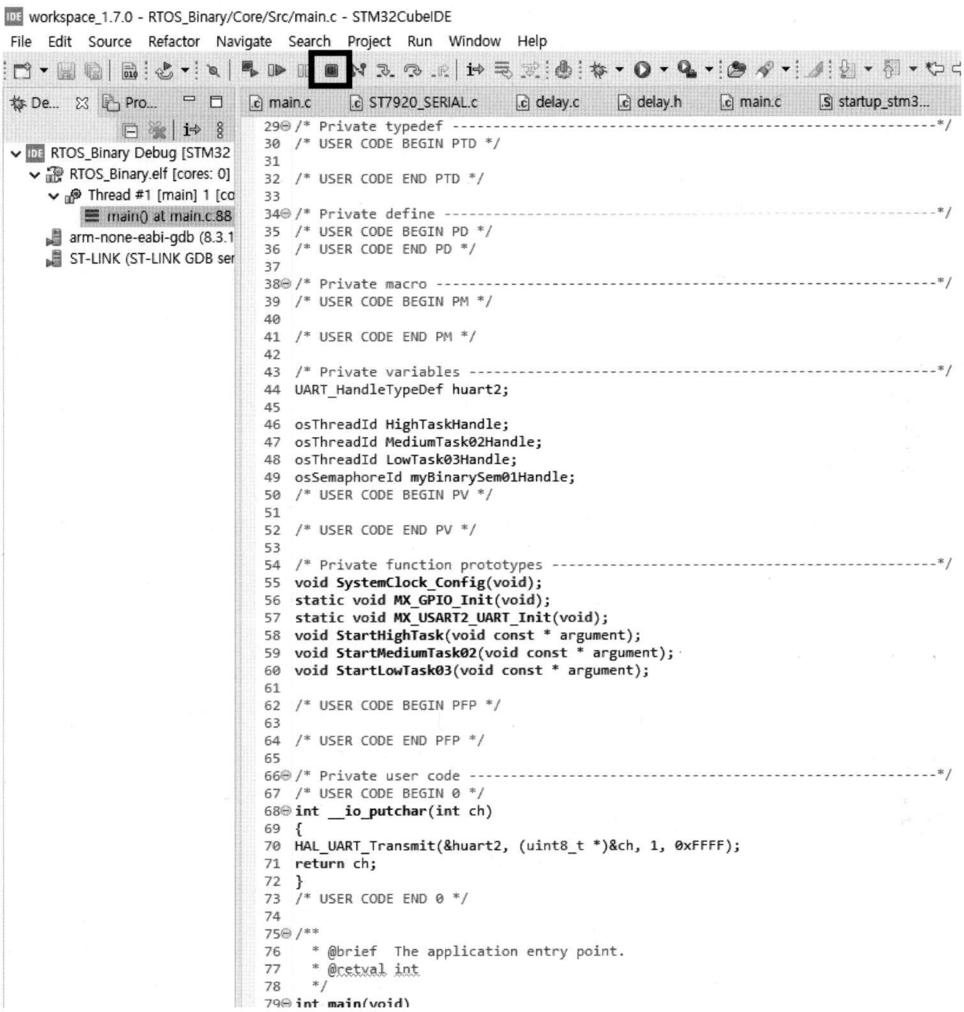

workspace_1.7.0 - RTOS_Binary/Core/Src/main.c - STM32CubeIDE
File Edit Source Refactor Navigate Search Project Run Window Help
De... Pro...
main.c ST7920_SERIAL.c delay.c delay.h main.c startup_stm3...
RTOS_Binary Debug [STM32
RTOS_Binary.elf [cores: 0]
Thread #1 [main] 1 [co
main() at main.c:88
arm-none-eabi-gdb (8.3.1
ST-LINK (ST-LINK GDB ser
/* Private typedef ---*/
/* USER CODE BEGIN PTD */
/* USER CODE END PTD */
/* Private define --*/
/* USER CODE BEGIN PD */
/* USER CODE END PD */
/* Private macro ---*/
/* USER CODE BEGIN PM */
/* USER CODE END PM */
/* Private variables ---*/
UART_HandleTypeDef huart2;
osThreadId HighTaskHandle;
osThreadId MediumTask02Handle;
osThreadId LowTask03Handle;
osSemaphoreId myBinarySem01Handle;
/* USER CODE BEGIN PV */
/* USER CODE END PV */
/* Private function prototypes ---*/
void SystemClock_Config(void);
static void MX_GPIO_Init(void);
static void MX_USART2_UART_Init(void);
void StartHighTask(void const * argument);
void StartMediumTask02(void const * argument);
void StartLowTask03(void const * argument);
/* USER CODE BEGIN PFP */
/* USER CODE END PFP */
/* Private user code ---*/
/* USER CODE BEGIN 0 */
int __io_putchar(int ch)
{
HAL_UART_Transmit(&huart2, (uint8_t *)&ch, 1, 0xFFFF);
return ch;
}
/* USER CODE END 0 */
/**
* @brief The application entry point.
* @retval int
*/
int main(void)

[그림 11.19] 디버그 모드 프로그램 실행

5.6. 프로그램 실행 확인

LED 확인 및 시리얼 통신 [그림 11.20] 데이터를 확인할 수 있다.

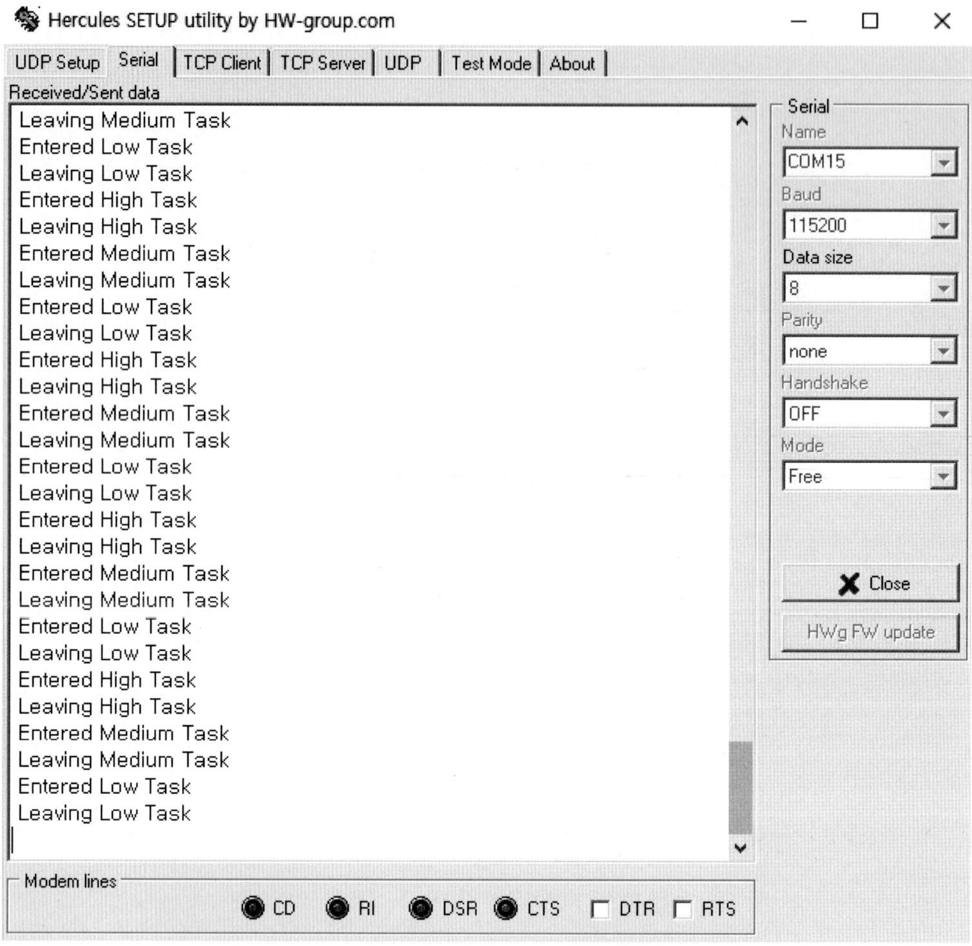

[그림 11.20] Task 우선순위에 따라 출력

5.7. Medium Task에 스위치 입력 부가 및 실행

보드에 user B1 sw[PC13]을 [그림 11.21]과 같이 입력으로 지정하고 [그림 11.22]와 같이 프로그램 내용을 수정하고 실행해 본다.

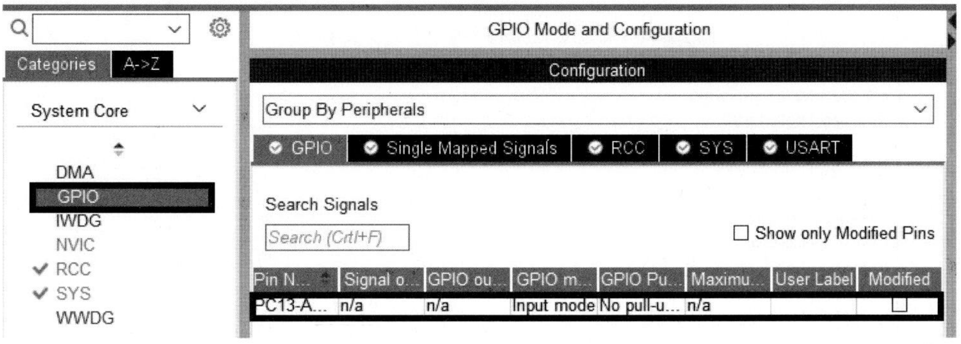

[그림 11.21] User sw 입력 모드 설정

```
void StartMediumTask02(void const * argument)
{
  /* USER CODE BEGIN StartMediumTask02 */
  /* Infinite loop */
  for(;;)
  {
      printf(" Entered Medium Task\r\n");
      while(HAL_GPIO_ReadPin(GPIOC,GPIO_PIN_13));
      printf(" Leaving Medium Task\r\n");
      osDelay(200);
  }
  /* USER CODE END StartMediumTask02 */
}
```

[그림 11.22] User sw 입력 모드 설정

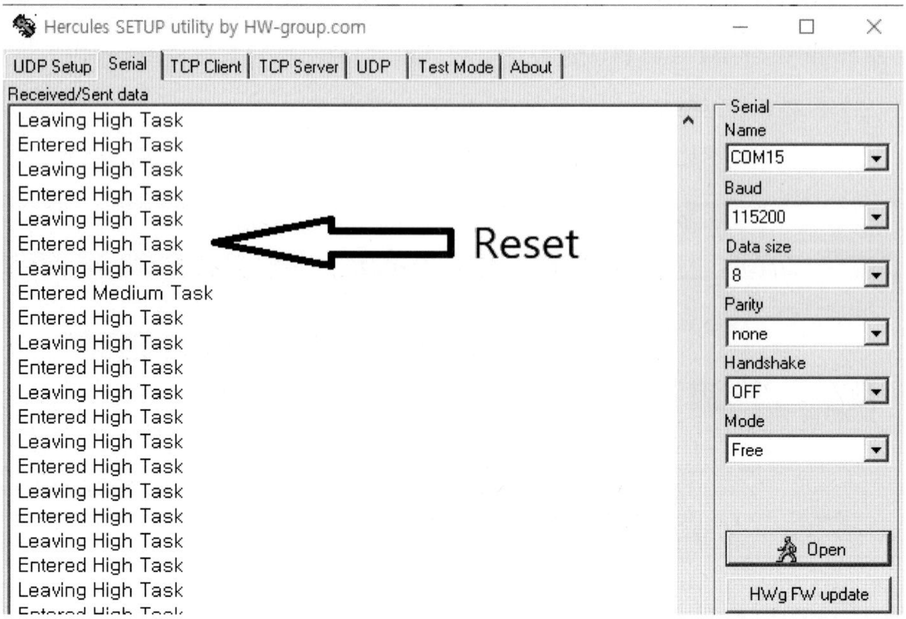

[그림 11.23] Reset/SW 이후 Task 동작

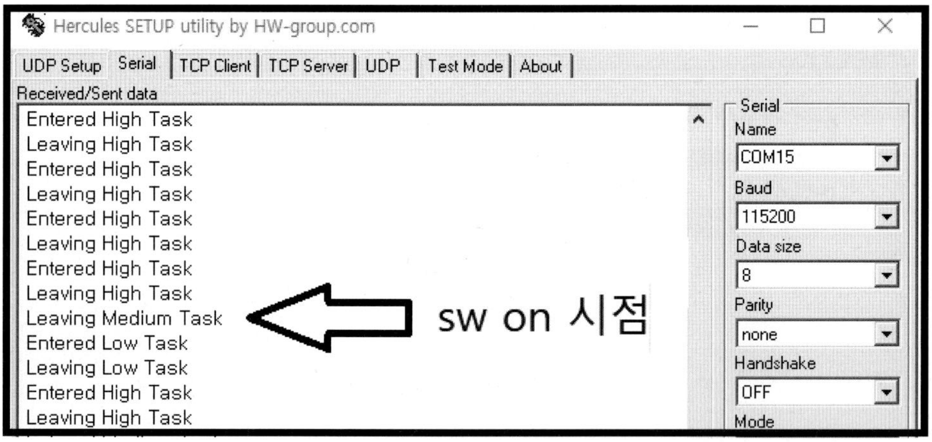

[그림 11.24] SW 동작에 따른 Task 동작

[그림 11.23]~[그림 11.24]를 보면 우선순위가 높은 순위대로 동작하기 때문인데 Reset 이후에 우선순위가 가장 높은 순서대로 다음과 같이(Entered High Task->leaving High Task ->Entered Medium Task) 동작하며 스위치 입력 대기 중 신

호를 만나서 입력 신호가 0.2초 이후가 지나면 우선순위가 가장 높은 Task로 이동하여 동작하게 된다. 만약 스위치가 On/Off 되면 스위치 이후의 동작, 즉 leaving Medium Task-〉 Entered Low Task-〉leaving Low Task-〉이 동작되고 이후 동작은 Reset 후의 동작으로 돌아간다. 결과적으로 위의 프로그램으로는 우리가 기대하는 동작을 얻기 위해서는 최적의 시간을 기다려 그때마다 스위치를 on/off 하면 되지만 수동으로는 불가능하다. 따라서 우리가 기대하는 동작을 얻기 위해 Binary Semiphore 개념은 도입한다.

5.8. Binary Semaphore 개념 도입

프로그램을 다음 [그림 11.25]~[그림 11.27]과 같이 수정한다. 결과는 [그림 11.28]~[그림 11.30]과 같다.

```
void StartHighTask(void const * argument)
{
  /* USER CODE BEGIN 5 */
  /* Infinite loop */
  for(;;)
  {
    printf(" Entered HighTask and Waiting for Semaphore\r\n");
    osSemaphoreWait(myBinarySem01Handle, osWaitForever);
    printf("Semaphore acquired by HighTask\r\n");
    printf(" Leaving HighTask and Semaphore\r\n");
    osSemaphoreRelease(myBinarySem01Handle);
    osDelay(200);
  }
  /* USER CODE END 5 */
}
```

[그림 11.25] High Task 프로그램

```
void StartMediumTask02(void const * argument)
{
  /* USER CODE BEGIN StartMediumTask02 */
  /* Infinite loop */
  for(;;)
  {
      printf(" Entered MediumTask and Waiting for Semaphore\r\n");
      osSemaphoreWait(myBinarySem01Handle, osWaitForever);
      printf("Semaphore acquired by MediumTask\r\n");
      while(HAL_GPIO_ReadPin(GPIOC,GPIO_PIN_13));
      printf(" Leaving MediumTask and Semaphore\r\n");
      osSemaphoreRelease(myBinarySem01Handle);
      osDelay(200);
  }
  /* USER CODE END StartMediumTask02 */
}
```

[그림 11.26] Medium 프로그램

```
void StartLowTask03(void const * argument)
{
  /* USER CODE BEGIN StartLowTask03 */
  /* Infinite loop */
  for(;;)
  {
      printf(" Entered LowTask\r\n");
      printf(" Leaving LowTask\r\n");
      osDelay(200);
  }
  /* USER CODE END StartLowTask03 */
}
```

[그림 11.27] LowTask 프로그램

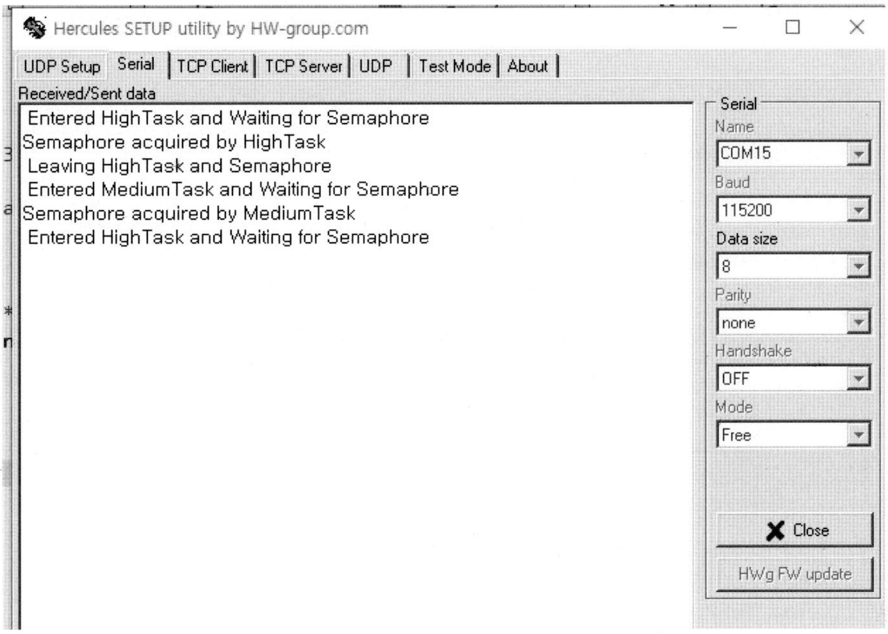

[그림 11.28] Reset 이후 Task 동작

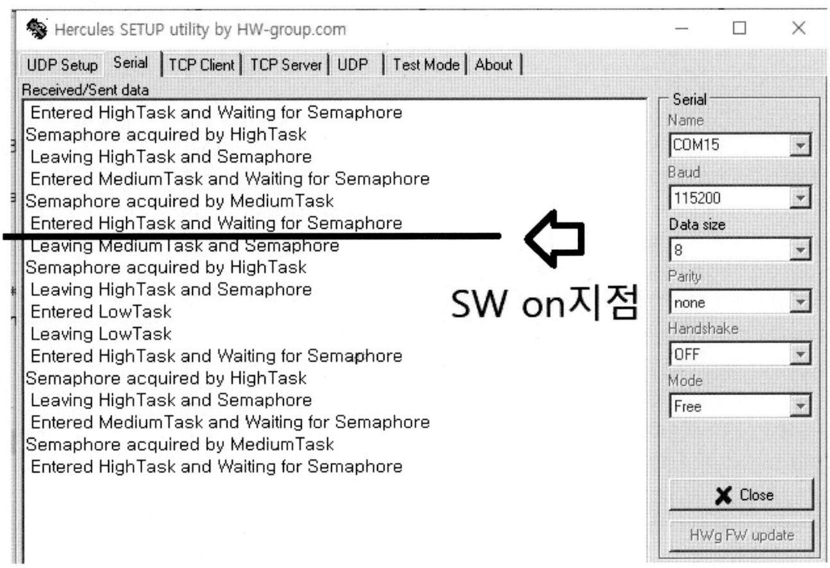

[그림 11.29] sw on/offt 이후 Task 동작

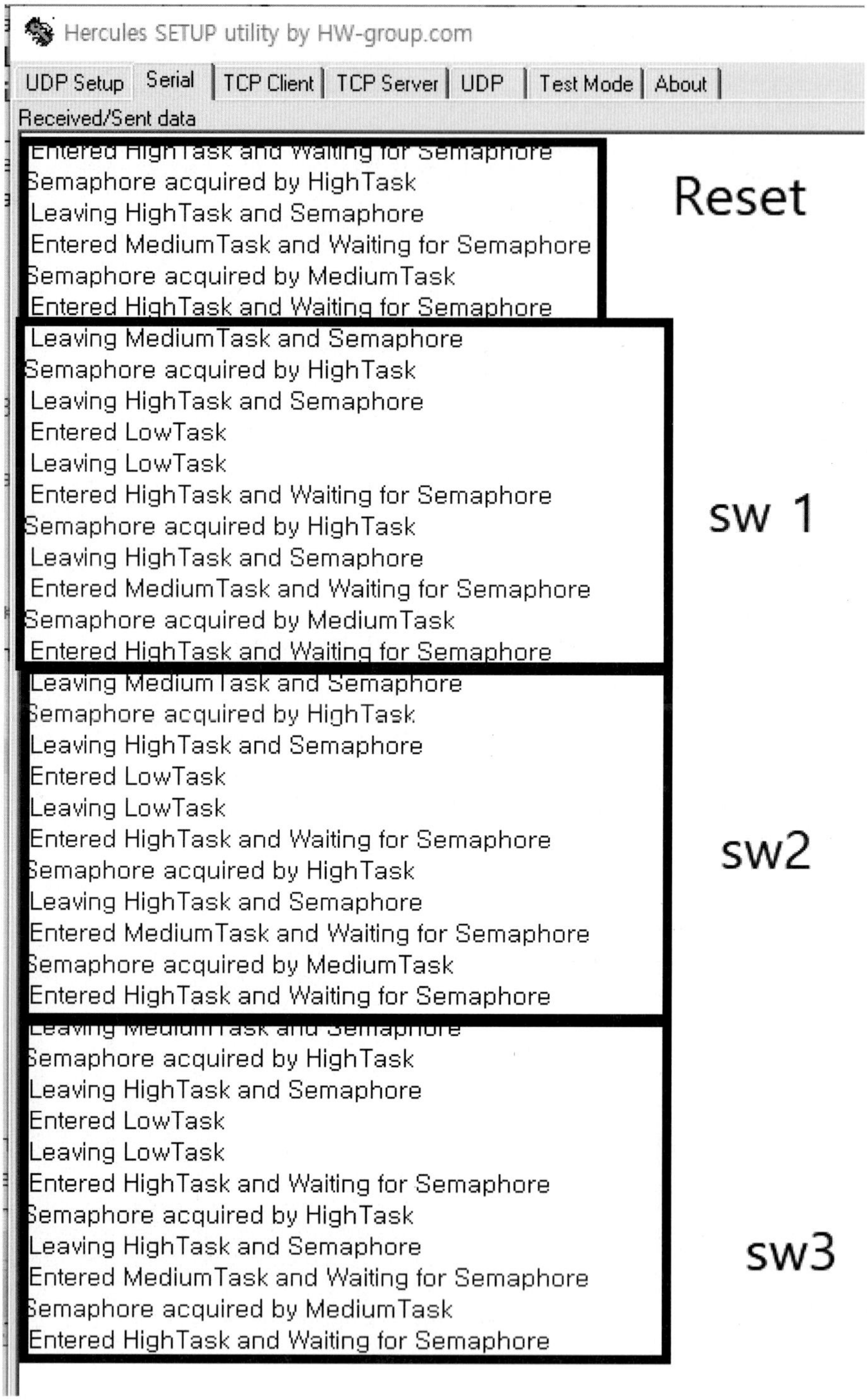

[그림 11.30] 실제 Task 동작

5.9. Task 동기화 예제

다음은 Thread Flags 함수를 사용하여 Task 동기화를 해보도록 한다. 관련된 함수는 다음과 같다. 이 함수들은 CMSIS_V2에서 제공한다. 따라서 아래 함수들을 이용하려면 FreeRTOS의 CMSIS_V2를 선택해야 한다.

```
//  ==== Thread Flags Functions ====

/// Set the specified Thread Flags of a thread.
/// \param[in]     thread_id     thread ID obtained by \ref osThreadNew or
\ref osThreadGetId.
/// \param[in]     flags         specifies the flags of the thread that
shall be set.
/// \return thread flags after setting or error code if highest bit set.
uint32_t osThreadFlagsSet (osThreadId_t thread_id, uint32_t flags);

/// Clear the specified Thread Flags of current running thread.
/// \param[in]     flags         specifies the flags of the thread that
shall be cleared.
/// \return thread flags before clearing or error code if highest bit set.
uint32_t osThreadFlagsClear (uint32_t flags);

/// Get the current Thread Flags of current running thread.
/// \return current thread flags.
uint32_t osThreadFlagsGet (void);

/// Wait for one or more Thread Flags of the current running thread to
become signaled.
/// \param[in]     flags         specifies the flags to wait for.
/// \param[in]     options       specifies flags options (osFlagsXxxx).
/// \param[in]     timeout       \ref CMSIS_RTOS_TimeOutValue or 0 in case
of no time-out.
/// \return thread flags before clearing or error code if highest bit set.
uint32_t osThreadFlagsWait (uint32_t flags, uint32_t options, uint32_t
timeout);
```

동기화를 하지 않는 프로그램을 [그림 11.31]과 같이 할 경우 LED 출력 및 시리얼 통신 출력이 일정하지 않다.

```
/* USER CODE END Header_StartLEDTask1 */
void StartLEDTask1(void *argument)
{
  /* init code for LWIP */
  MX_LWIP_Init();
  /* USER CODE BEGIN 5 */
  /* Infinite loop */
  for(;;)
  {
    HAL_GPIO_WritePin(GPIOB, GPIO_PIN_0, GPIO_PIN_RESET);
    printf("LED1 Off Task\r\n");
    osDelay(1000);
  }
  /* USER CODE END 5 */
}

/* USER CODE BEGIN Header_StartLEDTask2 */
/**
* @brief Function implementing the LEDTask2 thread.
* @param argument: Not used
* @retval None
*/
/* USER CODE END Header_StartLEDTask2 */
void StartLEDTask2(void *argument)
{
  /* USER CODE BEGIN StartLEDTask2 */
  /* Infinite loop */
  for(;;)
  {
      HAL_GPIO_WritePin(GPIOB, GPIO_PIN_0, GPIO_PIN_SET);
      printf("LED1 On Task\r\n");
      osDelay(1000);
  }
  /* USER CODE END StartLEDTask2 */
}
```

[그림 11.31] 비동기화 프로그램

다음 프로그램은 Thread Flag 함수를 이용 동기화를 했을 경우의 프로그램과 동작을 나타낸다[그림 11.32].

```
void StartLEDTask1(void *argument)
{
  /* USER CODE BEGIN 5 */

  osThreadFlagsSet(LEDTask2Handle, 0x0002);

  /* Infinite loop */
  for(;;)
  {
    osThreadFlagsWait(0x0001, osFlagsWaitAny, osWaitForever);

    HAL_GPIO_WritePin(GPIOB, GPIO_PIN_0, GPIO_PIN_RESET);
    printf("LED1 Off Task\r\n");
    osDelay(1000);

    osThreadFlagsSet(LEDTask2Handle, 0x0002);
  }
  /* USER CODE END 5 */
}

/* USER CODE BEGIN Header_StartLEDTask2 */
/**
 * @brief Function implementing the LEDTask2 thread.
 * @param argument: Not used
 * @retval None
 */
/* USER CODE END Header_StartLEDTask2 */
void StartLEDTask2(void *argument)
{
  /* USER CODE BEGIN StartLEDTask2 */

  /* Infinite loop */
  for(;;)
  {
    osThreadFlagsWait(0x0002, osFlagsWaitAny, osWaitForever);

    HAL_GPIO_WritePin(GPIOB, GPIO_PIN_0, GPIO_PIN_SET);
    printf("LED1 On Task\r\n");
    osDelay(1000);

    osThreadFlagsSet(LEDTask1Handle, 0x0001);
  }
  /* USER CODE END StartLEDTask2 */
}
```

[그림 11.32] 동기화 Task 프로그램

6. Software Timer 활용하기

FreeRTOS의 Software timer 중 One-Shot Timer와 Auto-reload timer를 활용하는 방법을 다룬다. One-Shot Timer는 그의 Callback 함수가 단 한 번만 실행되고 수동으로 재시동을 할 수 있고, 스스로 자동적으로 재시동을 할 수 없다. 반대로 Auto-reload timer는 그의 Callback 함수가 실행된 후 자동적으로 재시동이 되고 Callback 실행이 주기적으로 이루어진다.

6.1. FreeRTOS 설정

기본 클럭 설정은 기존과 동일하고 FreeRTOS Software timer 설정은 [그림 11.33]과 같이 한다.

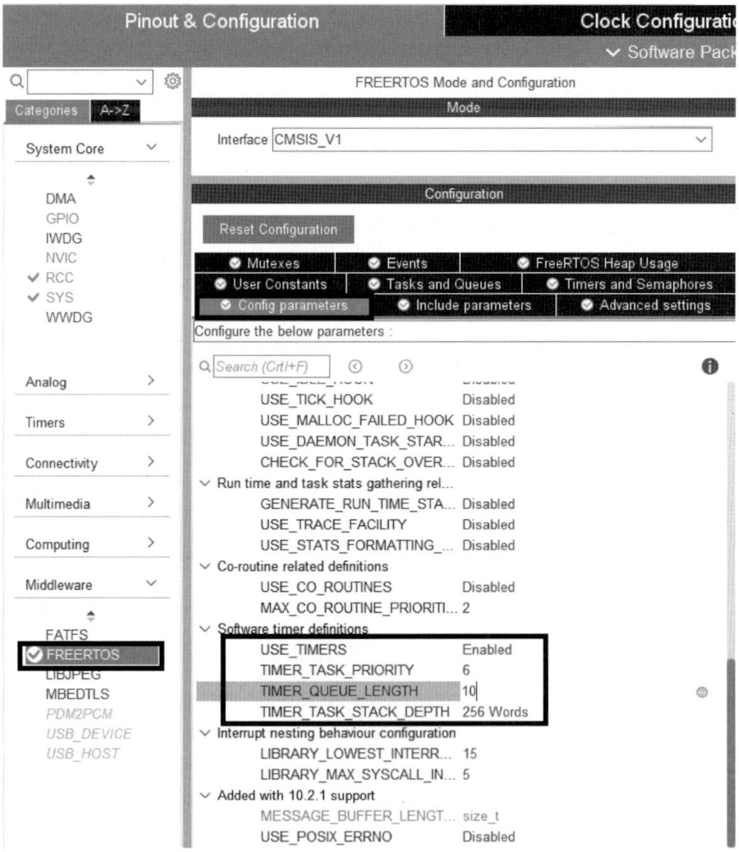

[그림 11.33] RTOS 설정

[그림 11.34]와 같이 Task 2개를 생성한다.

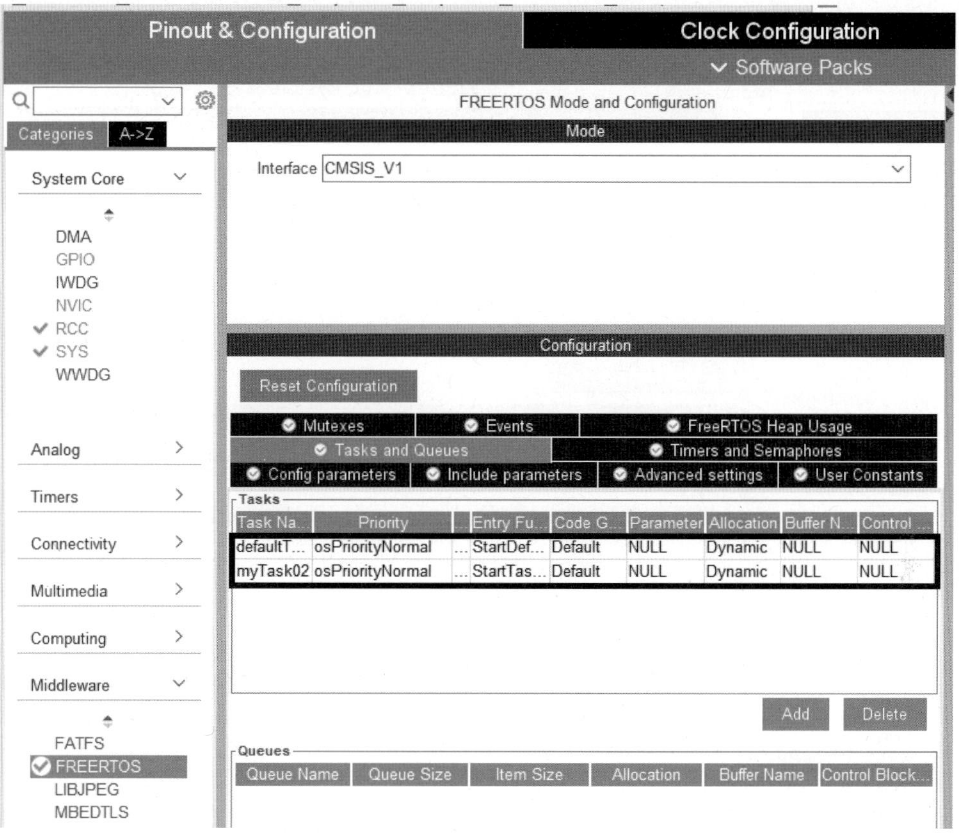

[그림 11.34] RTOS 설정

one shot timer와 Periodic Timer를 [그림 11.35]와 같이 생성한다.

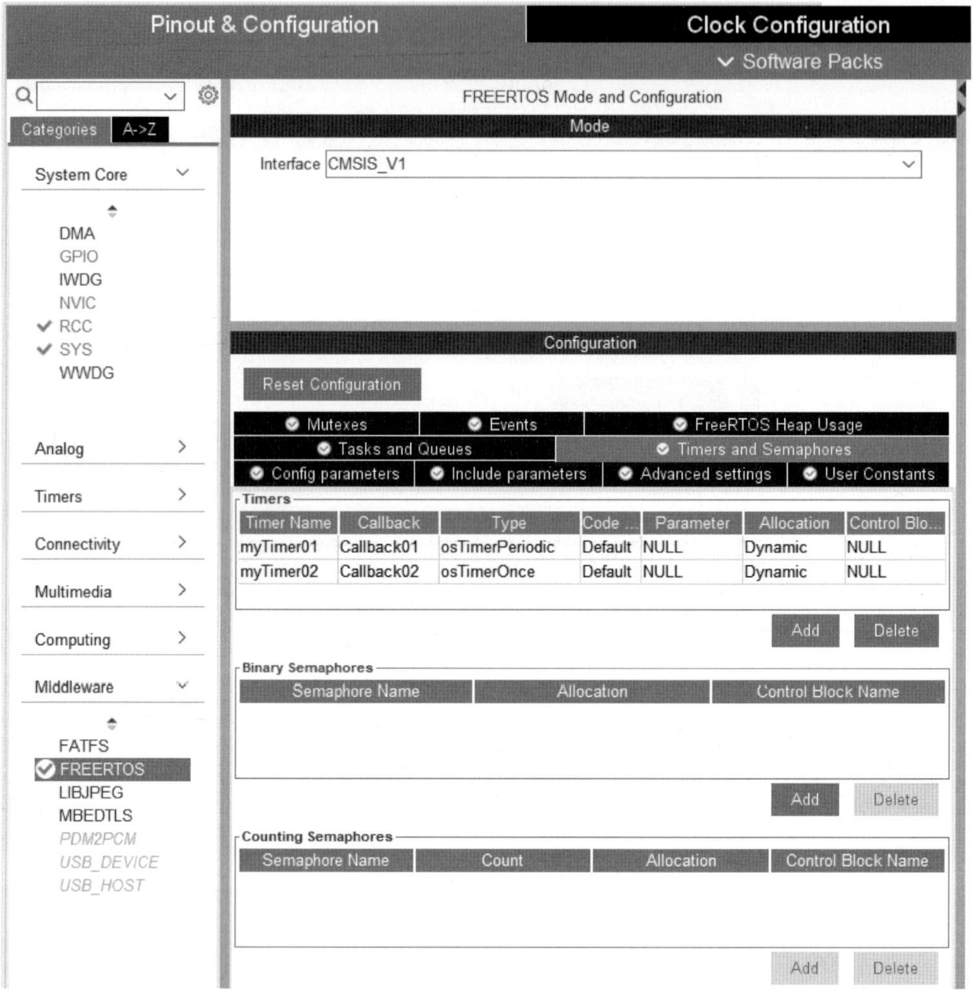

[그림 11.35] Timer 생성

6.2. Generate Code 생성

[그림 11.36]은 코드를 생성하기 위한 화면이다. 그림과 같이 클릭해서 프로그램을 생성한다.

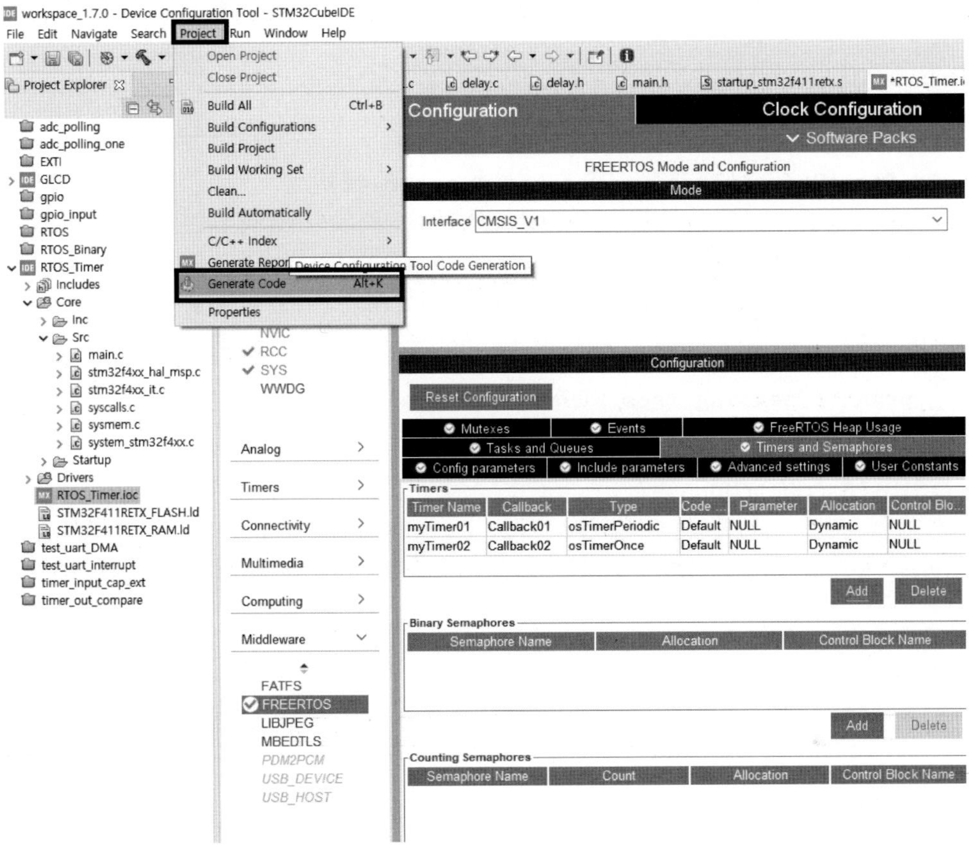

[그림 11.36] 프로그램 코드를 생성

6.3. Source Code 작성하기

Task 및 Timer의 프로그램은 [그림 11.37]과 [그림 11.38]과 같다. Callback01에 1초 간격으로 동작하고 Callback02에 2초를 설정한다. Callback01는 주기적으로 동작하고, Callback02는 One-shot timer로 스위치를 누르면 동작하고 2초간 LED가 On 되었다가 Off 된다. 스위츠를 누를 때마다 동작한다.

```
void StartDefaultTask(void const * argument)
{
  /* USER CODE BEGIN 5 */
    osTimerStart(myTimer01Handle, 1000);
  /* Infinite loop */
  for(;;)
  {
    printf("Sending from StartDefaultTask\r\n");
    osDelay(200);
  }
  /* USER CODE END 5 */
}

/* USER CODE BEGIN Header_StartTask02 */
/**
* @brief Function implementing the myTask02 thread.
* @param argument: Not used
* @retval None
*/
/* USER CODE END Header_StartTask02 */
void StartTask02(void const * argument)
{
  /* USER CODE BEGIN StartTask02 */
  /* Infinite loop */
  for(;;)
  {
    if(HAL_GPIO_ReadPin(GPIOC, GPIO_PIN_13)){
        HAL_GPIO_WritePin(GPIOA, GPIO_PIN_6, 1);
        printf("Sending from StartTask02\r\n");
        osTimerStart(myTimer01Handle, 1000);
    }
    osDelay(20);
  }
  /* USER CODE END StartTask02 */
}
```

[그림 11.37] Tasks 내용

```
/* Callback01 function */
void Callback01(void const * argument)
{
  /* USER CODE BEGIN Callback01 */
  HAL_GPIO_TogglePin(GPIOA, GPIO_PIN_7);
  printf("Sending from Callback01\r\n");

  /* USER CODE END Callback01 */
}

/* Callback02 function */
void Callback02(void const * argument)
{
  /* USER CODE BEGIN Callback02 */
     HAL_GPIO_WritePin(GPIOA, GPIO_PIN_6,0);
     printf("Sending from Callback02\r\n");
  /* USER CODE END Callback02 */
}
```

[그림 11.38] Timer Callback 함수 프로그램

6.4. 디버그 및 다운로드하기

메인 함수에 프로그램이 작성되었다면 다음으로 디버그는 [그림 11.39]의 를 클릭하거나 <u>단축키 F11</u>을 누르면 디버깅과 프로그램 다운로드가 완료된다.

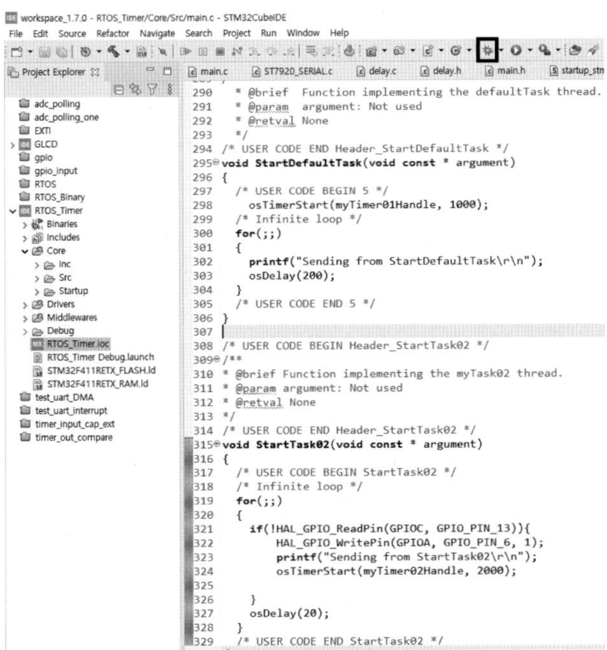

workspace_1.7.0 - RTOS_Timer/Core/Src/main.c - STM32CubeIDE
File Edit Source Refactor Navigate Search Project Run Window Help
Project Explorer
main.c ST7920_SERIAL.c delay.c delay.h main.h startup_stm
adc_polling
adc_polling_one
EXTI
GLCD
gpio
gpio_input
RTOS
RTOS_Binary
RTOS_Timer
Binaries
Includes
Core
Inc
Src
Startup
Drivers
Middlewares
Debug
RTOS_Timer.ioc
RTOS_Timer Debug.launch
STM32F411RETX_FLASH.ld
STM32F411RETX_RAM.ld
test_uart_DMA
test_uart_interrupt
timer_input_cap_ext
timer_out_compare
290 * @brief Function implementing the defaultTask thread.
291 * @param argument: Not used
292 * @retval None
293 */
294 /* USER CODE END Header_StartDefaultTask */
295 void StartDefaultTask(void const * argument)
296 {
297 /* USER CODE BEGIN 5 */
298 osTimerStart(myTimer01Handle, 1000);
299 /* Infinite loop */
300 for(;;)
301 {
302 printf("Sending from StartDefaultTask\r\n");
303 osDelay(200);
304 }
305 /* USER CODE END 5 */
306 }
307
308 /* USER CODE BEGIN Header_StartTask02 */
309 /**
310 * @brief Function implementing the myTask02 thread.
311 * @param argument: Not used
312 * @retval None
313 */
314 /* USER CODE END Header_StartTask02 */
315 void StartTask02(void const * argument)
316 {
317 /* USER CODE BEGIN StartTask02 */
318 /* Infinite loop */
319 for(;;)
320 {
321 if(!HAL_GPIO_ReadPin(GPIOC, GPIO_PIN_13)){
322 HAL_GPIO_WritePin(GPIOA, GPIO_PIN_6, 1);
323 printf("Sending from StartTask02\r\n");
324 osTimerStart(myTimer02Handle, 2000);
325
326 }
327 osDelay(20);
328 }
329 /* USER CODE END StartTask02 */

[그림 11.39] 프로그램 디버그

6.5. 결과 확인

결과는 PA7은 주기적으로 토글하고 PA6은 스위치를 누르면 켜지고 2초 이후에 Off 됨을 확인할 수 있다. [그림 11.40],[그림 11.41]은 스위치 on/off 전·후를 나타내고 있다.

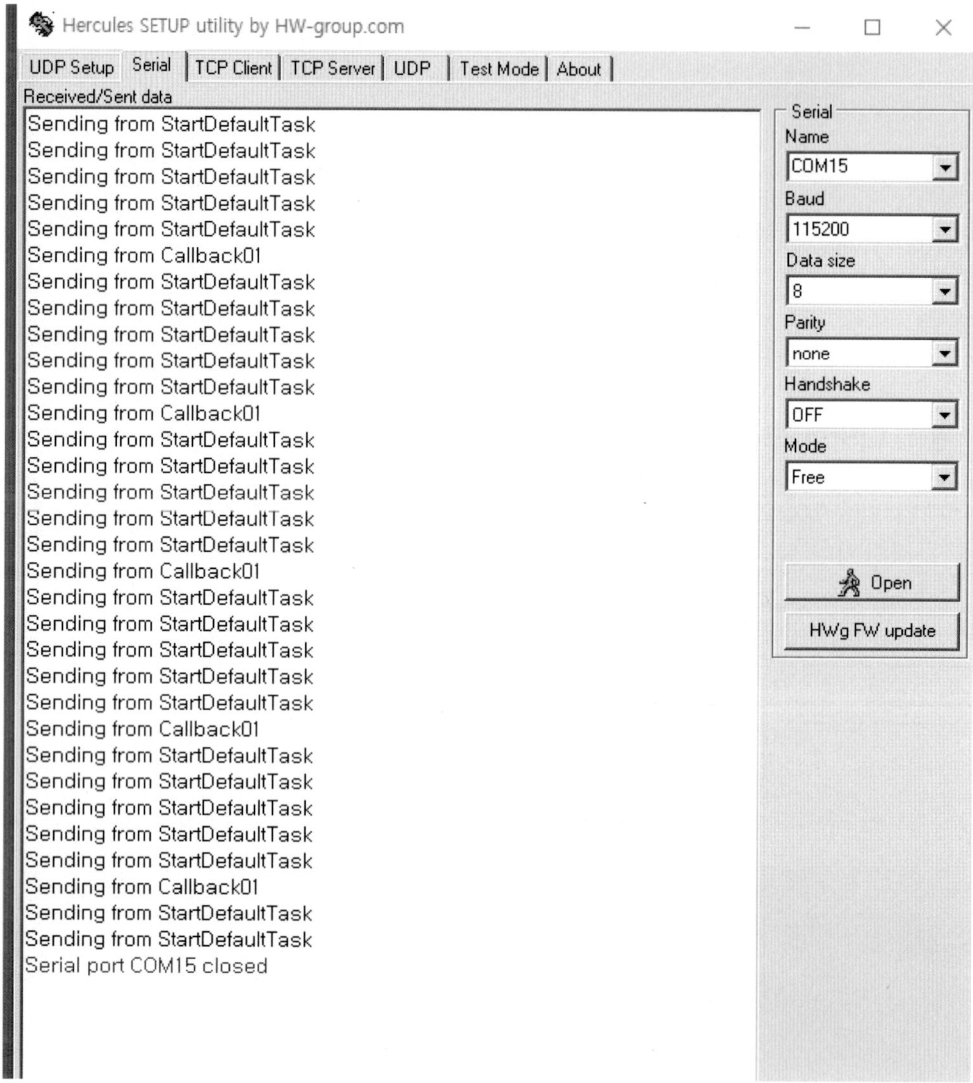

[그림 11.40] 스위를 On/off 전

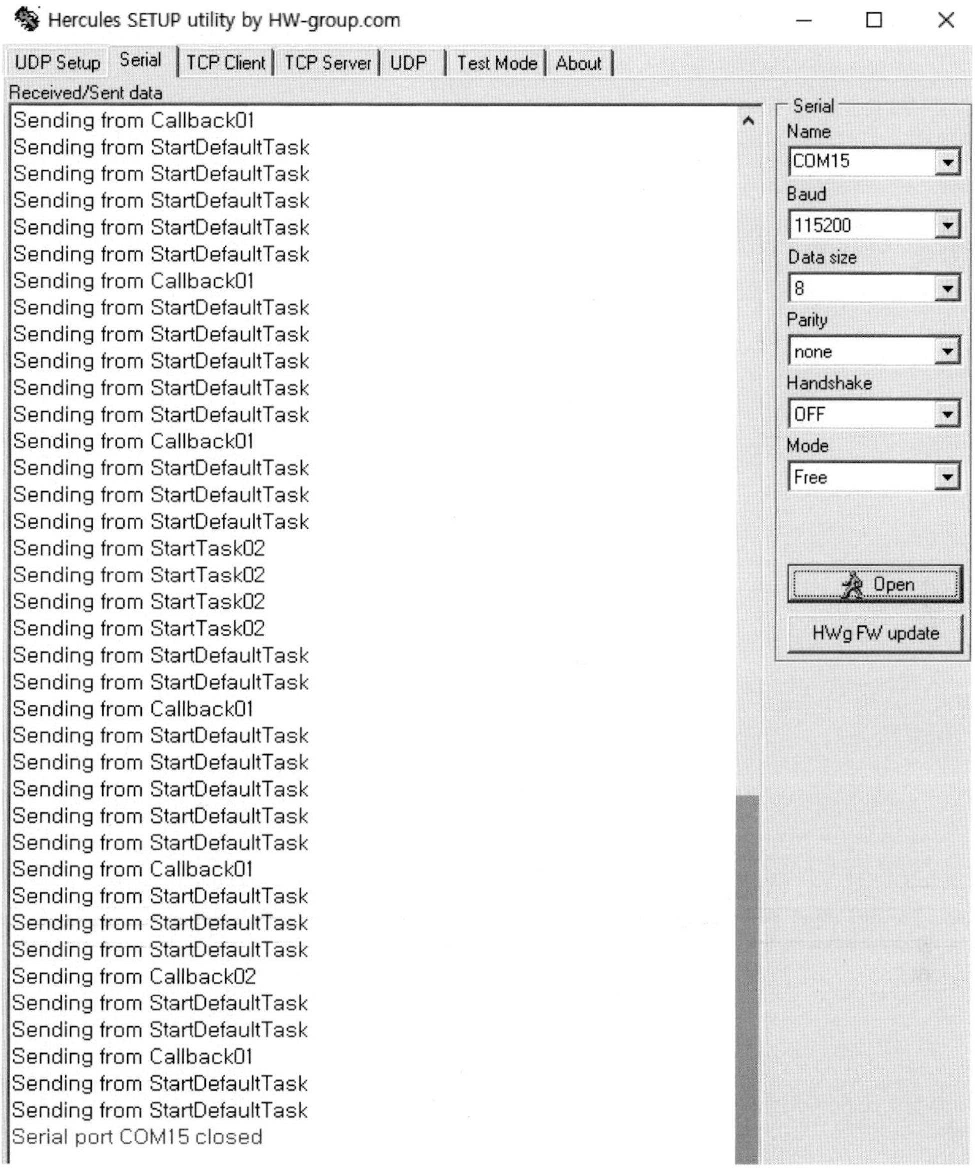

[그림 11.41] 스위를 On/off 후

CHAPTER 12

Application (초음파, 압력, 로드셀, 서미스터)

CHAPTER 12

Cortex-M4 기초 활용(ARM Core STM32)

Application (초음파, 압력, 로드셀, 서미스터)

1. 개요

본 장에서는 초음파 센서, 압력 센서, 가속도 센서를 MCU 측면에서 지금까지 언급한 다양한 방법들을 활용하여 다룬다.

2. 클럭 및 SYS 설정

[그림 12.1]~[그림 12.2]는 클럭이 설정된 경우를 나타낸다. 메인 클럭은 외부 클럭(Crystal/Ceramic Resonator)과 내부 클럭(HSI RC(BYPASS Clock Source) 둘 중의 하나를 이용하여야 한다. 내부 클럭보다는 외부 클럭이 정확하며, 본 보드에서는 Crystal/Ceramic Resonator는 연결되어 있지 않아 사용하지 않으며 ST-LINK MCO 출력 8MHz를 사용한다.

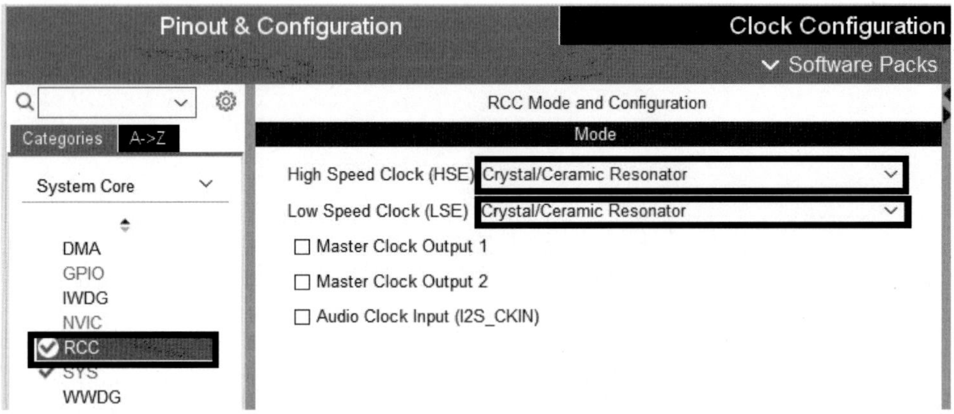

[그림 12.1] NUCLEO-F411RE 클럭 설정

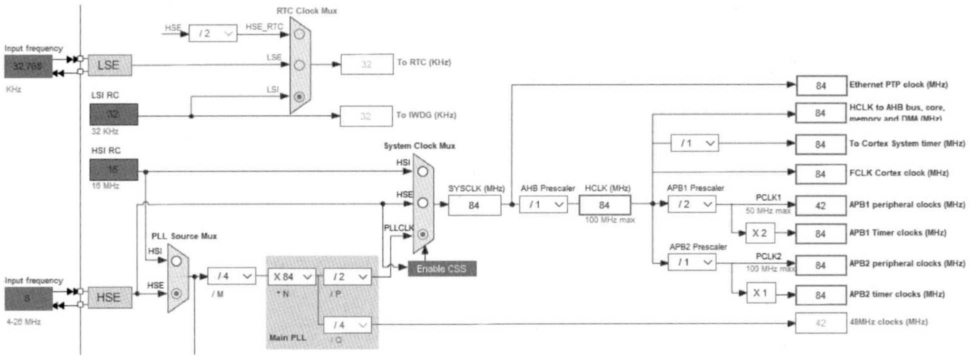

[그림 12.2] NUCLEO-F411RE 클럭 설정

[그림 12.3]은 Timebase 소스를 타이머 1로 설정한다. 일반적으로 Systick보다는 정확하다.

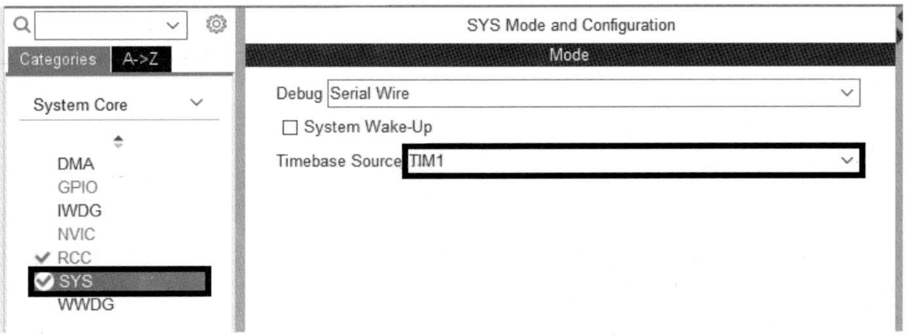

[그림 12.3] Timebase Source 설정

3. 초음파 센서 활용하기

3.1. 초음파 센서 개요

HC-SR04 초음파 센서 모듈의 동작은 Trig 단자에 10us 펄스 폭의 파형[그림 12.4]를 발생시키면 모듈에서 40kHz 8개의 버스트 신호가 전송되고 에코 신호가 발생된다. 이 에코 신호는 펄스 폭과 거리에 비례하는 물체의 거리를 나타낸다. 따라서 보낸 트리거 신호와 받은 에코 신호 사이의 시간 간격을 계산하면 된다. Trigger Port:PC0(출력 포트 지정), Echo Port:PC1(입력 포트 지정) 측정 시간의 단위는 us이다. 거리는 Cm로 환산한다. T: 에코의 펄스 폭[us]

왕복 거리= V*t= 340(m/s) *T(**측정 시간**)(s)= 34,000(cm/s) *T(s)

=34,000(cm/s) *T(us)* 1/1000000 = 0.034*T[Cm]

Distance(**물체의 거리**) =**왕복 거리**/2= 0.017*T[Cm]

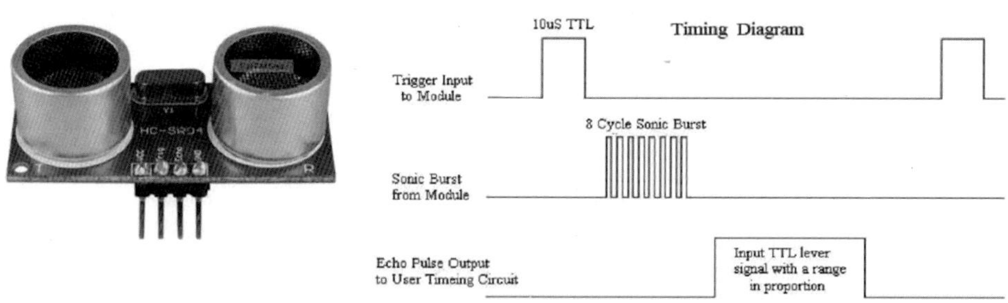

[그림 12.4] HC-SR04 타임 다이아그램

3.2. 타이머 설정

클럭 및 SYS 시리얼 통신 설정은 이전과 동일하게 설정하면 된다. 초음파 센서의 트리거 신호를 발생하기 위해서는 u(s)단위 딜레이가 필요하며 펄스 폭을 측정하기 위해서는 1u(s) 간격으로 작동하는 카운터가 필요하다. [그림 12.5]는 타이머 주파수를 fs=1MHz(1us)로 설정하였다.

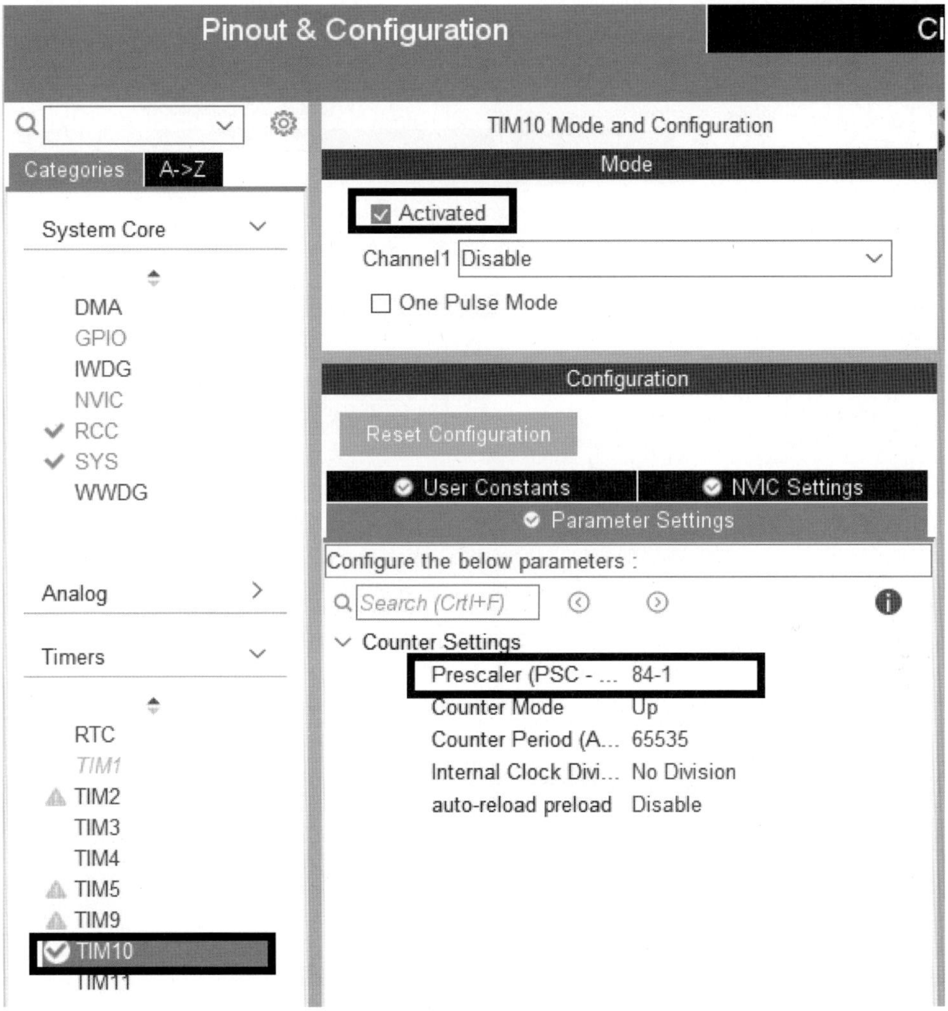

[그림 12.5] 타이머 10 설정

3.3. Generate Code 생성

[그림 12.6]은 코드를 생성하기 위한 화면이다. 그림과 같이 클릭해서 프로그램을 생성한다.

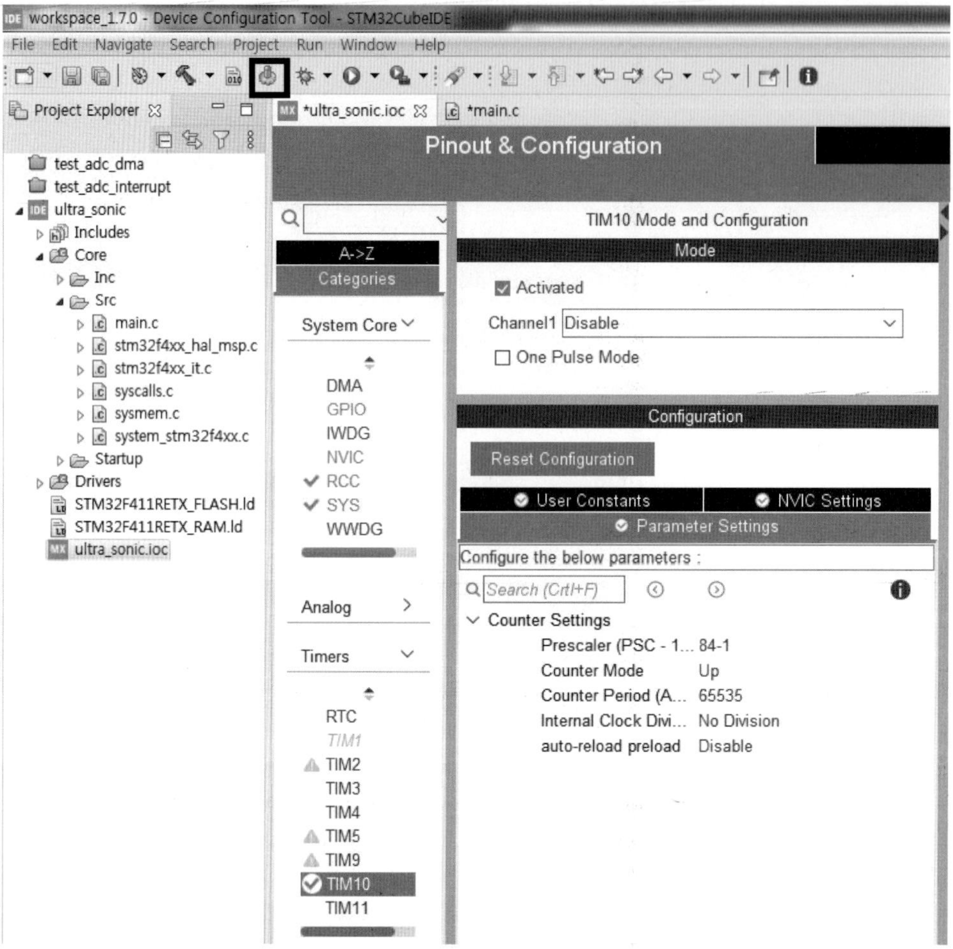

[그림 12.6] 프로그램 코드를 생성

3.4. Source Code 작성하기

딜레이 함수를 만들어서 클럭을 발생시키고 펄스 폭을 측정하면 된다(Polling 방식). [그림 12.7]은 메인 프로그램이다.

```
uint16_t counter_time, sensor_time;
float distance;
void delay_us(uint16_t delay){
    __HAL_TIM_SET_COUNTER(&htim10,0);
    while(__HAL_TIM_GET_COUNTER(&htim10)<delay);
}

uint16_t hc_sr04_read (void){
    counter_time=0;
    HAL_GPIO_WritePin(GPIOC, GPIO_PIN_0, 0);
    delay_us(2);
    HAL_GPIO_WritePin(GPIOC, GPIO_PIN_0, 1);
    delay_us(10);
    HAL_GPIO_WritePin(GPIOC, GPIO_PIN_0, 0);

    while (!HAL_GPIO_ReadPin(GPIOC, GPIO_PIN_1));
    while (HAL_GPIO_ReadPin(GPIOC, GPIO_PIN_1)){
          counter_time++;
          delay_us (1);
     }
    return counter_time;
}

HAL_TIM_Base_Start(&htim10);
/* USER CODE END 2 */

/* Infinite loop */
/* USER CODE BEGIN WHILE */
while (1)
{
  sensor_time = hc_sr04_read();
  distance =(float)sensor_time * 0.017;
  printf("distance=%f\r\n",distance);
  HAL_Delay(200);
  /* USER CODE END WHILE */
```

[그림 12.7] 메인 코드 내용

3.5. 디버그 및 다운로드하기

메인 함수에 프로그램이 작성되었다면 다음으로 디버그는 [그림 12.8]의 를 클릭하거나 단축키 F11을 누르면 디버깅과 프로그램 다운로드가 완료된다.

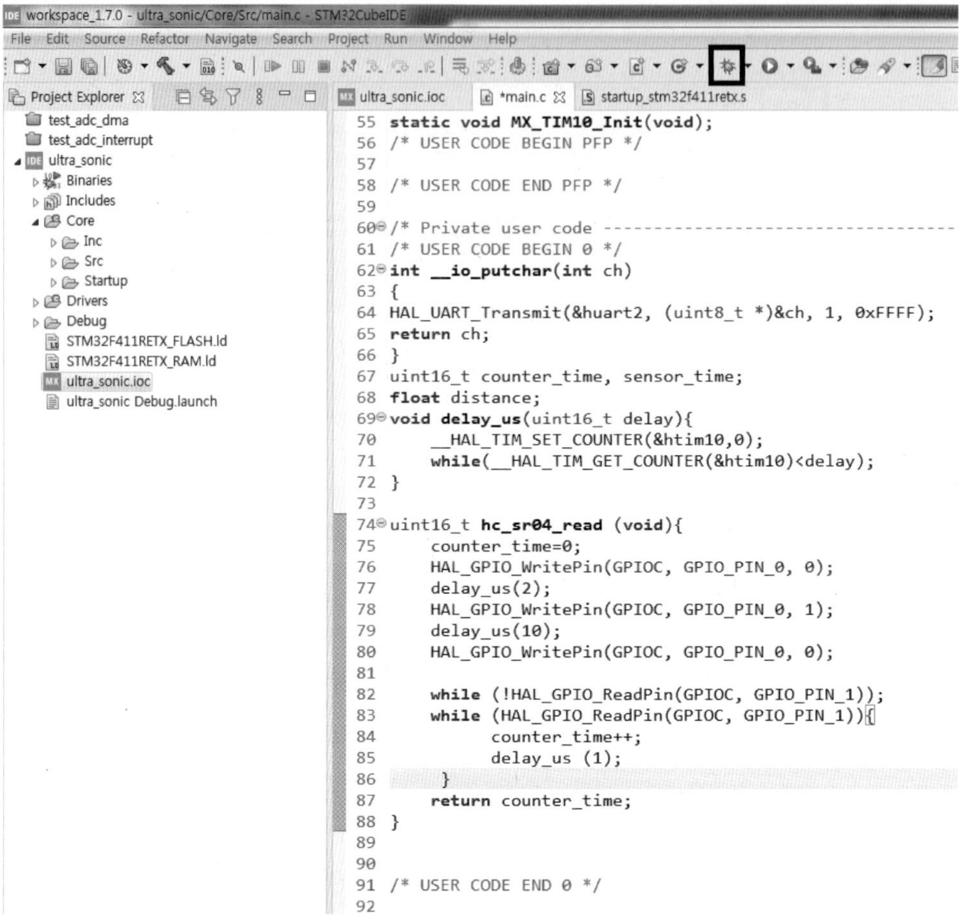

workspace_1.7.0 - ultra_sonic/Core/Src/main.c - STM32CubeIDE
File Edit Source Refactor Navigate Search Project Run Window Help
Project Explorer
test_adc_dma
test_adc_interrupt
ultra_sonic
Binaries
Includes
Core
Inc
Src
Startup
Drivers
Debug
STM32F411RETX_FLASH.ld
STM32F411RETX_RAM.ld
ultra_sonic.ioc
ultra_sonic Debug.launch
ultra_sonic.ioc
*main.c
startup_stm32f411retx.s
static void MX_TIM10_Init(void);
/* USER CODE BEGIN PFP */
/* USER CODE END PFP */
/* Private user code ---------------------------------
/* USER CODE BEGIN 0 */
int __io_putchar(int ch)
{
HAL_UART_Transmit(&huart2, (uint8_t *)&ch, 1, 0xFFFF);
return ch;
}
uint16_t counter_time, sensor_time;
float distance;
void delay_us(uint16_t delay){
__HAL_TIM_SET_COUNTER(&htim10,0);
while(__HAL_TIM_GET_COUNTER(&htim10)<delay);
}
uint16_t hc_sr04_read (void){
counter_time=0;
HAL_GPIO_WritePin(GPIOC, GPIO_PIN_0, 0);
delay_us(2);
HAL_GPIO_WritePin(GPIOC, GPIO_PIN_0, 1);
delay_us(10);
HAL_GPIO_WritePin(GPIOC, GPIO_PIN_0, 0);
while (!HAL_GPIO_ReadPin(GPIOC, GPIO_PIN_1));
while (HAL_GPIO_ReadPin(GPIOC, GPIO_PIN_1)){
counter_time++;
delay_us (1);
}
return counter_time;
}
/* USER CODE END 0 */

[그림 12.8] 프로그램 디버그

3.6. 보드 프로그램 실행하기

[그림 12.9] 상단 중앙에 아이콘 ■ 클릭하면 디버그 모드가 종료되어 보드의 프로그램이 실행된다.

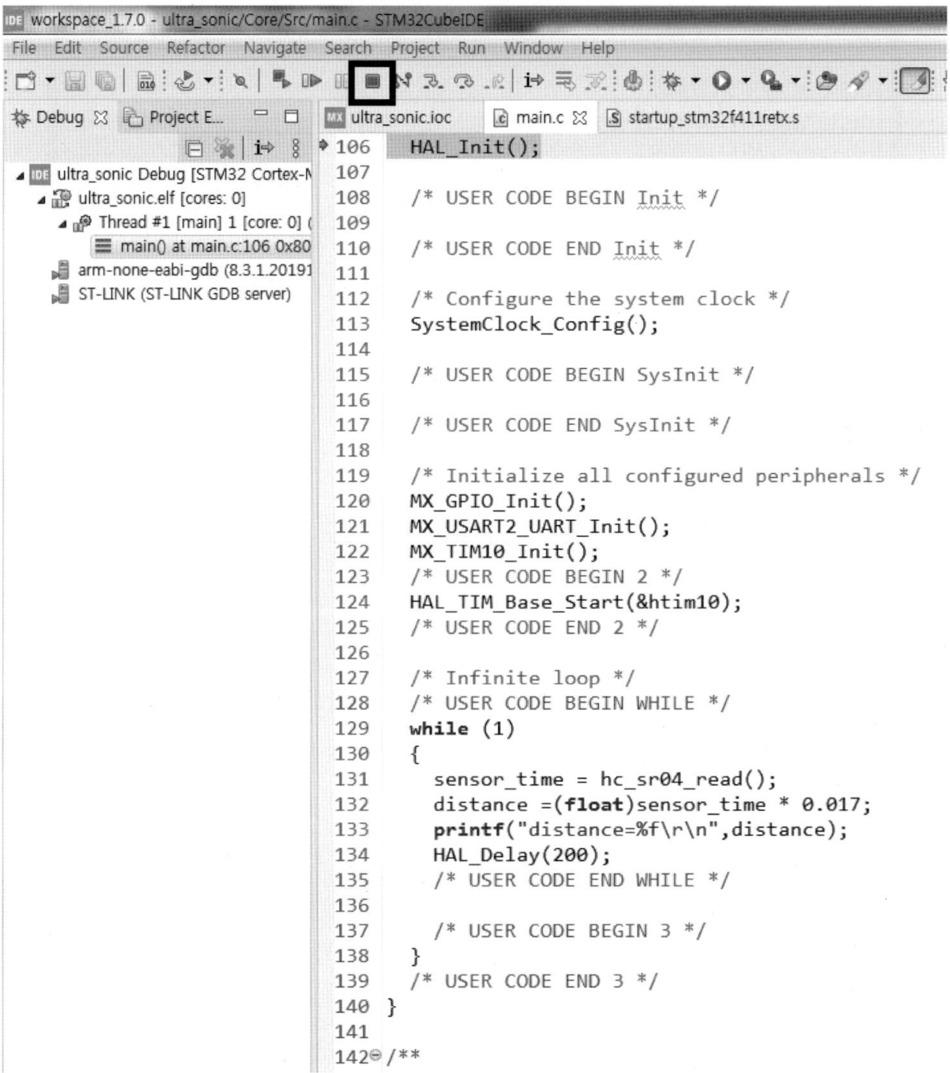
workspace_1.7.0 - ultra_sonic/Core/Src/main.c - STM32CubeIDE
File Edit Source Refactor Navigate Search Project Run Window Help
Debug
Project E...
ultra_sonic.ioc
main.c
startup_stm32f411retx.s
ultra_sonic Debug [STM32 Cortex-M
ultra_sonic.elf [cores: 0]
Thread #1 [main] 1 [core: 0]
main() at main.c:106 0x80
arm-none-eabi-gdb (8.3.1.20191
ST-LINK (ST-LINK GDB server)
HAL_Init();
/* USER CODE BEGIN Init */
/* USER CODE END Init */
/* Configure the system clock */
SystemClock_Config();
/* USER CODE BEGIN SysInit */
/* USER CODE END SysInit */
/* Initialize all configured peripherals */
MX_GPIO_Init();
MX_USART2_UART_Init();
MX_TIM10_Init();
/* USER CODE BEGIN 2 */
HAL_TIM_Base_Start(&htim10);
/* USER CODE END 2 */
/* Infinite loop */
/* USER CODE BEGIN WHILE */
while (1)
{
sensor_time = hc_sr04_read();
distance =(float)sensor_time * 0.017;
printf("distance=%f\r\n",distance);
HAL_Delay(200);
/* USER CODE END WHILE */
/* USER CODE BEGIN 3 */
}
/* USER CODE END 3 */
}
/**

[그림 12.9] 디버그 모드 프로그램 실행

3.7. 프로그램 실행 확인

LED 확인 및 시리얼 통신 [그림 12.10] 데이터를 확인할 수 있다.

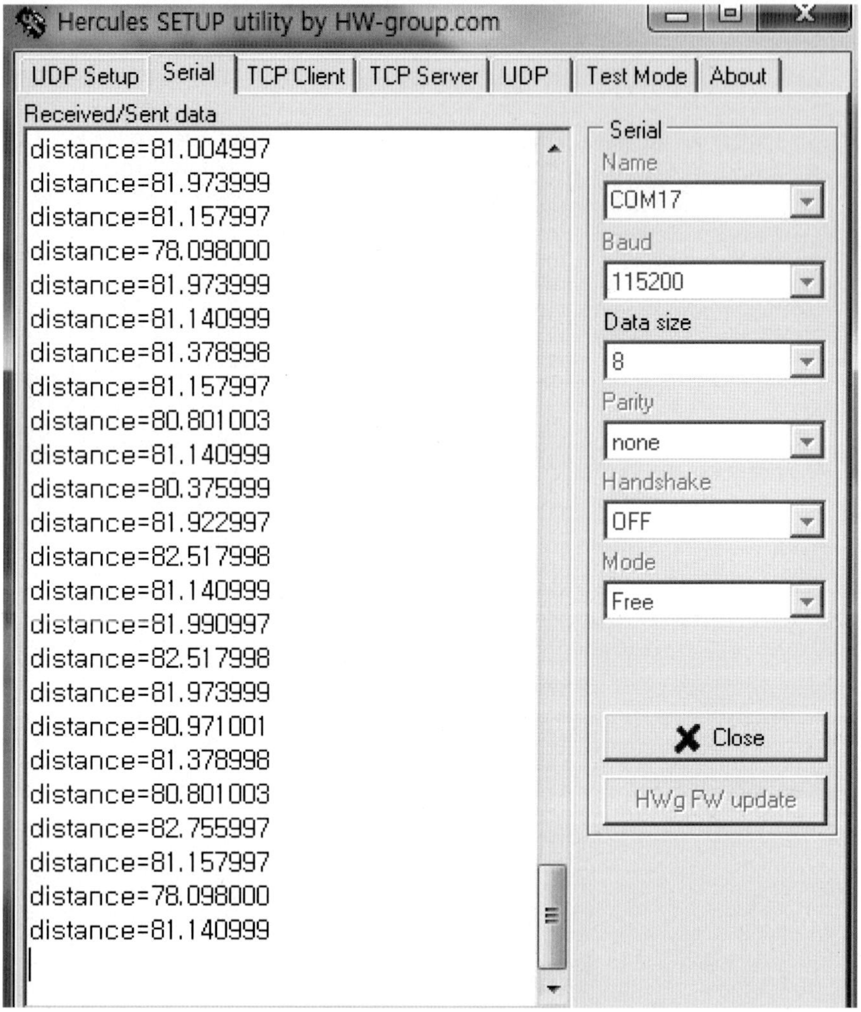

[그림 12.10] 거리 측정 결과

4. 압력 센서(BMP180) 활용하기

4.1. 압력 센서 개요

BOSCH의 BMP180 디지털 압력 센서는 I2C 인터페이스 방식이며 압력 범위는 300~1100hPa(9000~-500m 해수면 기준)이다. 또한, 정확도는 향상된 기능에서는 0.15m 정도이다. [그림 12.11]은 압력을 측정하기 위한 흐름도이다. 이를 실제 프로그램하는 과정은 22개의 데이터를 읽어 수식에 맞게 계산하는 과정 등이 있어 복잡하다. 따라서 아래 사이트를 방문 소스 및 헤더 파일을 다운하여 응용해 보기로 한다. 또한, 파일을 다운하여 디바이스에 맞도록 일부 수정이 필요하다.

https://gist.github.com/ggirjau/369d1520a21c7cf41413b00352446d61

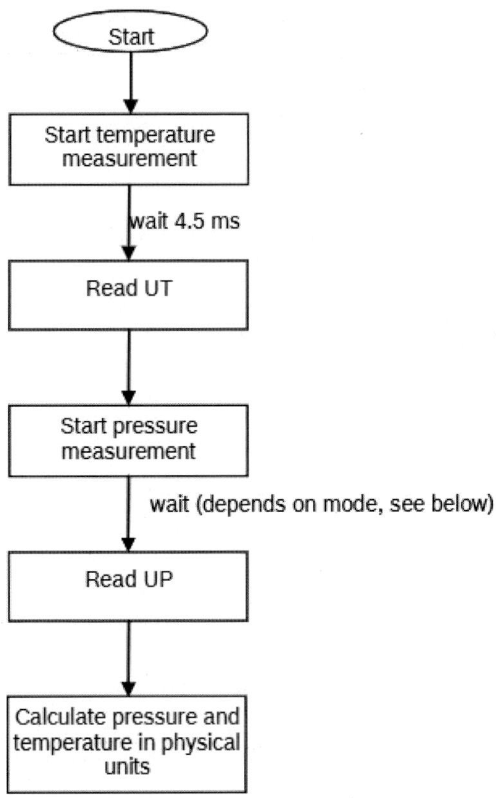

Start
Start temperature measurement
wait 4.5 ms
Read UT
Start pressure measurement
wait (depends on mode, see below)
Read UP
Calculate pressure and temperature in physical units

[그림 12.11] 압력 측정 흐름도

4.2. 압력 센서 설정

클럭 및 SYS 시리얼 통신 설정은 이전과 동일하게 설정하면 된다. 압력 센서는 I2C 인터페이스(3.3V GND SCL SDA, PB8,PB9)에 의해서 데이터를 읽기 때문에 별도의 하드웨어는 필요 없다. [그림 12.12]는 I2C 설정이다.

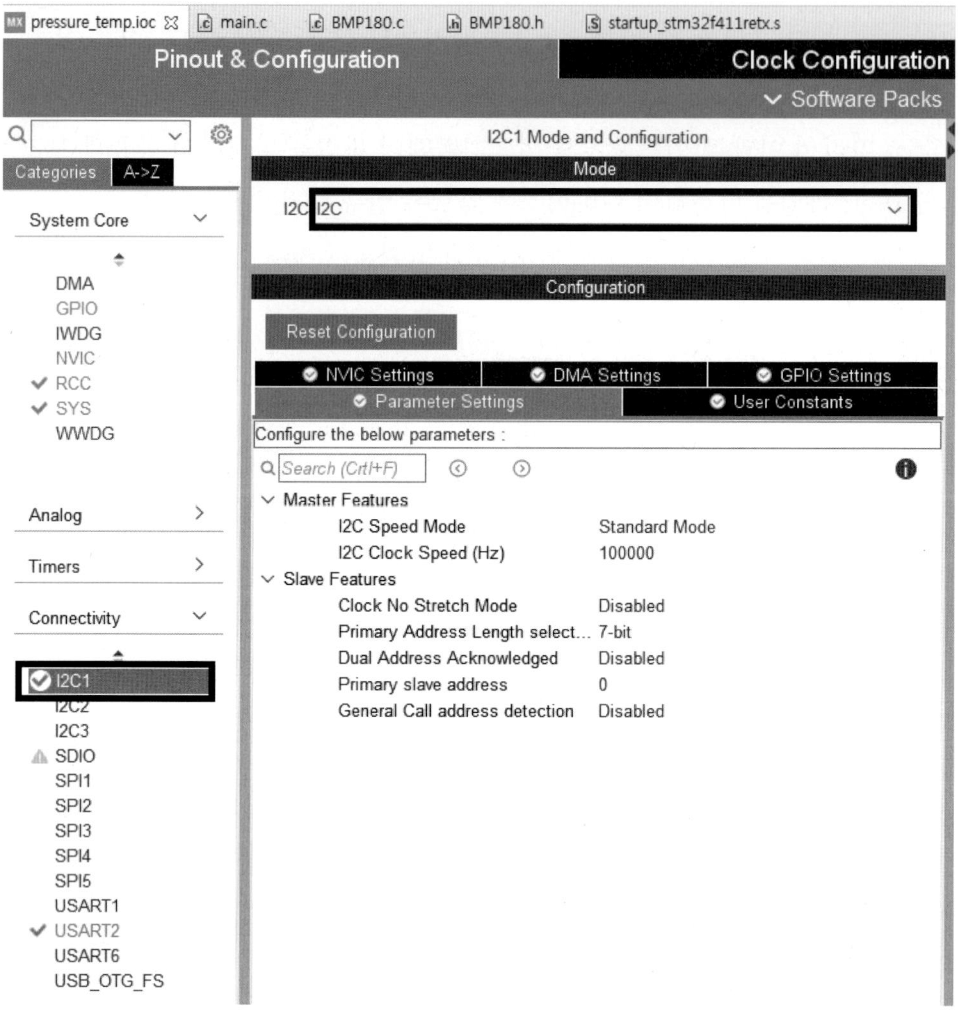

[그림 12.12] I2C 설정

4.3. Generate Code 생성

[그림 12.13]은 코드를 생성하기 위한 화면이다. 그림과 같이 클릭해서 프로그램을 생성한다.

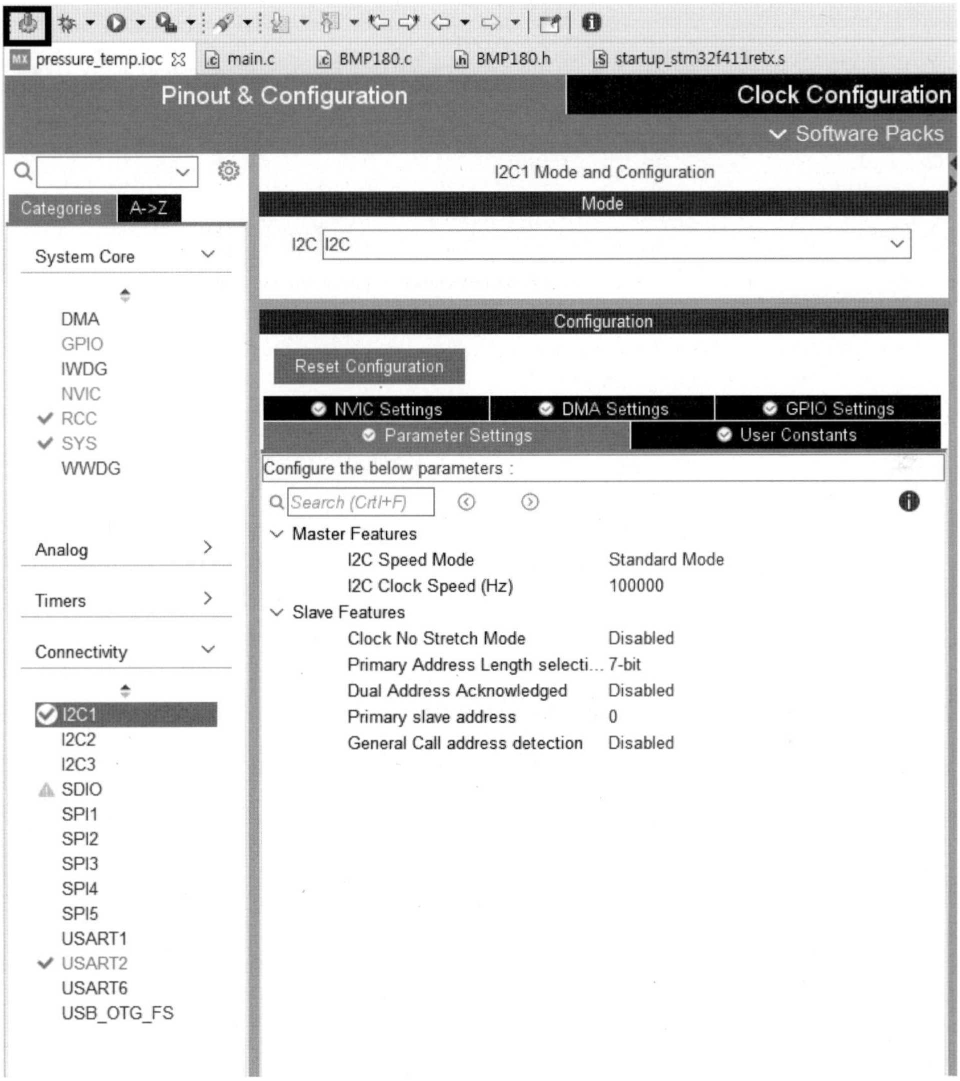

[그림 12.13] 프로그램 코드를 생성

4.4. Source Code 작성하기

내려받은 파일을 이용하기 위해서는 소스 파일 및 헤더 파일을 복사해서 추가하고 [그림 12.14], [그림 12.15]와 같이 프로그램하여 사용한다.

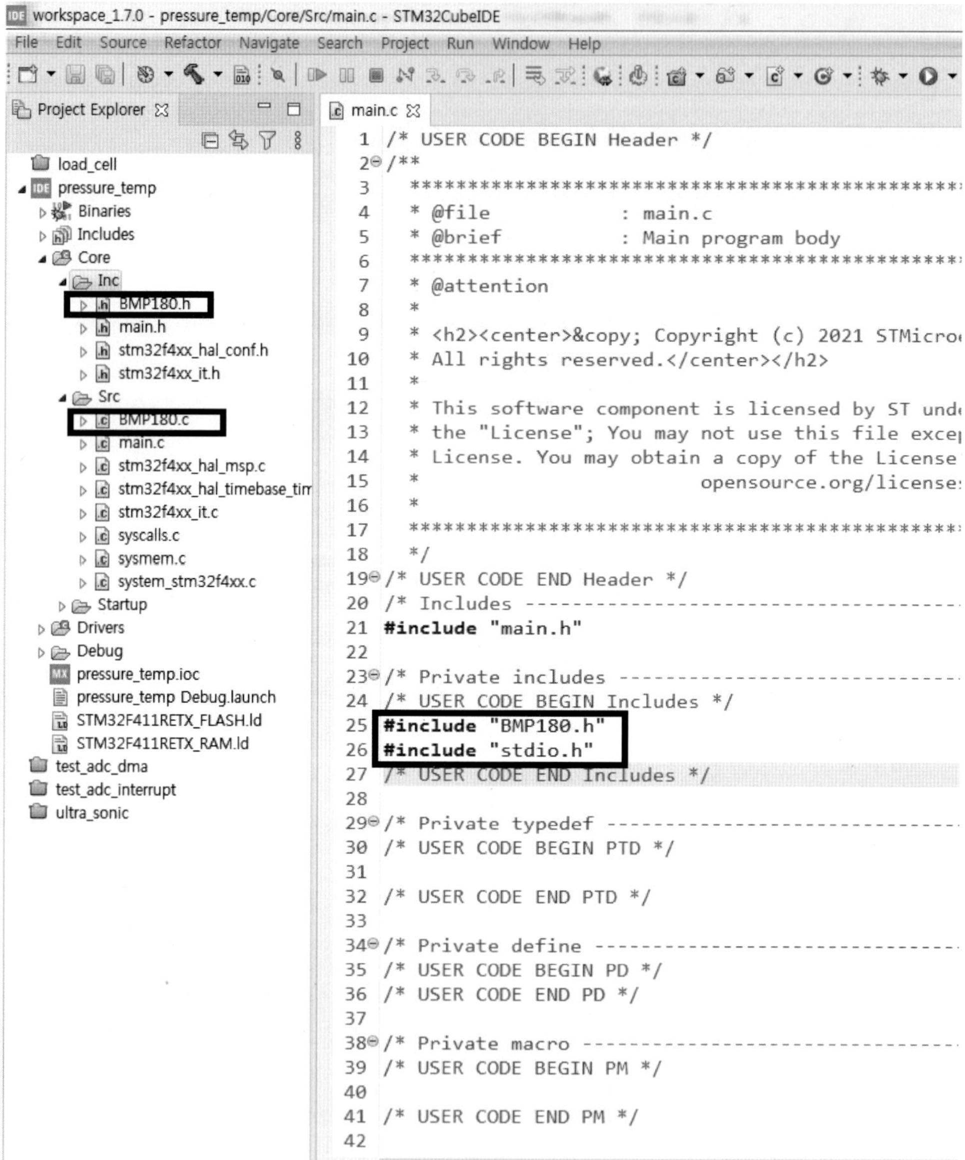

[그림 12.14] 메인 코드 내용

```
/* Private includes ---------------
/* USER CODE BEGIN Includes */
#include "BMP180.h"
#include "stdio.h"
/* USER CODE END Includes */
/* Private user code -------
/* USER CODE BEGIN 0 */
float Temperature = 0;
float Pressure = 0;
float Altitude = 0;
char Temperature1[10];
char Pressure1[10];
char Altitude1[10];
BMP180_Start();
/* USER CODE END 2 */

/* Infinite loop */
/* USER CODE BEGIN WHILE */
while (1)
{
    Temperature = BMP180_GetTemp();
    Pressure = BMP180_GetPress (0);
    Altitude = BMP180_GetAlt(0);
    printf("Temperature=%f,Pressure=%f,Altitude=%f\r\n",Temperature,Pressure/101325.0,Altitude);
    HAL_Delay(200);
  /* USER CODE END WHILE */

  /* USER CODE BEGIN 3 */
}
/* USER CODE END 3 */
```

[그림 12.15] 메인 코드 내용

4.5. 디버그 및 다운로드하기

메인 함수에 프로그램이 작성되었다면 다음으로 디버그는 [그림 12.16]의 를 클릭하거나 단축키 F11을 누르면 디버깅과 프로그램 다운로드가 완료된다.

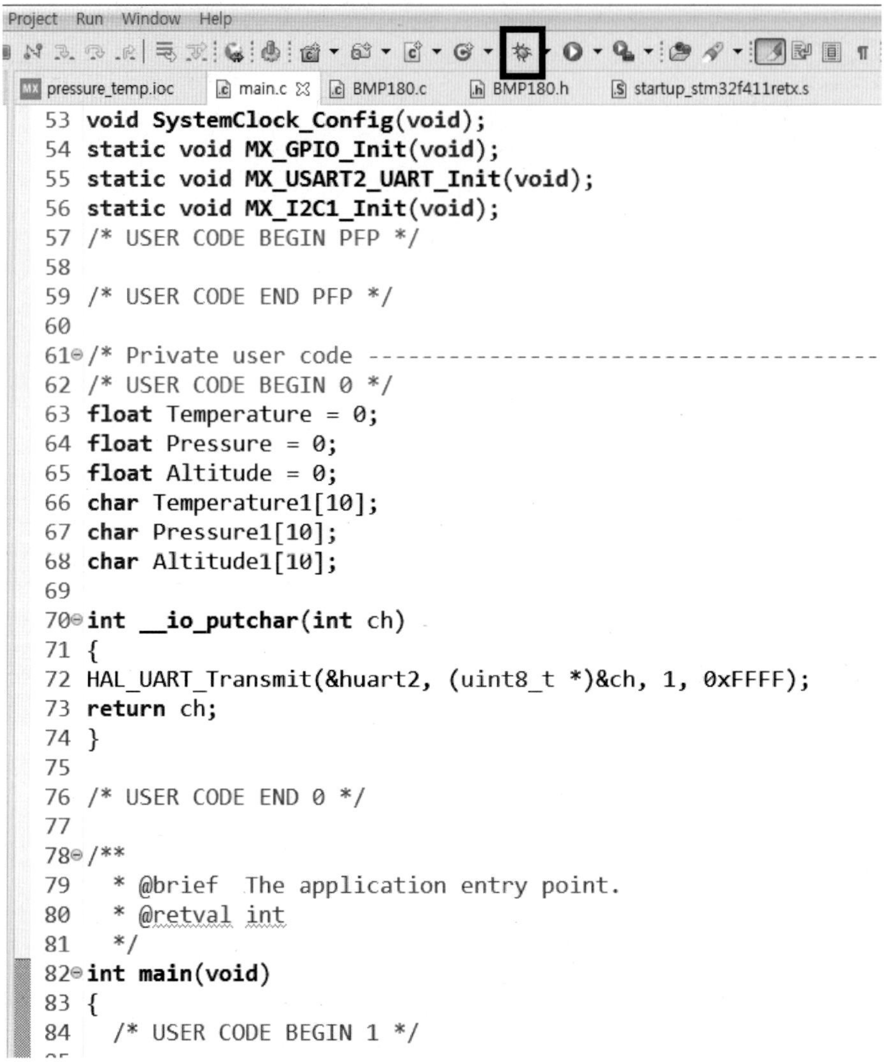
Project Run Window Help
pressure_temp.ioc
main.c
BMP180.c
BMP180.h
startup_stm32f411retx.s
void SystemClock_Config(void);
static void MX_GPIO_Init(void);
static void MX_USART2_UART_Init(void);
static void MX_I2C1_Init(void);
/* USER CODE BEGIN PFP */
/* USER CODE END PFP */
/* Private user code ----------------------------------
/* USER CODE BEGIN 0 */
float Temperature = 0;
float Pressure = 0;
float Altitude = 0;
char Temperature1[10];
char Pressure1[10];
char Altitude1[10];
int __io_putchar(int ch)
{
HAL_UART_Transmit(&huart2, (uint8_t *)&ch, 1, 0xFFFF);
return ch;
}
/* USER CODE END 0 */
/**
* @brief The application entry point.
* @retval int
*/
int main(void)
{
/* USER CODE BEGIN 1 */

[그림 12.16] 프로그램 디버그

4.6. 보드 프로그램 실행하기

[그림 12.17] 상단 중앙에 아이콘 클릭하면 디버그 모드가 종료되어 보드의 프로그램이 실행된다.

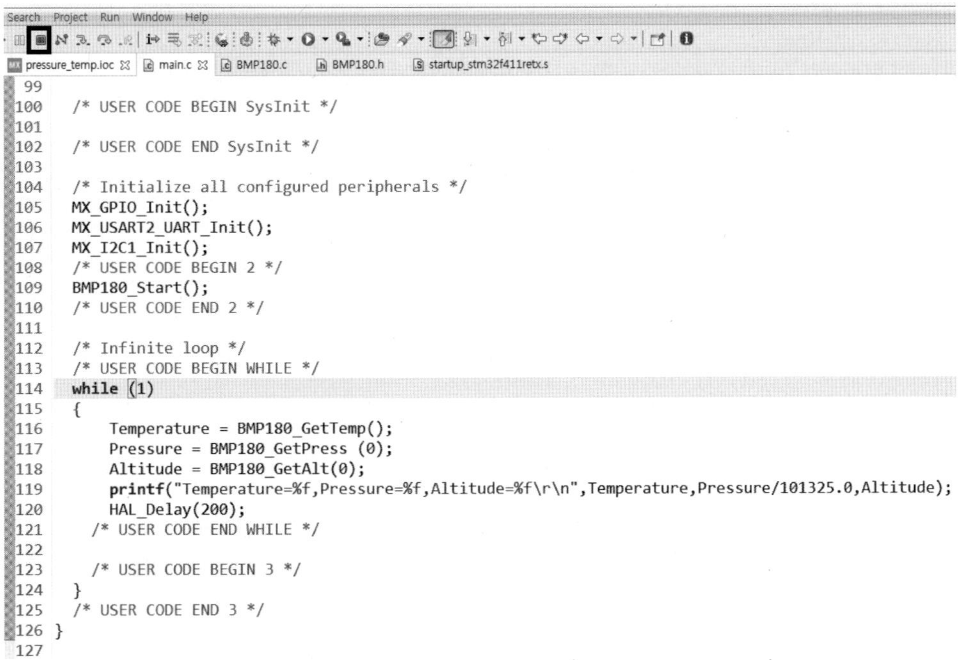
Search Project Run Window Help
pressure_temp.ioc
main.c
BMP180.c
BMP180.h
startup_stm32f411retx.s
/* USER CODE BEGIN SysInit */
/* USER CODE END SysInit */
/* Initialize all configured peripherals */
MX_GPIO_Init();
MX_USART2_UART_Init();
MX_I2C1_Init();
/* USER CODE BEGIN 2 */
BMP180_Start();
/* USER CODE END 2 */
/* Infinite loop */
/* USER CODE BEGIN WHILE */
while (1)
{
Temperature = BMP180_GetTemp();
Pressure = BMP180_GetPress (0);
Altitude = BMP180_GetAlt(0);
printf("Temperature=%f,Pressure=%f,Altitude=%f\r\n",Temperature,Pressure/101325.0,Altitude);
HAL_Delay(200);
/* USER CODE END WHILE */
/* USER CODE BEGIN 3 */
}
/* USER CODE END 3 */
}

[그림 12.17] 디버그 모드 프로그램 실행

4.7. 프로그램 실행 확인

시리얼 통신 [그림 12.18] 데이터를 확인할 수 있다.

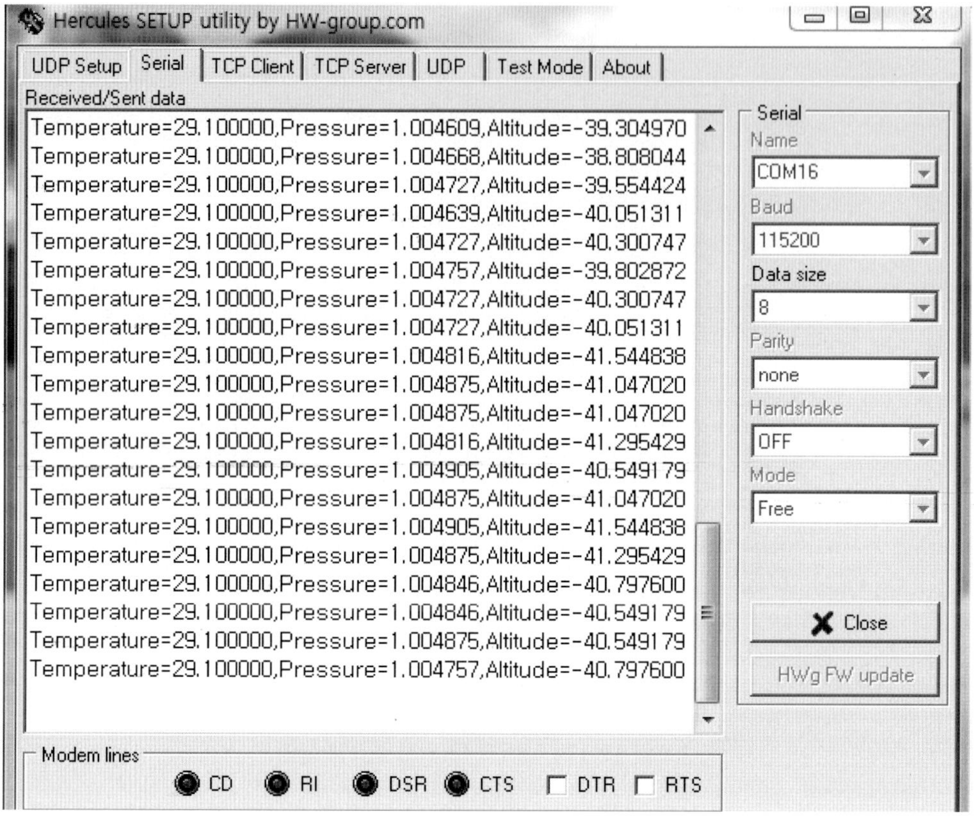

[그림 12.18] 온도 압력 고도 측정 결과

5. 로드 셀 (Load cell)(BCL-6L) 활용하기 1

5.1. 로드 셀 개요

로드 셀의 응용 분야 중 전자 저울과 같은 중량을 측정할 경우의 MCU와 관련된 부분을 살펴본다. 여기서 사용되는 센서의 종류는 빔형이며 [그림 12.19]와 같으며 실습에 사용되는 스펙은 아래와 같다.

〉〉실습용 로드셀 스펙〈〈: **최대 하중** 6kg, **볼트당** 2.0mv

제조사: **CAS ,MODEL: BCL-6L**

CAP: 6kgf,RO: 2.0mv/V

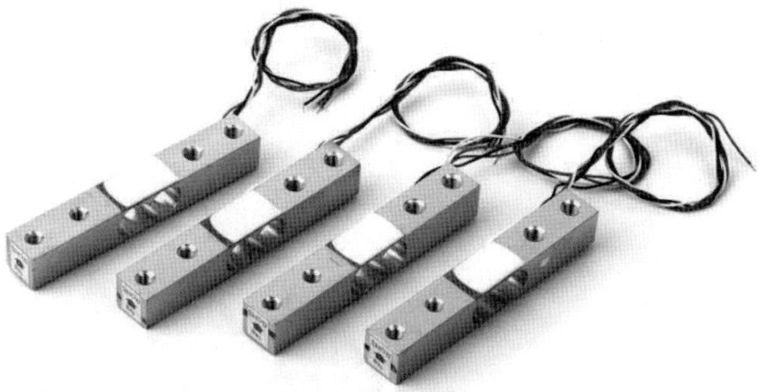

[그림 12.19] 빔형 로드 셀

실습에 사용되는 증폭기는 HX711 모듈이며 24비트 시리얼 통신 방식을 지원하며 [그림 12.20]과 같다. 이는 로드 셀의 신호(0~2.0mv*3.3)를 받아 128 또는 64배 증폭해서 채널 A나 B로 시리얼로 전송할 수 있다.

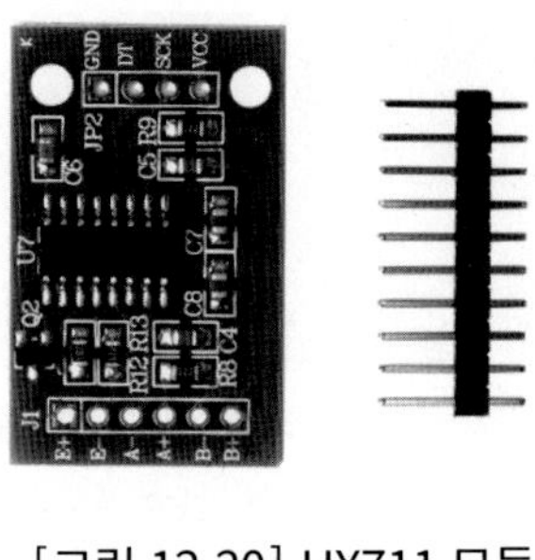

[그림 12.20] HX711 모듈

5.2. 센서 인터페이스 설정

[그림 12.21]은 실습을 위한 인터페이스이고 [그림 12.22]는 실습을 위한 로드셀의 설치 및 하중 측정 모습이며 AMP 채널은 A사용, 전압은 3.3V 사용, MCU 포트는 PORTC1은 클럭, PORTC0은 Data 입력 기능으로 설정한다.

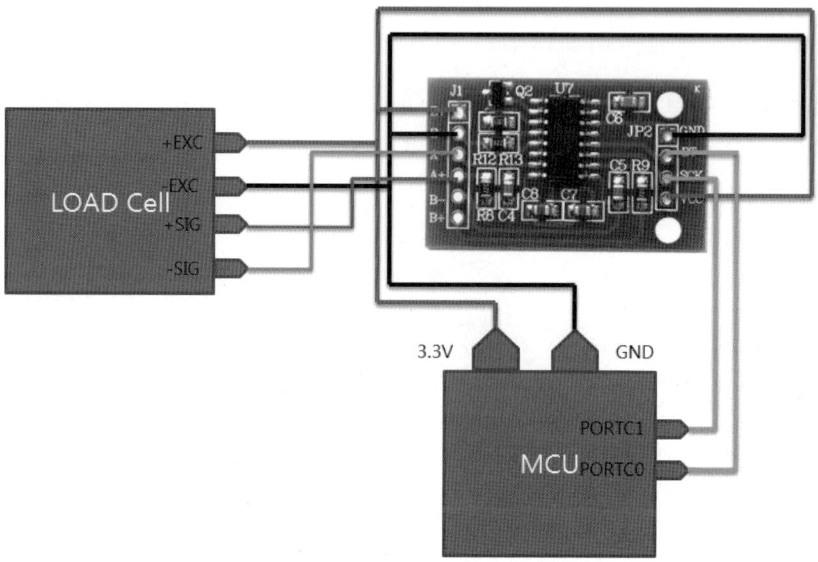

[그림 12.21] MCU, 센서 및 AMP 인터페이스

[그림 12.22] 셀에 100g, 1kg, 무부하, 205g

5.3. MCU 구성 설정

클럭 및 SYS 시리얼 통신 설정은 이전과 동일하게 설정하면 된다. 로드 셀 센서는 GPIO 인터페이스(3.3V GND PC1(SCK),PC0(SD))에 의해서 데이터를 읽기 때문에 별도의 하드웨어는 필요 없다. [그림 12.23]은 GPIO 설정이다.

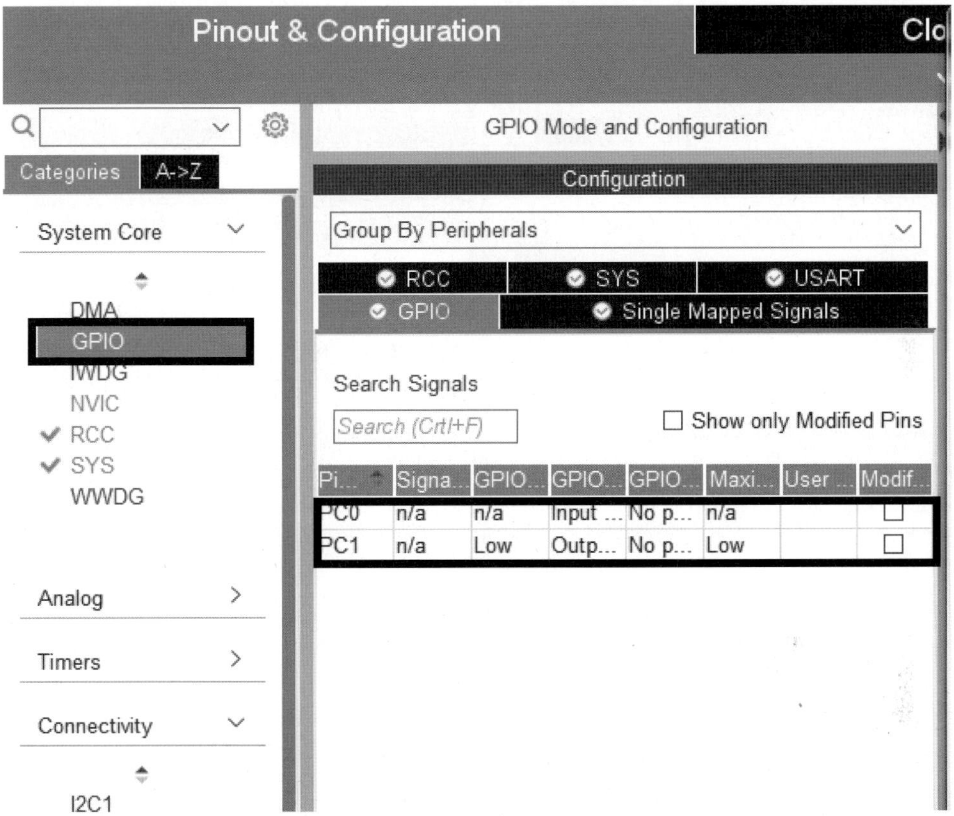

[그림 12.23] GPIO 설정

5.4. Generate Code 생성

[그림 12.24]는 코드를 생성하기 위한 화면이다. 그림의 [아이콘]을 클릭해서 프로그램을 생성한다.

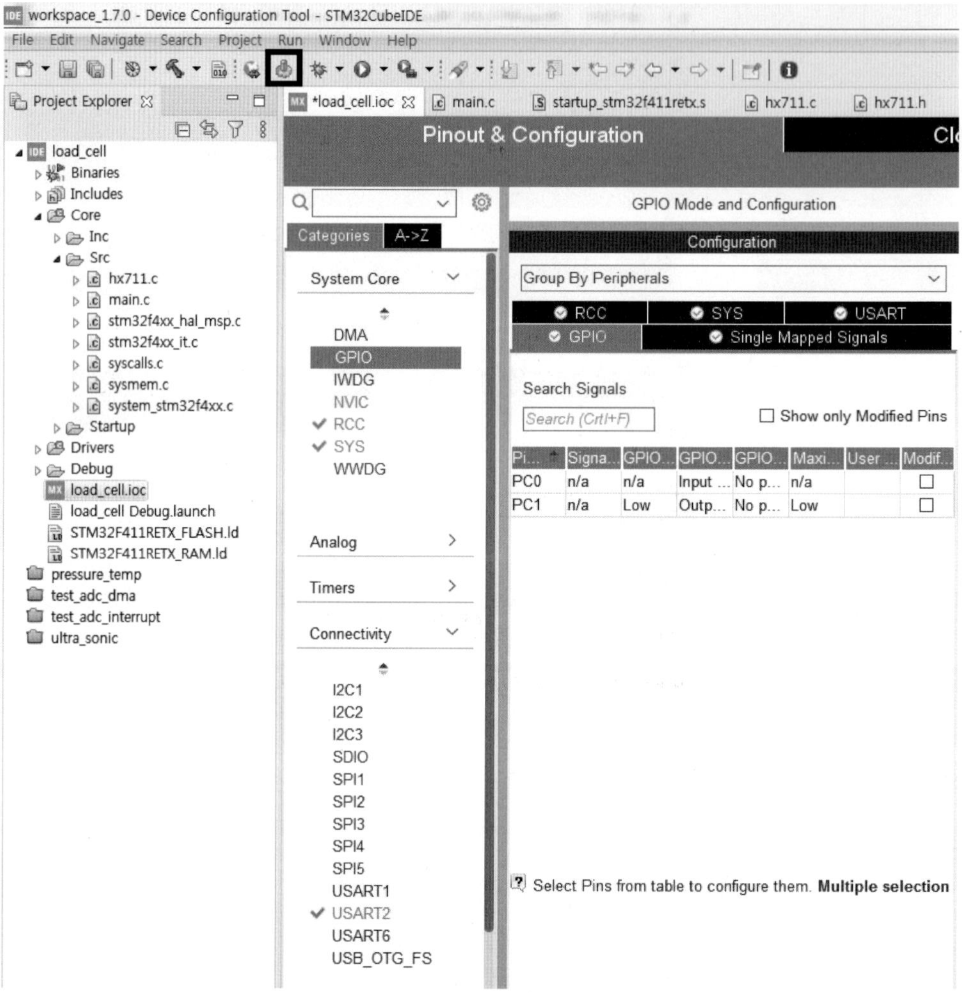

[그림 12.24] 프로그램 코드를 생성

5.5. Source Code 작성하기

소스 프로그램을 직접 작성하는 것도 중요하나 잘 작성된 프로그램을 활용하는 것도 중요하다. 그래서 아래의 사이트에서 관련 파일을 내려받아 파일을 이용하기 위해서는 소스 파일 및 헤더 파일을 복사해서 추가[그림 12.26]하고 [그림 12.27]과 같이 프로그램하여 사용한다.

https://m.blog.naver.com/PostView.naver?isHttpsRedirect=true&blogId=codelia77&logNo=221441025043//hx711_test_0113.zip

〉무게 측정방법(최대 중량이 6k 이므로 3kg 표준 분동 추천, 실습에는 1kg 사용)

1. **무 부하 일때의 출력 측정** : load_no_weight=189

2. 1kg **분동 출력측정**: load_onek_weight=1452.5;

3. **관련 수식**:

RO= 2mv/V, 3.3V **사용하므로** 6.6mv/3.3V, gain=128, 844.8mv/3.3V

6kg=〉 844.8mv/3.3V, ADC **해상도**= $\frac{3.3}{2^{24}}$ = 0.000000197

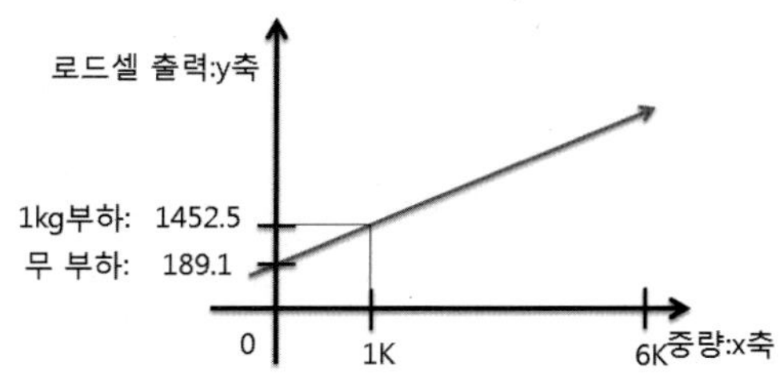

[그림 12.25] 중량과 로드 셀 출력 관계

y= (1452.5−189.1)*x/1000 +189.1

x=(y−189.1)*1000/(1452.5−189.1)

4. **프로그램**

```
Load_Raw = HX711_GetAvgValue(&Load_sensor,4)
Load_sensor = Load_Raw*0.000000197*6000/(0.8448)
Load_weight =
(Load_sensor-load_no_weight)*1k/(load_onek_weight-load_no_weight)
```

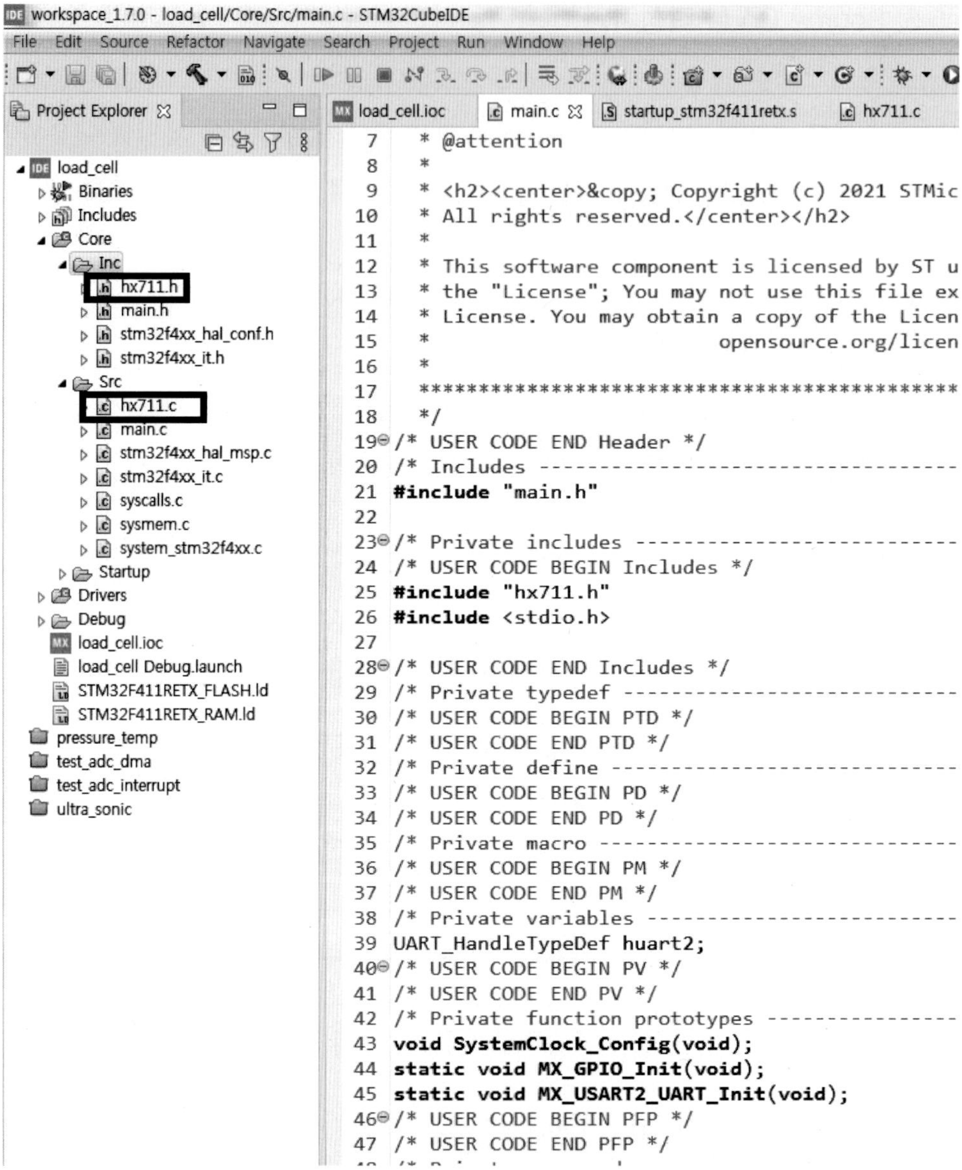

workspace_1.7.0 - load_cell/Core/Src/main.c - STM32CubeIDE
File Edit Source Refactor Navigate Search Project Run Window Help
Project Explorer
load_cell.ioc
main.c
startup_stm32f411retx.s
hx711.c
load_cell
Binaries
Includes
Core
Inc
hx711.h
main.h
stm32f4xx_hal_conf.h
stm32f4xx_it.h
Src
hx711.c
main.c
stm32f4xx_hal_msp.c
stm32f4xx_it.c
syscalls.c
sysmem.c
system_stm32f4xx.c
Startup
Drivers
Debug
load_cell.ioc
load_cell Debug.launch
STM32F411RETX_FLASH.ld
STM32F411RETX_RAM.ld
pressure_temp
test_adc_dma
test_adc_interrupt
ultra_sonic
* @attention
*
* <h2><center>© Copyright (c) 2021 STMic
* All rights reserved.</center></h2>
*
* This software component is licensed by ST u
* the "License"; You may not use this file ex
* License. You may obtain a copy of the Licen
* opensource.org/licen
*
****
*/
/* USER CODE END Header */
/* Includes ----------------------------------
#include "main.h"
/* Private includes --------------------------
/* USER CODE BEGIN Includes */
#include "hx711.h"
#include <stdio.h>
/* USER CODE END Includes */
/* Private typedef ---------------------------
/* USER CODE BEGIN PTD */
/* USER CODE END PTD */
/* Private define ----------------------------
/* USER CODE BEGIN PD */
/* USER CODE END PD */
/* Private macro -----------------------------
/* USER CODE BEGIN PM */
/* USER CODE END PM */
/* Private variables -------------------------
UART_HandleTypeDef huart2;
/* USER CODE BEGIN PV */
/* USER CODE END PV */
/* Private function prototypes ---------------
void SystemClock_Config(void);
static void MX_GPIO_Init(void);
static void MX_USART2_UART_Init(void);
/* USER CODE BEGIN PFP */
/* USER CODE END PFP */

[그림 12.26] 소스 코드 및 헤더 파일 추가

```
#include "hx711.h"
#include <stdio.h>

/* USER CODE BEGIN 0 */
HX711 Load_sensor;

int __io_putchar(int ch)
{
HAL_UART_Transmit(&huart2, (uint8_t *)&ch, 1, 0xFFFF);
return ch;
}
/* USER CODE END 0 */


/* USER CODE BEGIN 2 */
 Load_sensor.PD_SCK_PinType=GPIOC;
 Load_sensor.PD_SCK_PinNumber=GPIO_PIN_1;
 Load_sensor.DOUT_PinType=GPIOC;
 Load_sensor.DOUT_PinNumber=GPIO_PIN_0;
 Load_sensor.mode=0;
 Load_sensor.SCALE = 1.f;
 float load_no_weight=147.5;
 float load_onek_weight=1134.6;
/* USER CODE END 2 */

/* Infinite loop */
/* USER CODE BEGIN WHILE */
while (1)
{
    int Load_Raw = HX711_GetAvgValue(&Load_sensor,4);
    float Load_sensor=(float)Load_Raw*0.000000197*6000/(0.8448);
    float Load_weight= (Load_sensor-load_no_weight)/(load_onek_weight-load_no_weight);
    printf("Load_sensor=%f,Load_weight=%f[g]\r\n",Load_sensor,Load_weight*1000);
  /* USER CODE END WHILE */

  /* USER CODE BEGIN 3 */
}
/* USER CODE END 3 */
```

[그림 12.27] 메인 코드 내용

5.6. 디버그 및 다운로드하기

메인 함수에 프로그램이 작성되었다면 다음으로 디버그는 [그림 12.28]의 를 클릭하거나 단축키 F11을 누르면 디버깅과 프로그램 다운로드가 완료된다.

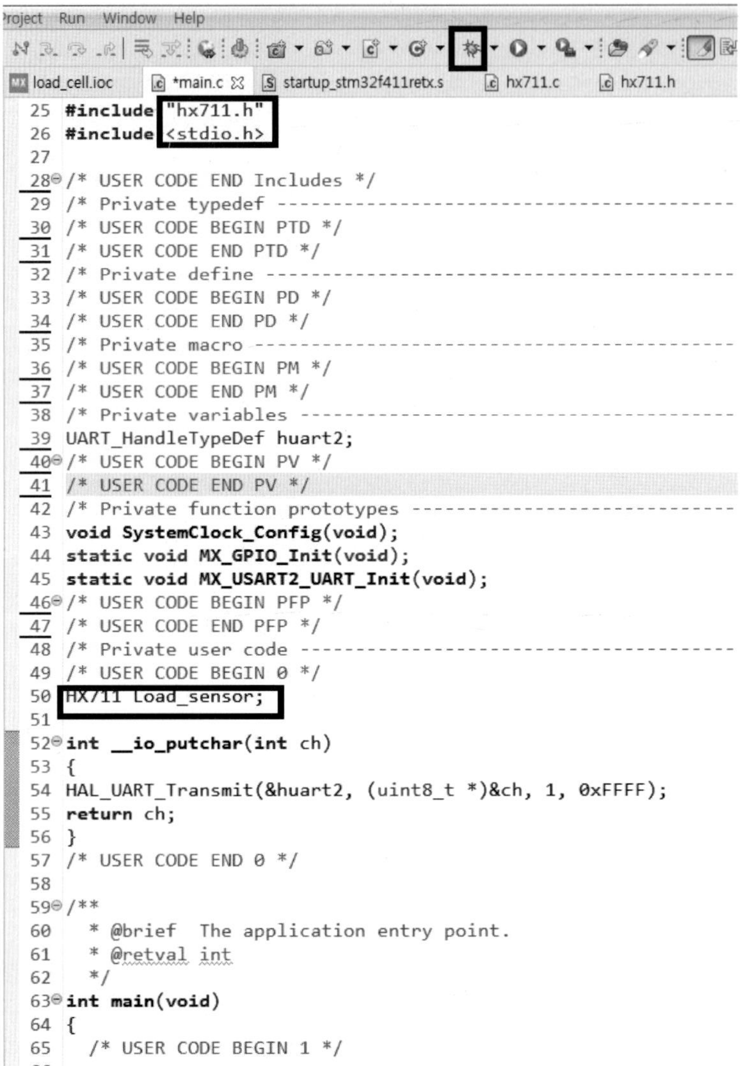

Project Run Window Help
load_cell.ioc
*main.c
startup_stm32f411retx.s
hx711.c
hx711.h
#include "hx711.h"
#include <stdio.h>
/* USER CODE END Includes */
/* Private typedef ---
/* USER CODE BEGIN PTD */
/* USER CODE END PTD */
/* Private define ---
/* USER CODE BEGIN PD */
/* USER CODE END PD */
/* Private macro ---
/* USER CODE BEGIN PM */
/* USER CODE END PM */
/* Private variables ---
UART_HandleTypeDef huart2;
/* USER CODE BEGIN PV */
/* USER CODE END PV */
/* Private function prototypes -----------------------------
void SystemClock_Config(void);
static void MX_GPIO_Init(void);
static void MX_USART2_UART_Init(void);
/* USER CODE BEGIN PFP */
/* USER CODE END PFP */
/* Private user code ---
/* USER CODE BEGIN 0 */
HX711 Load_sensor;
int __io_putchar(int ch)
{
HAL_UART_Transmit(&huart2, (uint8_t *)&ch, 1, 0xFFFF);
return ch;
}
/* USER CODE END 0 */
/**
* @brief The application entry point.
* @retval int
*/
int main(void)
{
/* USER CODE BEGIN 1 */

[그림 12.28] 메인 코드 내용

5.7. 프로그램 실행하기

[그림 12.29] 단 중앙에 아이콘 클릭하면 디버그 모드가 종료되어 보드의 프로그램이 실행된다.

Search Project Run Window Help

load_cell.ioc | main.c | startup_stm32f411retx.s | hx711.c | hx711.h

```
  /* USER CODE END Init */

  /* Configure the system clock */
  SystemClock_Config();

  /* USER CODE BEGIN SysInit */

  /* USER CODE END SysInit */

  /* Initialize all configured peripherals */
  MX_GPIO_Init();
  MX_USART2_UART_Init();
  /* USER CODE BEGIN 2 */
   Load_sensor.PD_SCK_PinType=GPIOC;
   Load_sensor.PD_SCK_PinNumber=GPIO_PIN_1;
   Load_sensor.DOUT_PinType=GPIOC;
   Load_sensor.DOUT_PinNumber=GPIO_PIN_0;
   Load_sensor.mode=0;
   Load_sensor.SCALE = 1.f;
   float load_no_weight=147.5;
   float load_onek_weight=1134.6;
  /* USER CODE END 2 */

  /* Infinite loop */
  /* USER CODE BEGIN WHILE */
  while (1)
  {
      int Load_Raw = HX711_GetAvgValue(&Load_sensor,4);
      float Load_sensor=(float)Load_Raw*0.000000197*6000/(0.8448);
      float Load_weight= (Load_sensor-load_no_weight)/(load_onek_weight-load_no_weight);
      printf("Load_sensor=%f,Load_weight=%f[g]\r\n",Load_sensor,Load_weight*1000);
    /* USER CODE END WHILE */

    /* USER CODE BEGIN 3 */
  }
  /* USER CODE END 3 */
}

/**
  * @brief System Clock Configuration
```

[그림 12.29] 버그 모드 프로그램 실행

5.8. 프로그램 실행 확인

시리얼 통신 [그림 12.30] 데이터를 확인할 수 있다.

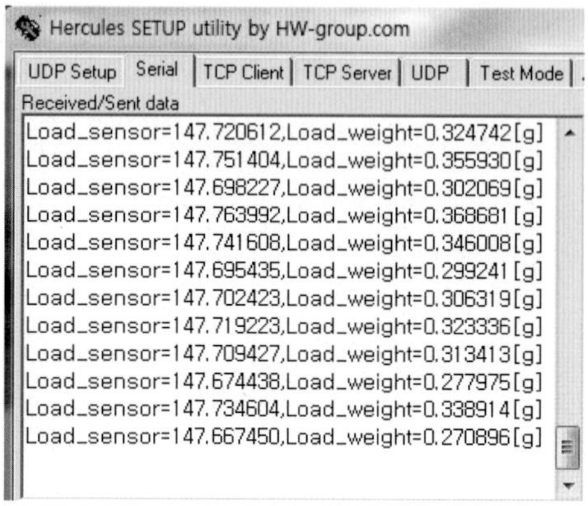

Hercules SETUP utility by HW-group.com
UDP Setup
Serial
TCP Client
TCP Server
UDP
Test Mode
Received/Sent data
Load_sensor=147.720612,Load_weight=0.324742[g]
Load_sensor=147.751404,Load_weight=0.355930[g]
Load_sensor=147.698227,Load_weight=0.302069[g]
Load_sensor=147.763992,Load_weight=0.368681[g]
Load_sensor=147.741608,Load_weight=0.346008[g]
Load_sensor=147.695435,Load_weight=0.299241[g]
Load_sensor=147.702423,Load_weight=0.306319[g]
Load_sensor=147.719223,Load_weight=0.323336[g]
Load_sensor=147.709427,Load_weight=0.313413[g]
Load_sensor=147.674438,Load_weight=0.277975[g]
Load_sensor=147.734604,Load_weight=0.338914[g]
Load_sensor=147.667450,Load_weight=0.270896[g]

무부하 측정

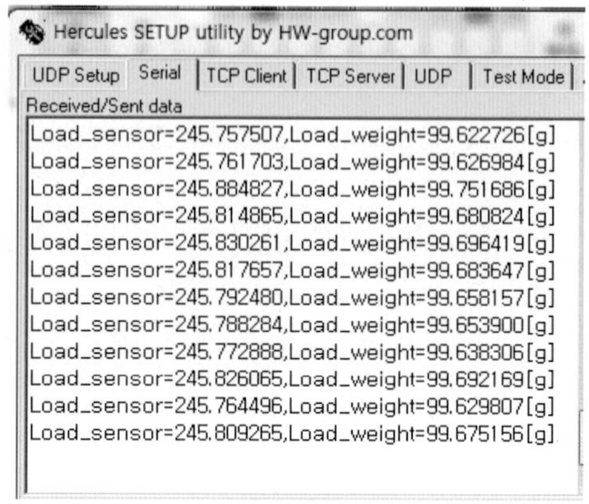

Hercules SETUP utility by HW-group.com
UDP Setup
Serial
TCP Client
TCP Server
UDP
Test Mode
Received/Sent data
Load_sensor=245.757507,Load_weight=99.622726[g]
Load_sensor=245.761703,Load_weight=99.626984[g]
Load_sensor=245.884827,Load_weight=99.751686[g]
Load_sensor=245.814865,Load_weight=99.680824[g]
Load_sensor=245.830261,Load_weight=99.696419[g]
Load_sensor=245.817657,Load_weight=99.683647[g]
Load_sensor=245.792480,Load_weight=99.658157[g]
Load_sensor=245.788284,Load_weight=99.653900[g]
Load_sensor=245.772888,Load_weight=99.638306[g]
Load_sensor=245.826065,Load_weight=99.692169[g]
Load_sensor=245.764496,Load_weight=99.629807[g]
Load_sensor=245.809265,Load_weight=99.675156[g]

100g 측정

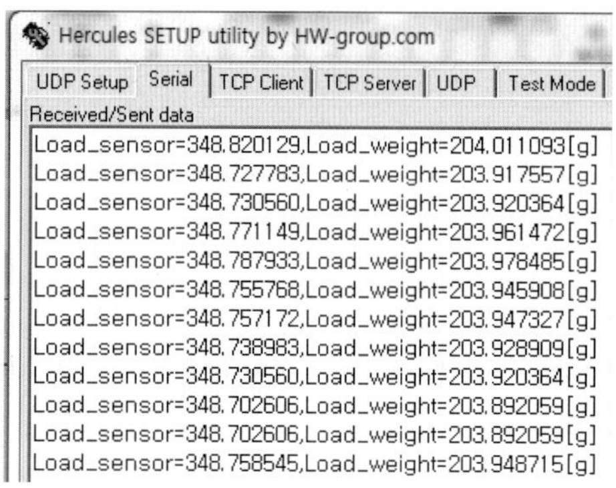

Hercules SETUP utility by HW-group.com
UDP Setup
Serial
TCP Client
TCP Server
UDP
Test Mode
Received/Sent data
Load_sensor=348.820129,Load_weight=204.011093[g]
Load_sensor=348.727783,Load_weight=203.917557[g]
Load_sensor=348.730560,Load_weight=203.920364[g]
Load_sensor=348.771149,Load_weight=203.961472[g]
Load_sensor=348.787933,Load_weight=203.978485[g]
Load_sensor=348.755768,Load_weight=203.945908[g]
Load_sensor=348.757172,Load_weight=203.947327[g]
Load_sensor=348.738983,Load_weight=203.928909[g]
Load_sensor=348.730560,Load_weight=203.920364[g]
Load_sensor=348.702606,Load_weight=203.892059[g]
Load_sensor=348.702606,Load_weight=203.892059[g]
Load_sensor=348.758545,Load_weight=203.948715[g]

205g 측정

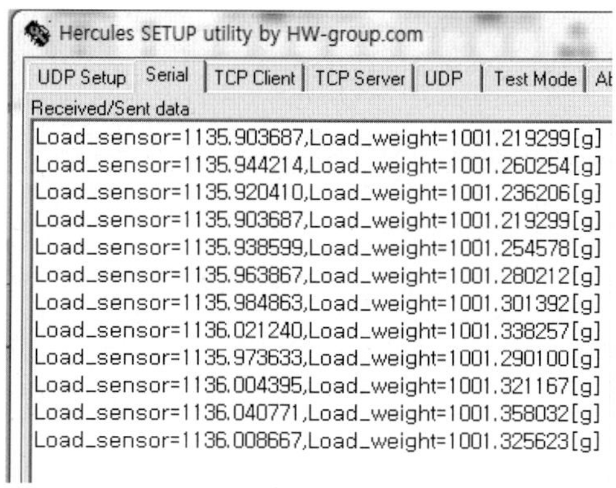

Hercules SETUP utility by HW-group.com
UDP Setup
Serial
TCP Client
TCP Server
UDP
Test Mode
Received/Sent data
Load_sensor=1135.903687,Load_weight=1001.219299[g]
Load_sensor=1135.944214,Load_weight=1001.260254[g]
Load_sensor=1135.920410,Load_weight=1001.236206[g]
Load_sensor=1135.903687,Load_weight=1001.219299[g]
Load_sensor=1135.938599,Load_weight=1001.254578[g]
Load_sensor=1135.963867,Load_weight=1001.280212[g]
Load_sensor=1135.984863,Load_weight=1001.301392[g]
Load_sensor=1136.021240,Load_weight=1001.338257[g]
Load_sensor=1135.973633,Load_weight=1001.290100[g]
Load_sensor=1136.004395,Load_weight=1001.321167[g]
Load_sensor=1136.040771,Load_weight=1001.358032[g]
Load_sensor=1136.008667,Load_weight=1001.325623[g]

1kg 측정

[그림 12.30] 0g ,100g, 205g, 1kg 측정 결과

6. 로드 셀 (Load cell)(BCL-6L) 활용하기 2

6.1. 개요

이번에는 HA711 Data sheet 뒤편에 있는 프로그램을 활용하여 센서를 구현해 보자. 모든 조건을 동일하게 한다. 포트 또한 PC1(출력 지정): CLK, PC0(입력 지정): SDA으로 동일하다. 단지 다음 조건만 다르다.

1. **무부하 일 때의 출력 측정**: load_no_weight=11884.8
2. **1kg 분동 출력 측정**: load_onek_weight=12872.2;

6.2. 프로그램

메인프로그램 함수는 [그림 12.31], 전체 프로그램은 [그림 12.32]와 같다.

```
uint32_t load_cell(void){
        uint32_t Count;
        uint8_t i;
        HAL_GPIO_WritePin(GPIOC, GPIO_PIN_1, 0);
        Count=0;
        while(HAL_GPIO_ReadPin(GPIOC, GPIO_PIN_0));
        for (i=0;i<24;i++){
            HAL_GPIO_WritePin(GPIOC, GPIO_PIN_1, 1);
            Count=Count<<1;
            HAL_GPIO_WritePin(GPIOC, GPIO_PIN_1, 0);
            if(HAL_GPIO_ReadPin(GPIOC, GPIO_PIN_0)) Count++;
        }
        HAL_GPIO_WritePin(GPIOC, GPIO_PIN_1, 1);
        Count=Count^0x800000;
        HAL_GPIO_WritePin(GPIOC, GPIO_PIN_1, 0);
        return(Count);
}
uint32_t load_cell_avr(uint8_t time){
    uint32_t sum=0;
    for(uint8_t i=0;i<time; i++){
        sum=sum+load_cell();
    }
    return sum/time;
}
```

[그림 12.31] 메인 프로그램 함수

```
uint32_t load_cell(void){
        uint32_t Count;
        uint8_t i;
        HAL_GPIO_WritePin(GPIOC, GPIO_PIN_1, 0);
        Count=0;
        while(HAL_GPIO_ReadPin(GPIOC, GPIO_PIN_0));
        for (i=0;i<24;i++){
            HAL_GPIO_WritePin(GPIOC, GPIO_PIN_1, 1);
            Count=Count<<1;
            HAL_GPIO_WritePin(GPIOC, GPIO_PIN_1, 0);
            if(HAL_GPIO_ReadPin(GPIOC, GPIO_PIN_0)) Count++;
        }
        HAL_GPIO_WritePin(GPIOC, GPIO_PIN_1, 1);
        Count=Count^0x800000;
        HAL_GPIO_WritePin(GPIOC, GPIO_PIN_1, 0);
        return(Count);
}
uint32_t load_cell_avr(uint8_t time){
    uint32_t sum=0;
    for(uint8_t i=0;i<time; i++){
        sum=sum+load_cell();
    }
    return sum/time;
}

float load_no_weight=11884.8;
float load_onek_weight=12872.5;

/* USER CODE END 2 */

/* Infinite loop */
/* USER CODE BEGIN WHILE */
while (1)
{
    int Load_Raw = load_cell_avr(4);
    float Load_sensor=(float)Load_Raw*0.000000197*6000/(0.8448);
    float Load_weight= (Load_sensor-load_no_weight)*1000/(load_onek_weight-load_no_weight);
    printf("Load_sensor=%f,Load_weight=%f[g]\r\n",Load_sensor,Load_weight);
  /* USER CODE END WHILE */

  /* USER CODE BEGIN 3 */
}
/* USER CODE END 3 */
```

[그림 12.32] 프로그램

6.3. 결과

[그림 12.33]은 측정 결과이며 만약 다양한 표준 중량이 있다면 더 정확한 측정을 할 수 있을 것이다. 여러분이 연구해 보길 바란다.

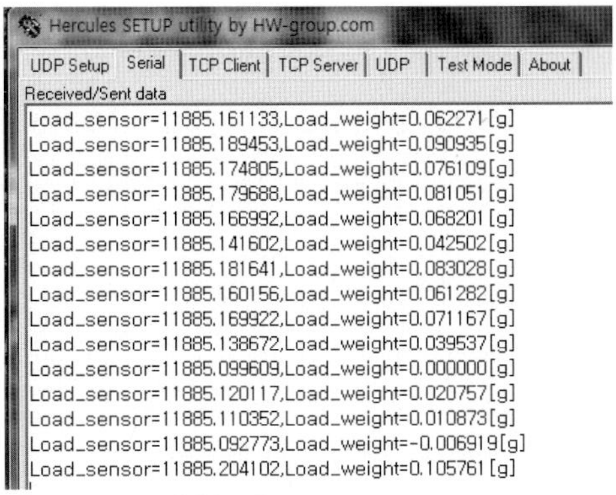

Hercules SETUP utility by HW-group.com
UDP Setup Serial TCP Client TCP Server UDP Test Mode About
Received/Sent data
Load_sensor=11885.161133,Load_weight=0.062271[g]
Load_sensor=11885.189453,Load_weight=0.090935[g]
Load_sensor=11885.174805,Load_weight=0.076109[g]
Load_sensor=11885.179688,Load_weight=0.081051[g]
Load_sensor=11885.166992,Load_weight=0.068201[g]
Load_sensor=11885.141602,Load_weight=0.042502[g]
Load_sensor=11885.181641,Load_weight=0.083028[g]
Load_sensor=11885.160156,Load_weight=0.061282[g]
Load_sensor=11885.169922,Load_weight=0.071167[g]
Load_sensor=11885.138672,Load_weight=0.039537[g]
Load_sensor=11885.099609,Load_weight=0.000000[g]
Load_sensor=11885.120117,Load_weight=0.020757[g]
Load_sensor=11885.110352,Load_weight=0.010873[g]
Load_sensor=11885.092773,Load_weight=-0.006919[g]
Load_sensor=11885.204102,Load_weight=0.105761[g]

무부하 측정

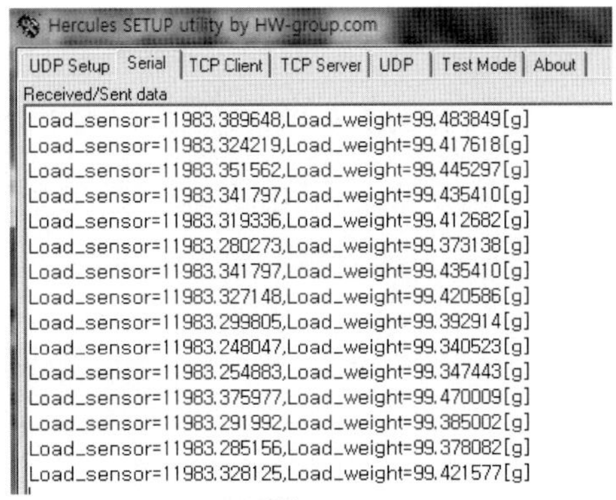

Hercules SETUP utility by HW-group.com
UDP Setup Serial TCP Client TCP Server UDP Test Mode About
Received/Sent data
Load_sensor=11983.389648,Load_weight=99.483849[g]
Load_sensor=11983.324219,Load_weight=99.417618[g]
Load_sensor=11983.351562,Load_weight=99.445297[g]
Load_sensor=11983.341797,Load_weight=99.435410[g]
Load_sensor=11983.319336,Load_weight=99.412682[g]
Load_sensor=11983.280273,Load_weight=99.373138[g]
Load_sensor=11983.341797,Load_weight=99.435410[g]
Load_sensor=11983.327148,Load_weight=99.420586[g]
Load_sensor=11983.299805,Load_weight=99.392914[g]
Load_sensor=11983.248047,Load_weight=99.340523[g]
Load_sensor=11983.254883,Load_weight=99.347443[g]
Load_sensor=11983.375977,Load_weight=99.470009[g]
Load_sensor=11983.291992,Load_weight=99.385002[g]
Load_sensor=11983.285156,Load_weight=99.378082[g]
Load_sensor=11983.328125,Load_weight=99.421577[g]

100g 측정

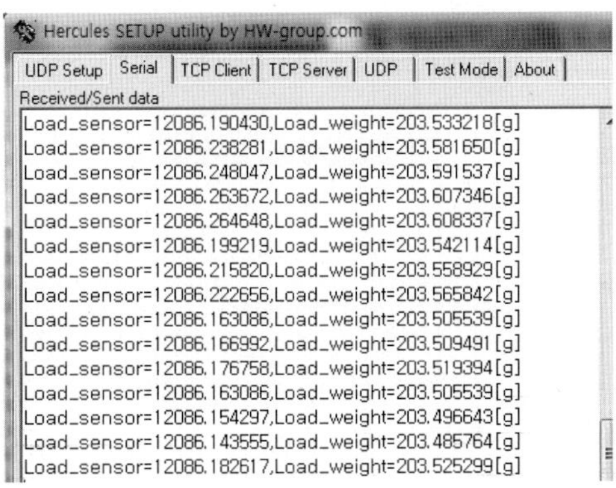

Hercules SETUP utility by HW-group.com
UDP Setup Serial TCP Client TCP Server UDP Test Mode About
Received/Sent data
Load_sensor=12086.190430,Load_weight=203.533218[g]
Load_sensor=12086.238281,Load_weight=203.581650[g]
Load_sensor=12086.248047,Load_weight=203.591537[g]
Load_sensor=12086.263672,Load_weight=203.607346[g]
Load_sensor=12086.264648,Load_weight=203.608337[g]
Load_sensor=12086.199219,Load_weight=203.542114[g]
Load_sensor=12086.215820,Load_weight=203.558929[g]
Load_sensor=12086.222656,Load_weight=203.565842[g]
Load_sensor=12086.163086,Load_weight=203.505539[g]
Load_sensor=12086.166992,Load_weight=203.509491[g]
Load_sensor=12086.176758,Load_weight=203.519394[g]
Load_sensor=12086.163086,Load_weight=203.505539[g]
Load_sensor=12086.154297,Load_weight=203.496643[g]
Load_sensor=12086.143555,Load_weight=203.485764[g]
Load_sensor=12086.182617,Load_weight=203.525299[g]

205g 측정

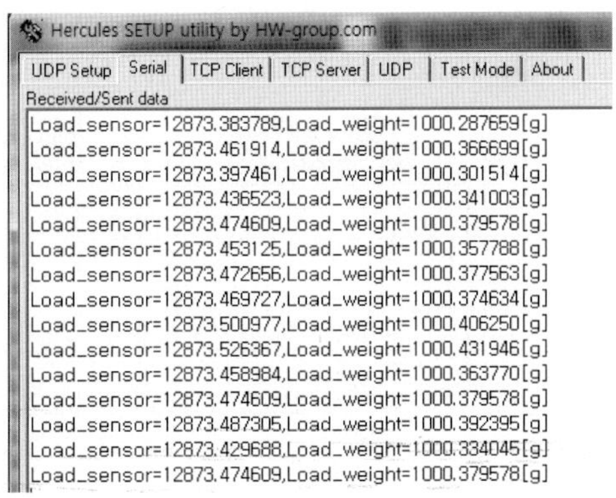

Hercules SETUP utility by HW-group.com
UDP Setup Serial TCP Client TCP Server UDP Test Mode About
Received/Sent data
Load_sensor=12873.383789,Load_weight=1000.287659[g]
Load_sensor=12873.461914,Load_weight=1000.366699[g]
Load_sensor=12873.397461,Load_weight=1000.301514[g]
Load_sensor=12873.436523,Load_weight=1000.341003[g]
Load_sensor=12873.474609,Load_weight=1000.379578[g]
Load_sensor=12873.453125,Load_weight=1000.357788[g]
Load_sensor=12873.472656,Load_weight=1000.377563[g]
Load_sensor=12873.469727,Load_weight=1000.374634[g]
Load_sensor=12873.500977,Load_weight=1000.406250[g]
Load_sensor=12873.526367,Load_weight=1000.431946[g]
Load_sensor=12873.458984,Load_weight=1000.363770[g]
Load_sensor=12873.474609,Load_weight=1000.379578[g]
Load_sensor=12873.487305,Load_weight=1000.392395[g]
Load_sensor=12873.429688,Load_weight=1000.334045[g]
Load_sensor=12873.474609,Load_weight=1000.379578[g]

1000g 측정

[그림 12.33] 프로그램

7. 서미스터 (Thermistor) 활용하기

7.1. 서미스터 개요

여기에서는 NTC Type 서미스터로 온도 측정을 하고자 한다. 일반적인 서미스터의 특성은 [그림 12.34]이고 [그림 12.35] [참고자료, TI, Temperature sensing with thermistors,2020,Feb]는 서미스터 인터페이스 회로 구성이다. 실습에는 R_{BIAS} 저항를 10k로, V_{VIAS} 전압은 3.3V로 설정하고 실습한다. [그림 12.35]의 (B)는 선형성을 높이기 위한 인터페이스이다.

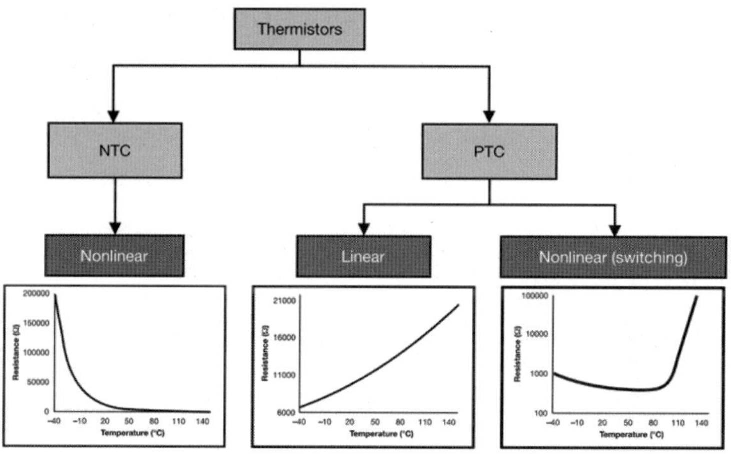

[그림 12.34] 서미스터 특성

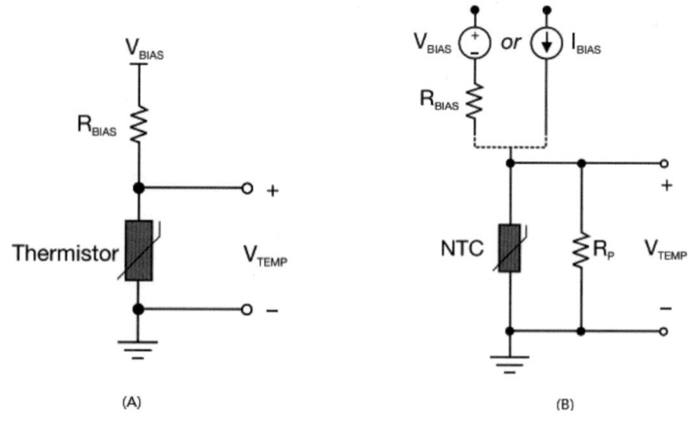

[그림 12.35] 서미스터 연결

7.2. 센서 인터페이스 설정

[그림 12.36]는 실습을 위한 인터페이스이고 [그림 12.37]은 실습 그림이다. 전압은 3.3V 사용, MCU ADC1 채널 0인 PA0로 설정한다.

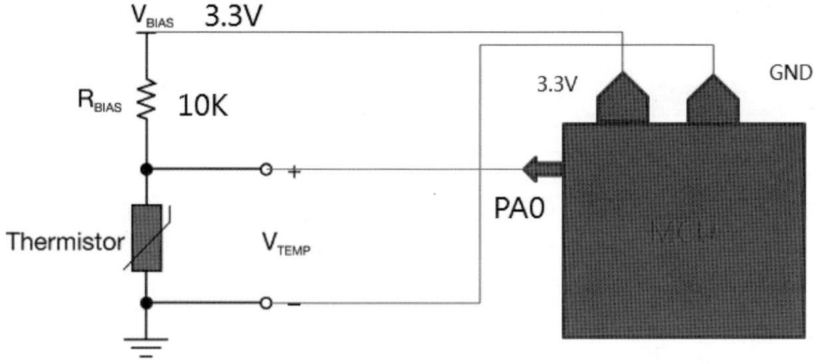

[그림 12.36] 서미스터 실습 회로

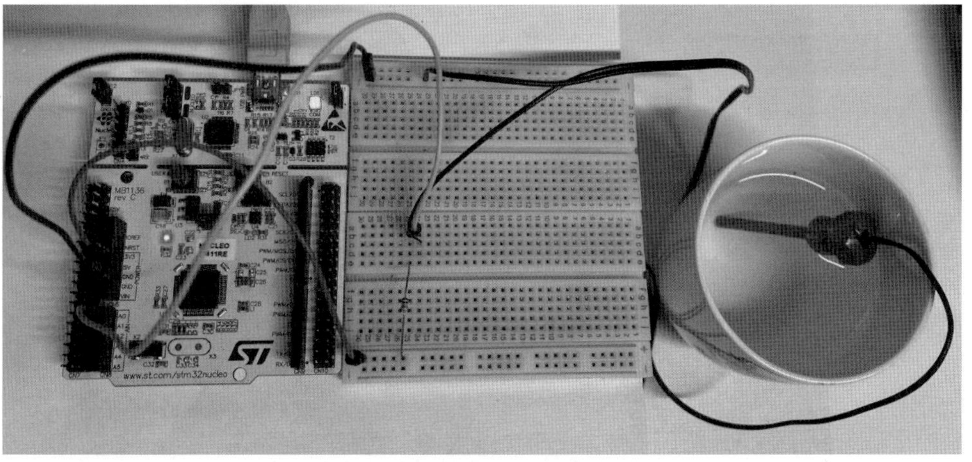

[그림 12.37] 실습 사진

7.3. MCU 구성 설정

클럭 및 SYS 시리얼 통신 설정은 이전과 동일하게 설정하면 된다. [그림 12.38]은 서미스터 양단의 전압을 읽기 위해서 ADC 채널 0을 DMA 방식으로 설정한다.

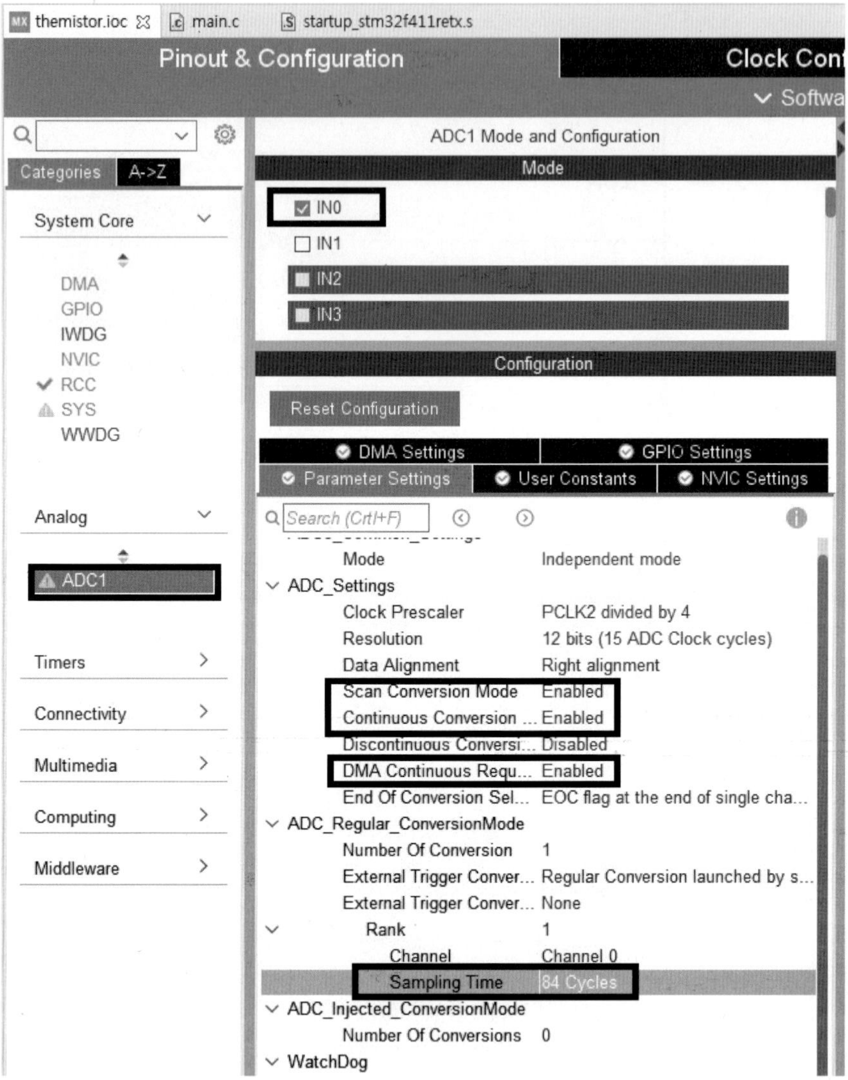

[그림 12.38] ADC DMA 방식 설정

7.4. Generate Code 생성

[그림 12.39]는 코드를 생성하기 위한 화면이다. 그림의 을 클릭해서 프로그램을 생성한다.

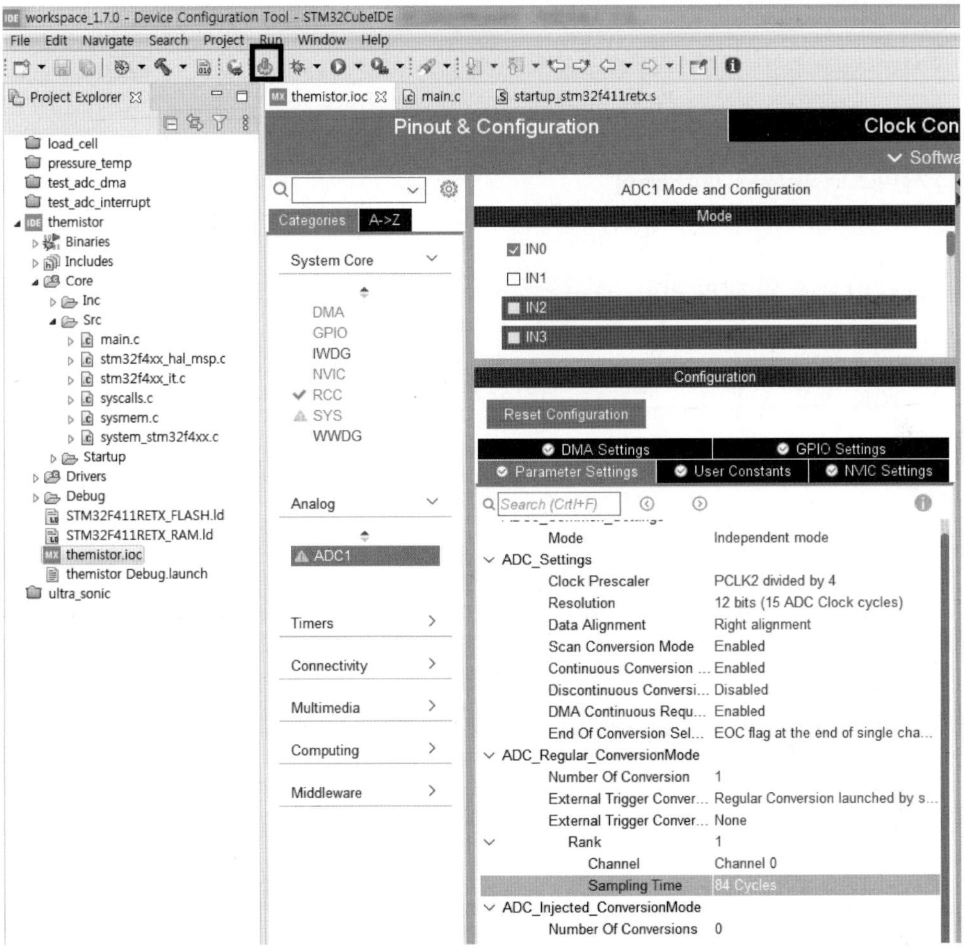

[그림 12.39] 프로그램 코드를 생성

7.5. Source Code 작성하기

[그림 12.36]에서 다음과 같은 관련 수식을 정리할 수 있다.

〉 $V_{TEMP} = \frac{R_t}{R_t + R_{10k}} * 3.3[V]$

〉서미스터 저항: $R_t = \frac{V_{TEMP} * R_{10k}}{3.3 - V_{TEMP}}$

〉서미스터의 온도 계산 공식

$R_t = R\ e^{(B(1/T1 - 1/T2))}$, B: 상수

Rt는 T1 온도일 때의 저항,

R은 T2 온도일 때의 저항

R=10K 경우, 보통 T2=273.15 +25(25도),

〉T1과 Rt 관계식 정리

$$T1 = \frac{1}{(\log(R_t/R)/B + 1/T2)}$$

섭씨 경우 :T1= T1−273.15 +0.5

4.**프로그램**

```
float thermistor_con_data(void)
{
   float Rt=0;
   float Rp=10000;
   float T2=273.0+25;
   float Bx=3380;//3380
   float Ka=273.15;
   float vol=0,temp;
   vol=(float)therm_data*(3.3/4096);
   Rt=10000*vol/(3.3-vol);
   temp=1/(1/T2+log(Rt/Rp)/Bx)-Ka+0.5;
   return temp;
}
```

```
float therm_avr(uint8_t time){
	float sum=0.0;
	for(uint8_t i=0;i<time;i++){
		sum=sum+thermistor_con_data();
}
	return sum/time;
}
```

```
#include "stdio.h"
#include "math.h"

uint32_t therm_data;
float therm_con_data;

float thermistor_con_data(void)
{
   float Rt=0;
   float Rp=10000;
   float T2=273.0+25;
   float Bx=3380;//3380 343
   float Ka=273.0;
   float vol=0,temp;
   vol=(float)therm_data*(3.3/4096);
   Rt=10000*vol/(3.3-vol); // bottom side    // Rt=(3.3-vol)*10000/vol; // up side
   temp=1/(1/T2+log(Rt/Rp)/Bx)-Ka+0.5;
   return temp;
}

float therm_avr(uint8_t time){
    float sum=0.0;
    for(uint8_t i=0;i<time;i++){
        sum=sum+thermistor_con_data();
    }
    return sum/time;
}
HAL_ADC_Start_DMA(&hadc1, &therm_data, 1);
/* USER CODE END 2 */

/* Infinite loop */
/* USER CODE BEGIN WHILE */
while (1)
{
  therm_con_data=therm_avr(50);
  printf("thermistor_temp=%f\r\n",therm_con_data);
  HAL_Delay(50);
  /* USER CODE END WHILE */

  /* USER CODE BEGIN 3 */
}
/* USER CODE END 3 */
```

[그림 12.40] 메인 코드 내용

7.6. 디버그 및 다운로드하기

메인 함수에 프로그램이 작성되었다면 다음으로 디버그는 [그림 12.41]의 를 클릭하거나 단축키 F11을 누르면 디버깅과 프로그램 다운로드가 완료된다.

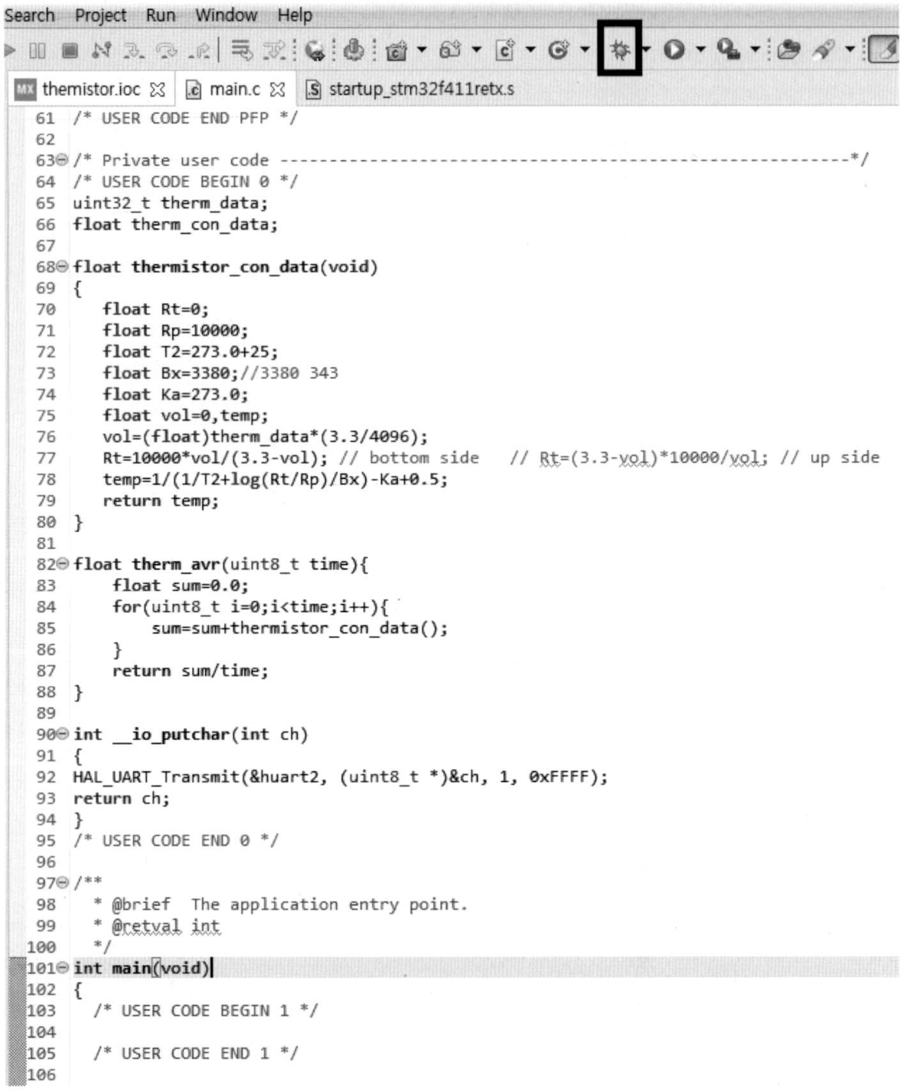

Search Project Run Window Help
themistor.ioc
main.c
startup_stm32f411retx.s
/* USER CODE END PFP */
/* Private user code ---*/
/* USER CODE BEGIN 0 */
uint32_t therm_data;
float therm_con_data;
float thermistor_con_data(void)
{
float Rt=0;
float Rp=10000;
float T2=273.0+25;
float Bx=3380;//3380 343
float Ka=273.0;
float vol=0,temp;
vol=(float)therm_data*(3.3/4096);
Rt=10000*vol/(3.3-vol); // bottom side // Rt=(3.3-vol)*10000/vol; // up side
temp=1/(1/T2+log(Rt/Rp)/Bx)-Ka+0.5;
return temp;
}
float therm_avr(uint8_t time){
float sum=0.0;
for(uint8_t i=0;i<time;i++){
sum=sum+thermistor_con_data();
}
return sum/time;
}
int __io_putchar(int ch)
{
HAL_UART_Transmit(&huart2, (uint8_t *)&ch, 1, 0xFFFF);
return ch;
}
/* USER CODE END 0 */
/**
* @brief The application entry point.
* @retval int
*/
int main(void)
{
/* USER CODE BEGIN 1 */
/* USER CODE END 1 */

[그림 12.41] 프로그램 디버그

7.7. 프로그램 실행하기

[그림 12.42] 상단 중앙에 아이콘 ▣ 클릭하면 디버그 모드가 종료되어 보드의 프로그램이 실행된다.

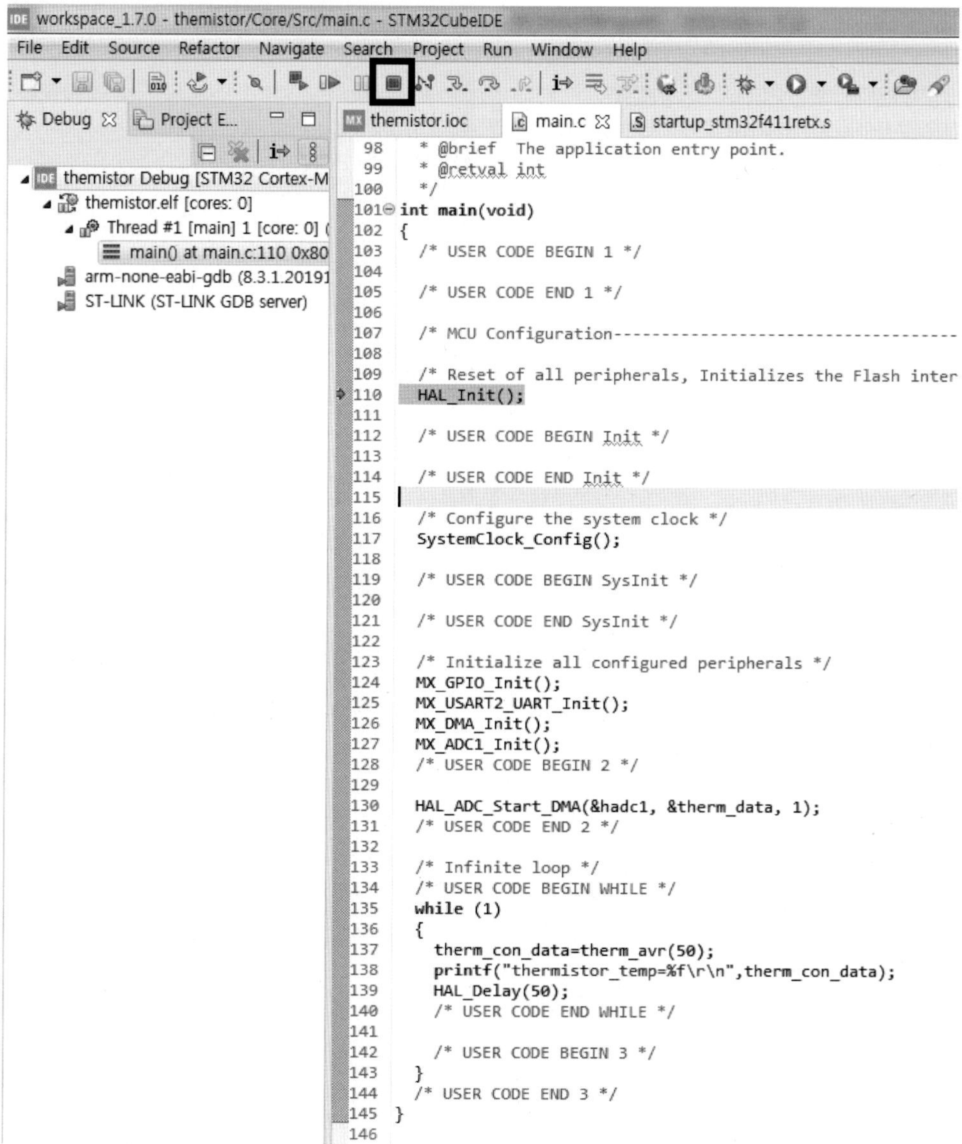

workspace_1.7.0 - themistor/Core/Src/main.c - STM32CubeIDE
File Edit Source Refactor Navigate Search Project Run Window Help
Debug
Project E...
themistor.ioc
main.c
startup_stm32f411retx.s
themistor Debug [STM32 Cortex-M
themistor.elf [cores: 0]
Thread #1 [main] 1 [core: 0]
main() at main.c:110 0x80
arm-none-eabi-gdb (8.3.1.20191
ST-LINK (ST-LINK GDB server)
* @brief The application entry point.
* @retval int
*/
int main(void)
{
/* USER CODE BEGIN 1 */
/* USER CODE END 1 */
/* MCU Configuration----------------------------------
/* Reset of all peripherals, Initializes the Flash inter
HAL_Init();
/* USER CODE BEGIN Init */
/* USER CODE END Init */
/* Configure the system clock */
SystemClock_Config();
/* USER CODE BEGIN SysInit */
/* USER CODE END SysInit */
/* Initialize all configured peripherals */
MX_GPIO_Init();
MX_USART2_UART_Init();
MX_DMA_Init();
MX_ADC1_Init();
/* USER CODE BEGIN 2 */
HAL_ADC_Start_DMA(&hadc1, &therm_data, 1);
/* USER CODE END 2 */
/* Infinite loop */
/* USER CODE BEGIN WHILE */
while (1)
{
therm_con_data=therm_avr(50);
printf("thermistor_temp=%f\r\n",therm_con_data);
HAL_Delay(50);
/* USER CODE END WHILE */
/* USER CODE BEGIN 3 */
}
/* USER CODE END 3 */
}

[그림 12.42] 버그 모드 프로그램 실행

7.8. 프로그램 실행확인

시리얼통신[그림 12.43] 데이터를 확인할 수 있다.

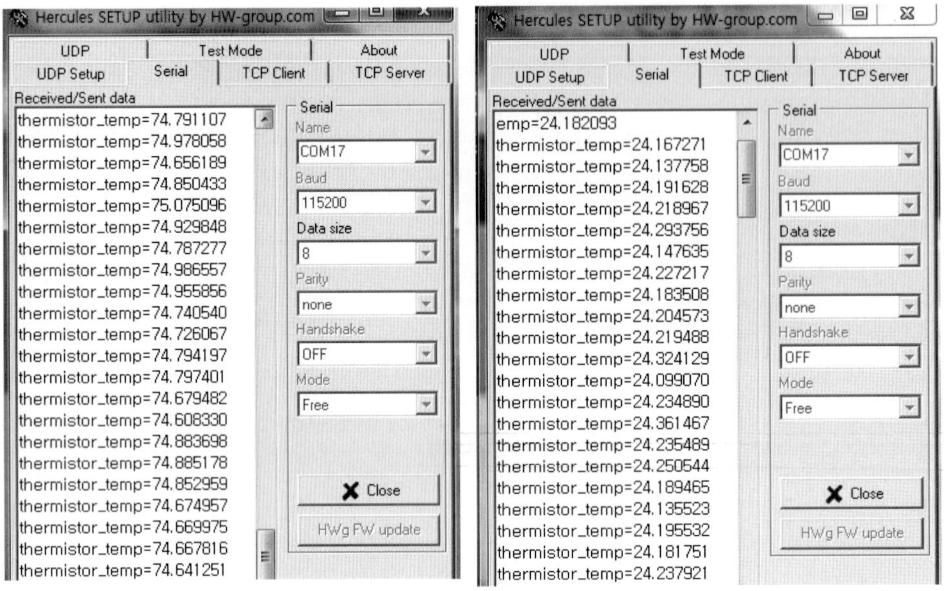

[그림 12.43] 측정 결과

[참고 문헌 및 참고 자료]

1. STM32F405xx STM32F407xx datasheet:DocID022152 Rev 8
2. TNK0023 Tech Note :Doc ID TNK0023 Rev 1.0
3. Audio and waveform generation using the DAC in STM32 products(AN3126 Application note:AN3126 Rev 4)
4. STM32F0, STM32F1, STM32F2, STM32F4, STM32L1 series,STM32F30x, STM32F3x8, STM32F373 lines timer overview(AN4013 Application note: DocID022500 Rev 4)
5. STM32F405/415, STM32F407/417, STM32F427/437 and STM32F429/439 advanced Arm®-based 32-bit MCUs(RM0090 Reference manual:RM0090 Rev 18)
6. Discovery kit with STM32F407VG MCU(UM1472 User manual: UM1472 Rev 7)
7. STM32 cross-series timer overview(AN4013 Application note: AN4013 Rev 8)
8. STM32 Nucleo-32 boards (MB1180)(UM1956 User manual:UM1956 Rev5)
9. General-purpose timer cookbook for STM32 microcontrollers(AN4776 Application note:AN4776 Rev 3)
10 .STM32 Nucleo-144 boards (MB1137)(UM1974 User manual: UM1974 Rev 8)
11. STM32 microcontroller debug toolbox(AN4989 Application note:DocID030181 Rev 2)
12. NUCLEO-XXXXKX Data brief(en.DM00214578)
13. STM32 NUCLEO BOARDS(en.flstm32nucleo:www.st.com/stm32nucleo)
14. STM32G0-ADC(Analog-to Digital Converter Revision 1.0)
15. NUCLEO-XXXXRX NUCLEO-XXXXRX-P Data brief(nucleo-64)
16. NUCLEO-XXXXZX NUCLEO-XXXXZX-P NUCLEO-XXXXZX-Q Data brief(nucleo-144)
17. STM32 Nucleo 보드 (소개)(TNK0020 Tech Note:Doc ID TNK0020 Rev 1.0)
18. https://www.st.com/content/st_com/en/stm32cube-ecosystem.html
19. https://www.youtube.com/watch?v=KWePALYozeo
20. https://www.youtube.com/watch?v=jGYQBLauvQo
21. https://www.youtube.com/watch?v=Ar5lmpiDJCo
22. https://www.youtube.com/watch?v=_Cc_zqBGWSM
23. https://www.youtube.com/watch?v=xQPOxOfhZlY
24. https://www.youtube.com/watch?v=b_MdsF9cFio

Cortex-M4
이론과 실무(기초 활용편)

ARM Core STM 32 마스터하기

2021년 9월 25일 1판 1쇄 인 쇄
2021년 10월 1일 1판 1쇄 발 행

지 은 이 : 정금섭
펴 낸 이 : 박정태

펴 낸 곳 : **광 문 각**

10881
경기도 파주시 파주출판문화도시 광인사길 161
광문각 B/D 4층
등 록 : 1991. 5. 31 제12-484호
전 화(代) : 031) 955-8787
팩 스 : 031) 955-3730
E - mail : kwangmk7@hanmail.net
홈페이지 : www.kwangmoonkag.co.kr

ISBN : 978-89-7093-569-0 93560

값: 23,000원

※ 교재와 관련된 자료는 광문각 홈페이지(www.kwangmoonkag.com) 자료실에서 다운로드 할 수 있습니다.